COMPUTATIONAL MATHEMATICS SERIES

CRYPTANALYSIS of NUMBER THEORETIC CIPHERS

COMPUTATIONAL MATHEMATICS SERIES

Series Editor Mike J. Atallah

Published Titles

Inside the FFT Black Box: Serial and Parallel Fast Fourier Transform Algorithms
Eleanor Chu and Alan George

Mathematics of Quantum Computation
Ranee K. Brylinski and Goong Chen

Fuzzy Automata and Languages: Theory and Applications
John N. Mordeson and Davender S. Malik

Cryptanalysis of Number Theoretic Ciphers
Samuel S. Wagstaff, Jr.

COMPUTATIONAL MATHEMATICS SERIES

CRYPTANALYSIS of NUMBER THEORETIC CIPHERS

Samuel S. Wagstaff, Jr.

CHAPMAN & HALL/CRC

A CRC Press Company

Boca Raton   London   New York   Washington, D.C.

Library of Congress Cataloging-in-Publication Data

Wagstaff, Samuel S.
  Cryptanalysis of number theorectic ciphers / Samuel S. Wagstafff, Jr.
    p. cm. — (Computational mathematics)
    ISBN 1-58488-153-4
    1. Computer security. 2. Cryptography. 3. Number theory. I. Title. II. Computational
  mathematics series.

QA76.9.A25 W33 2002
005.8—dc21   2002034919

This book contains information obtained from authentic and highly regarded sources. Reprinted material is quoted with permission, and sources are indicated. A wide variety of references are listed. Reasonable efforts have been made to publish reliable data and information, but the author and the publisher cannot assume responsibility for the validity of all materials or for the consequences of their use.

Neither this book nor any part may be reproduced or transmitted in any form or by any means, electronic or mechanical, including photocopying, microfilming, and recording, or by any information storage or retrieval system, without prior permission in writing from the publisher.

The consent of CRC Press LLC does not extend to copying for general distribution, for promotion, for creating new works, or for resale. Specific permission must be obtained in writing from CRC Press LLC for such copying.

Direct all inquiries to CRC Press LLC, 2000 N.W. Corporate Blvd., Boca Raton, Florida 33431.

Trademark Notice: Product or corporate names may be trademarks or registered trademarks, and are used only for identification and explanation, without intent to infringe.

Visit the CRC Press Web site at www.crcpress.com

Preface

This work has its origins in a cryptography course taught by the author many times during the past twenty years in the Computer Science Department at Purdue University.

Part I gives the mathematical background for cryptography as well as some definitions and simple examples from cryptography. The cryptographic definitions appear in the first chapter.

The second chapter treats some topics from elementary probability theory which are needed most for cryptanalysis.

Chapters 3 through 7 give a standard first course in elementary number theory, but with a slant toward computation and with the needs of cryptography always in mind. Thus, Chapter 3, on divisibility, also tells how to perform arithmetic with large integers and Chapter 4, which is about primes, discusses the probability that a "random" large integer will have only small prime factors. This topic is rarely discussed in the chapter on primes in an elementary number theory book, but is needed to estimate the difficulty of breaking certain ciphers.

Chapter 5 introduces congruences, which are used in many modern cryptographic algorithms. Chapter 6 proves Fermat's little theorem and Euler's generalization of it. These important results are used throughout the rest of the book. This chapter also introduces primitive roots and discrete logarithms, which are needed for many ciphers and protocols.

Chapter 7 deals with the solution of quadratic congruences. We do not prove the quadratic reciprocity law, but do explain its importance in computation. We state this law in a form useful for programming rather than in the slick concise way found in many number theory texts.

Chapter 8 introduces information theory and gives examples of some obsolete ciphers.

Chapter 9 offers a selection of topics from modern algebra that are used in later chapters to make and break various ciphers.

Chapters 10 through 13 treat the complementary problems of factoring large integers and identifying large primes. Many cryptographic algorithms begin by choosing large primes. Some ciphers and protocols can be broken by factoring a large integer. Slow but nevertheless important factoring methods are the topic of Chapter 10. In Chapter 11, the reader learns how to tell whether a large integer is probably prime, how to give a rigorous proof that a large number is prime, and how to construct large primes that have an easy rigorous proof of primality. Chapter 12 deals with the important elliptic curve groups used in prime proofs, in factoring integers, and directly in ciphers and protocols. The fastest known factoring algorithms are described in Chapter 13.

Chapter 14 discusses the best ways to break certain ciphers by computing "discrete logarithms." We describe several good methods for choosing random numbers in Chapter 15. Cryptographic algorithms that need secret random integers can be compromised if the numbers are not sufficiently random.

Part II describes a selection of cryptographic algorithms, most of which use number theory. Chapter 16 presents some single-key ciphers, in which all keys are supposed to remain secret. Rijndael, the new Advanced Encryption Standard, is the fastest of these ciphers. The Pollig-Hellman ciphers are slower, but enjoy special properties which make them useful in certain protocols. Chapter 17 introduces public-key ciphers, including those of Rivest, Shamir and Adleman, Massey-Omura, ElGamal, and Rabin-Williams.

Methods of signing messages electronically are presented in Chapter 18. Chapter 19 explains ways for two users to exchange keys in a secure manner, so that no one else can discover these keys by eavesdropping on their messages, and so that the users can be sure that they are talking to each other and not to an impersonator.

In Chapter 20 we describe simple protocols for playing games, sharing secrets, signing documents without seeing them, and establishing one's identity. The protocols in Chapter 21 are more complicated, and include signing contracts over the Internet, holding an election over the Internet and using digital "cash" to purchase goods. Chapter 22 explains two complete cryptographic systems, Kerberos for user authentication and Pretty Good Privacy for secure electronic mail.

Some attacks on the cryptographic algorithms are discussed as the algorithms are presented in Part II. In Part III, we collect together some general methods of attack on the cryptographic algorithms of Part II and assess their effectiveness.

Chapter 23 treats direct attacks in which the attacker has no contact with the victim and the victim does nothing wrong. These attacks involve a direct assault on a secret key. They are analogous to the attacker breaking into the victim's house when he is away and taking his money.

In the attack techniques of Chapter 24, either the victim or his computer makes an error which allows an attacker to learn a secret key. These methods are similar to an attacker entering a victim's house and taking his money when the victim left the door unlocked or the lock is broken.

In the attacks of Chapter 25, the attacker interacts with the victim and either steals a secret key or makes the victim do something he wishes he had not done. These attacks are like being mugged or raped.

The second and third parts of the book give copious references to the theorems in the first part, so that the reader can learn more about why the cryptography works and the nature of the attacks on it.

More than 200 interesting exercises test the reader's understanding of the text. The exercises range in difficulty from nearly trivial to quite challenging. We hope you enjoy the antics of Alice, Bob, and their gang.

The prerequisites for reading this book are calculus and linear algebra. From calculus, you should know how to differentiate, integrate, and find extrema. You should be familiar with the logarithm and exponential functions and with Newton's method for finding zeros of functions. You should know that sums may be approximated by integrals. You should know the rudiments of set theory, intersections, unions, and subsets. From linear algebra, you should be familiar with matrices and know how to solve a system of linear equations in several unknowns. For complete understanding of this book, you should also be familiar with proof by mathematical induction.

Throughout the book, we use the notation $\lfloor x \rfloor$ and $\lceil x \rceil$ to mean the largest integer $\leq x$ and the smallest integer $\geq x$, respectively.

When a and b are integers with $b > 0$, we write $a \bmod b$ for the (nonnegative) remainder when a is divided by b. It is always in the range $0 \leq a \bmod b < b$. Since $\lfloor a/b \rfloor$ is the integer part of the quotient when a is divided by b we always have

$$a = b\lfloor a/b \rfloor + (a \bmod b).$$

When n is a positive integer, we define "n factorial" to be the product $n! = 1 \cdot 2 \cdot 3 \cdots (n-1)n$. Also define $0! = 1$. When $0 \leq i \leq n$ are integers, define the **binomial coefficient**, "n choose i," to be

$$\binom{n}{i} = \frac{n!}{i!(n-i)!}.$$

The name comes from the **binomial theorem**, which says that if n is a nonnegative integer, then

$$(a+b)^n = \sum_{i=0}^{n} \binom{n}{i} a^i b^{n-i}.$$

We write $\log x$ to mean the logarithm of x to an unspecified base, $\log_b x$ for the logarithm of x to base b, and $\ln x = \log_e x$ for the natural logarithm of x. We write $\exp(x)$ for e^x when x is a complicated expression.

Explicit algorithms are written in a simple pseudocode which should be clear to anyone familiar with a modern computer language like C or Java. We use `0x123ABC` for the hexadecimal number "123ABC," just as many computer languages would do.

From computer science, you should know that one can sort a list of n items with $n \log n$ comparisons.

We should explain the notion of "amortization," which appears in several algorithms. If one special instruction in a block of repeated instructions is performed only once in every k repetitions, then the time needed to execute the block once may be estimated as $1/k$ times the time for the special instruction plus the time needed for the other instructions. We say the time for the special instruction is *amortized* over the time for the block of instructions.

I thank Abhilasha Bhargav for drawing some exquisite graphs.

I am grateful to Mikail Atallah, Richard Crandall, Joe Doob, Jason Gower, Darren King, Peter Montgomery, Stephen Samuels, and Chaogui Zhang for providing insightful comments on earlier versions of parts of this book or other information that made the book better.

I wish to thank the hundreds of students who took my cryptography class during the past twenty years for testing the exercises.

Finally, I thank the Center for Education and Research in Information Assurance and Security, CERIAS, its sponsors and its director, Professor Eugene Spafford, for support while this book was being written.

Sam Wagstaff
Purdue University CERIAS
West Lafayette, Indiana
ssw@cerias.purdue.edu

I dedicate this work to my parents,
Helen and Sam,
who gave me wings and let me fly.

Contents

III Methods of Attack 279

Part I

Mathematical Foundations
of Cryptanalysis

Chapter 1

Terminology of Cryptography

This chapter introduces the basic facts of cryptography. Refer to Denning [36] for more basic information on cryptography.

1.1 *Notation*

Cryptography is the study of secret writing. A **cipher** is a way of hiding ordinary text, called **plaintext**, by transforming it into **ciphertext**. This process is called **enciphering** or **encryption** of the plaintext into ciphertext. The reverse process is called **deciphering** or **decryption**. The following figure illustrates this terminology.

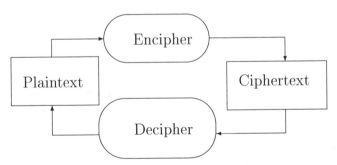

Ciphers are divided into two categories: substitution and transposition ciphers. **Substitution ciphers** replace letters or larger blocks with substitutes, usually of the same length. In a **simple substitution cipher**, the same alphabet is used for plaintext and ciphertext, and a fixed permutation of this alphabet gives the substitution rule. As an example, suppose the letters of the alphabet are arranged in a circle (with A following Z) and a message is encrypted by replacing each plaintext letter by the fifth letter after it in the circle. Thus, the message "SECRET" would be enciphered as "XJHWJY." Decryption is performed by replacing each ciphertext letter by the fifth letter before it in the alphabet circle. This type of cipher, in which the alphabet is rotated, is called a **Caesar cipher** because Julius Caesar used it.

The letters in the ciphertext of a **transposition cipher** are the same letters, with the same frequencies, as the letters in the plaintext, but they are rearranged. A simple example of a transposition cipher uses a matrix. The plaintext and ciphertext are broken into blocks with length equal to the number of entries in the matrix. A message is enciphered by writing each block of plaintext into the matrix by rows and reading a ciphertext block out of the matrix by columns backwards. For example, suppose we use a 2×3 matrix to encipher the message "SECRET," which has only one block. We form the matrix

$$\begin{array}{ccc} S & E & C \\ R & E & T \end{array}$$

and read the ciphertext "CTEESR." Decryption is performed by writing each ciphertext block into the matrix by columns backwards and reading the plaintext block by rows.

Product ciphers are created by the composition of several ciphers whose types alternate between substitution and transposition. Substitution and transposition ciphers each have certain weaknesses which may be overcome by composing them in this alternating fashion. To give a simple example, compose the two ciphers above, using the Caesar cipher first. The plaintext "SECRET" is first changed into "XJHWJY." This is written into the matrix

$$\begin{array}{ccc} X & J & H \\ W & J & Y \end{array}$$

and the ciphertext is "HYJJXW." The Data Encryption Standard, DES, and Rijndael are two well known examples of product ciphers.

Both encryption and decryption are controlled by **keys**. The key for a transposition cipher is the fixed permutation of the letters in a block. The key for a simple substitution cipher is the fixed permutation of the alphabet. In the simple case of the Caesar cipher, the key is the amount of shift of the alphabet.

1.2   Types of Attacks

Suppose E and D are the prototype encryption and decryption methods before the key is specified. When the key is K, the encryption and decryption functions, obtained by specifying K in E and D, are often written E_K and D_K, respectively. Thus we would write $C = E_K(M)$ to mean that C is the ciphertext obtained when the plaintext M is enciphered with key K. Likewise, $M = D_K(C)$ means C was deciphered using the key K to give M.

A fundamental property shared by all ciphers is that $D_K(E_K(M)) = M$ for every M. This equation says that if you encipher a message M with key K, then you can decipher it, using the same key, and recover M.

In general, a cipher must be easy to use, the functions E_K and D_K must be fast for all keys K, and the security of the cipher should depend only on the secrecy of the keys and not on the secrecy of the methods E and D. This last requirement is needed because the methods may be public or spies may acquire them, but keys can and should be changed frequently. In some ciphers, all keys of a given length are equally good. But in other ciphers, one must choose a key having special properties in order for it to work or be secure. The first requirement above implies that it should be easy to select a key with any required properties.

Cryptanalysis is the study of attacks on ciphers. Methods of attack may be classified into several general types based on what information is known and what is unknown to the cryptanalyst.

In a **ciphertext-only attack**, only the ciphertext is known, although often the language of the plaintext and the type of cipher are also known. The goal of the cryptanalyst is to find the plaintext and the key. This is the most difficult type of attack. Sometimes the cryptanalyst has only a string of bits to work with.

In a **known-plaintext attack**, the cryptanalyst is given some ciphertext and the corresponding plaintext. For example, it may be known that all messages from Alice to Bob begin with a standard header. In this case the first part of each ciphertext can be deciphered because it is always the same. The goal is to find the key so that other ciphertext may be deciphered.

In a **chosen-plaintext attack**, the cryptanalyst may specify some plaintext, perhaps even a meaningless message, and somehow learn the corresponding ciphertext. This feat may be accomplished by tricking the cipher machine operator into enciphering a given message or by capturing a cipher chip with an unreadable key etched into it, for example. The goal is to find the key.

Public-key ciphers, described in the next section, give rise to a **chosen-ciphertext attack**, in which the cryptanalyst may specify some ciphertext and learn the corresponding plaintext. Again the goal is to find the key. Although the plaintext obtained might not be a meaningful message it may still aid in finding the key.

A good cipher should resist all of these kinds of attack. Specifically, it should be computationally infeasible for a cryptanalyst to do any of the following, no matter how much ciphertext is given.

1. Find M given C.

2. Find D_K given C or C and the corresponding M.

3. Construct C so that $D_K(C)$ is any meaningful message.

4. Find E_K given C or C and the corresponding M.

The first two requirements ensure the secrecy of the cipher and messages enciphered with it. Requirements 1 and 2 say that a ciphertext-only attack should be hard. Requirement 2 says that a known-plaintext attack should be hard.

The last two requirements ensure the authenticity of messages enciphered with the cipher. Requirement 4 says that no attacker can discover the enciphering function, use it to encipher a phony message (such as a bank transfer), and have the recipient accept it as authentic. Requirement 3 says that no attacker can create ciphertext which would decipher into a meaningful plaintext, although this plaintext may be strange and even unknown to the attacker. Both say that if an active attacker replaces one ciphertext with another, the change will almost certainly be detected.

1.3   Public Key Ciphers

Until the 1970's, cryptographers assumed that if one knew an enciphering function E_K, including its key K, then one could easily deduce the corresponding deciphering function D_K. All ciphers invented until then were of this type. In 1976, Diffie and Hellman [40] proposed a new type of cipher, called **public-key encryption**, for which this drawback did not hold. Until [40] appeared, it was generally assumed that this drawback could not be removed. Each user of the new cipher would have an enciphering function, which would be made public, and a deciphering function, which would be kept secret. When Alice wanted to communicate secretly with Bob, she would find Bob's public enciphering function in a directory, encrypt her message using that function, and send it to Bob. Bob could decrypt it because he knew his secret deciphering function. But no eavesdropper could read the message because he would not know Bob's secret deciphering function.

In actual public-key ciphers both methods E and D are public and the same for every user. However, the two functions for one user have different keys. The encryption key is public and the decryption key is secret. It is not computationally feasible to deduce the decryption key from the encryption key, or *vice versa*. This difference in keys led Simmons to this classification of ciphers: If the same key is used for both encryption and decryption, the cipher is called a **one-key** or **symmetric cipher**. If two different keys are used, and neither can be deduced easily from the other, the cipher is called a **two-key** or **asymmetric cipher**. Symmetric ciphers are also called **conventional** or **secret key** ciphers. All public-key ciphers are asymmetric ciphers.

If Alice uses a public-key cipher, her encryption and decryption functions would be written E_A and D_A. Here the "A" denotes "Alice," not a key as it would for be a symmetric cipher. These functions would have the property that $D_A(E_A(M)) = M$ for every plaintext M. Likewise, Bob would have the public enciphering function E_B and a secret deciphering function D_B.

Recall that symmetric ciphers provide authentication in addition to secrecy.

If Alice and Bob share a key K for a one-key cipher, then no one else will know K. If Bob receives an encrypted message allegedly from Alice and successfully decrypts it using key K, then Bob can be sure that the message really came from Alice because only Bob and Alice know K.

Public-key ciphers, as described above, provide secrecy but not authentication. Suppose Alice, Bob, and others use public-key ciphers. Alice sends a message M to Bob as $E_B(M)$ and Bob deciphers it by applying D_B. However, anyone who knows Bob's public enciphering function E_B could write a message signed, "from Alice," encipher it with E_B and send it to Bob. When Bob receives an encrypted message allegedly from Alice and successfully decrypts it using D_B, he cannot be sure it came from Alice.

Fortunately, it is easy to add authentication to public-key ciphers. Alice can "sign" her message to Bob by applying her secret deciphering function to it. She would send the signed ciphertext $C = E_B(D_A(M))$ to Bob, along with a plaintext note saying that this message came from Alice. Bob would apply D_B to C and obtain $D_A(M)$. Then Bob would locate Alice's public key E_A from a secure source and apply E_A to $D_A(M)$ and obtain M. Note that we require $D_B(E_B(M)) = M$ for *every* M, not just meaningful M. Several public-key ciphers enjoy this property.

If the plaintext M were not secret, but Alice wanted Bob to be certain that it came from her, then Alice could merely send $D_A(M)$ to Bob, along with a plaintext note saying it was from her. Then Bob, or anyone else for that matter, could apply the public E_A to $D_A(M)$ and read M. In this case, Alice has signed M but not hidden it. Public-key ciphers separate authentication from secrecy.

Diffie and Hellman gave no example of the public-key ciphers they proposed. The first example was given two years later by Rivest, Shamir and Adleman [97] and is called the RSA cryptosystem. Many more public-key ciphers have been invented since then.

1.4 *Block and Stream Ciphers*

Ciphers are classified according to how the key is used to encipher the plaintext M. **Block ciphers** break M into blocks $M_1, M_2, \ldots$ of equal length and encipher each block with the same key, so that the ciphertext is $E_K(M) = E_K(M_1)E_K(M_2)\ldots$. The transposition cipher which uses a matrix and the Caesar cipher are examples of block ciphers. The block lengths are 1 letter for the Caesar cipher and the number of letters that fit in the matrix for the transposition cipher. DES is a block cipher with a block length of 64 bits. Usually the block length is several letters.

Stream ciphers have a key expressed as a **key stream** $K = k_1 k_2 k_3 \ldots$. These ciphers break the plaintext M into pieces, $M = m_1 m_2 m_3 \ldots$, which may be letters or bits, and encipher the i-th piece m_i with the i-th piece k_i of the key, so that the ciphertext is $E_K(M) = E_{k_1}(m_1)E_{k_2}(m_2)E_{k_3}(m_3)\ldots$.

A stream cipher is called **periodic** if its key repeats after d pieces, for some fixed d.

Example 1.1

Let E_k denote encryption by a Caesar cipher with the alphabet rotated by k letters, where $1 \leq k \leq 26$. Let the key K be a sequence of five integers $k_1, \ldots, k_5$, each between 1 and 26. We can create a periodic stream cipher with period $d = 5$ by using the five Caesar ciphers $E_{k_1}, \ldots, E_{k_5}$ in a round-robin fashion to encipher the successive letters of the plaintext. If $M = m_1 m_2 m_3 \ldots$, then the ciphertext is

$$E_K(M) = E_{k_1}(m_1) E_{k_2}(m_2) \ldots E_{k_5}(m_5) E_{k_1}(m_6) E_{k_2}(m_7) \ldots.$$

It is convenient and practical to use the k-th letter of the alphabet to represent the number k between 1 and 26. Then one must remember only a five-letter word for the key, rather than five numbers. This cipher (with the key expressed in letters) is called a **Vigenère cipher**.

The key stream of a stream cipher need not be periodic. If it is not periodic, then it should be as long as the plaintext. A nonperiodic key stream may be created in two ways: If it is generated in some fashion independent of M, the cipher is called a **synchronous stream cipher**. But if the key stream is computed from the ciphertext already produced, the cipher is called a **self-synchronous stream cipher**.

Example 1.2

To give a simple example of a synchronous stream cipher, let us build on the Caesar cipher E_k with variable key letter k of the previous example. Suppose the sender and receiver agree on a standard text and a position in that text. Let k_i be the i-th letter of the standard text, beginning at the position. Then encipher $M = m_1 m_2 m_3 \ldots$ as $E_K(M) = E_{k_1}(m_1) E_{k_2}(m_2) E_{k_3}(m_3) \ldots$. This is called a **running key cipher**.

Many stream ciphers exclusive-or the plaintext and key to form the ciphertext. The **exclusive-or** operation $\oplus$ is defined on bits by

$$0 \oplus 0 = 1 \oplus 1 = 0, \text{ and } 0 \oplus 1 = 1 \oplus 0 = 1.$$

Note that $x \oplus y = y \oplus x$ and $x \oplus (y \oplus z) = (x \oplus y) \oplus z$ for all x, y and z. Note also that $x \oplus x = 0$ and $x \oplus 0 = x$ for all x. A useful property of exclusive-or, which follows from the ones just stated, is that $x \oplus y \oplus y = x$ for all x and y. We also write $X \oplus Y$ for the bitwise exclusive-or of the bit strings X and Y having the same length. If $M = m_1 m_2 m_3 \ldots$ and the key stream is $k_1 k_2 k_3 \ldots$, and the pieces m_i and k_i all have the same length in bits, then one can define the ciphertext $C = c_1 c_2 c_3 \ldots$ by $c_i = m_i \oplus k_i$ for each i. It follows from the useful property above that the deciphering rule is the same as for enciphering:

"Exclusive-or with the key." In symbols, this means $m_i = c_i \oplus k_i$. It is true because $(m_i \oplus k_i) \oplus k_i = m_i$.

In other synchronous stream ciphers, the key stream may be produced by a random number generating procedure or by special hardware. Sometimes a block cipher is used to generate the key stream. One way this may be done is to encipher a given block repeatedly. Suppose E_K is a block cipher with key K. Pick a random block B_0 and repeatedly encipher it to generate $B_i = E_K(B_{i-1})$ for $i \geq 1$. Let k_i be the first eight bits of B_i (or however many bits are needed). Use k_i as the key stream for the synchronous stream cipher. The key to the stream cipher consists of the initial block B_0 and the key K for the block cipher. This method of using a block cipher to produce a key stream is called **output-block feedback mode**.

Another way to use a block cipher to produce a key stream is **counter mode**. In it, one chooses a random number R which fits in one block of the block cipher E_K. Then k_i is the low-order eight bits (or some other selected bits) of $E_K(R + i)$. The key to the stream cipher is the random number R and the key K.

Here is a trivial example of a self-synchronous stream cipher to illustrate the idea. Suppose $M = m_1 m_2 m_3 \ldots$, where the pieces m_i are bytes or characters. Choose an initial key byte k_1 and encipher m_1 as $c_1 = m_1 \oplus k_1$. Now define $k_i = c_{i-1}$ for $i > 1$ and encipher with the rule $c_i = m_i \oplus k_i = m_i \oplus c_{i-1}$. The point is that the (nonrepeating) key stream is generated from the previous ciphertext. Of course, if a cryptanalyst knew the enciphering rule (but not the key), he could easily decipher all but the first byte of the message by computing $m_i = c_i \oplus c_{i-1}$ for $i > 1$.

One can design a slightly better self-synchronous stream cipher by disguising the previous ciphertext piece c_{i-1} before using it to encipher the next plaintext piece. For example, one may let $k_i = E_K(c_{i-1})$, where E_K is a block cipher. Then the key for the self-synchronous stream cipher consists of the initial key piece k_1 and the key K for the block cipher E. Of course, K is far more important than k_1, because any cryptanalyst who discovers K can read all but the first message piece m_1.

Synchronous stream ciphers are simpler than self-synchronous stream ciphers. However, the latter have some advantages over the former. If ciphertext is broadcast by radio, and interference changes a few bits, then users of a synchronous stream cipher would have to resynchronize before correct deciphering could resume. But if a self-synchronous stream cipher were used, then deciphering could continue with just a few characters lost. If ciphertext is stored in a file, then the file can be deciphered only from the beginning when a synchronous stream cipher is used. But a file enciphered with a self-synchronous stream cipher can be read from any starting point—just start reading a few characters before the desired portion to get the correct initial keys—and one could even change the end of the file without disturbing the rest of it.

1.5   Protocols

Cryptographic ideas are not limited to just enciphering and deciphering text. We have already mentioned that digital signatures provide authentication. We will discuss many **protocols**, which are dialogs using cryptographic techniques to accomplish some purpose. Here are a few examples.

The Diffie-Hellman key exchange protocol allows two users to choose a common key for a symmetric cipher while someone eavesdrops on them. The eavesdropper does not learn their key from the messages sent between them.

Electronic cash is a system for purchasing goods electronically that has many of the same properties as cash. It is not a credit card. You withdraw electronic cash from your bank account. You can send it securely through computer networks. You can spend it with a merchant without communication with the bank during the transaction. Neither the bank nor the merchant will know who you are. You can spend the money only once.

A zero-knowledge proof is a dialog between two people, the Prover and the Verifier, in which the Prover convinces the Verifier that she knows a certain secret, but without revealing to the Verifier (or to an eavesdropper) any part of the secret. After the protocol concludes, neither the Verifier nor an eavesdropper could masquerade as the Prover and convince someone else that they know the secret.

Other protocols allow users to toss coins, play poker, vote or sign contracts over computer networks. The goal of these protocols is to make the exchange just as fair as if the people were together in the same room using coins, cards, ballots or pens to do these things.

1.6   Exercises

1. In the discussion of "signatures" using public-key ciphers, we said that if Alice signed a message to Bob, he could verify that it came from her by obtaining her public key E_A from a "secure source." Design a system that provides a secure source for public keys. Keep in mind that network addresses can be spoofed, so that if Bob tries to get E_A from Alice's home page he may instead receive a phony E_A from Irene the Impersonator, who actually sent the signed message. If E_A were signed by someone Bob trusted, then he could be certain it is authentic. But suppose Bob doesn't know Alice or anyone who knows her. Design a system that provides a chain of signatures of public keys which could be checked and which provides a secure source for anyone's public key.

2. Your friend has a secret file enciphered with a synchronous stream cipher using exclusive-or as the cipher function. You would like to read the file. You have made a copy of the ciphertext. One day your friend mentions that the "L" key on his terminal sticks, so that sometimes he gets "LL" when he meant "L." He says this happened once when he originally

typed the secret file, but that he has just corrected it. You run to your computer and make a new copy of the ciphertext. It is one character shorter than the copy you made earlier. Since the two files agree for the first twelve characters, and differ thereafter, you suspect that the same key stream was used to encipher the file both times. Assuming this is the case, how much of the file can you decipher? Give an explicit algorithm for deciphering the part you can decipher.

3. Derive the decryption function and comment on cryptographic security of the following encryption scheme: Let a key $= k_0, \ldots, k_{b-1}$ have b bits, and let the bits of plaintext be $m_0, m_1, m_2, \ldots$. The encryption function produces ciphertext bits $c_0, c_1, c_2, \ldots$, where $c_i = c_{i-1} \oplus k_{i \bmod b} \oplus m_i$ and c_{-1} is understood to be 0. What type of cipher is this?

4. You are the Chief Information Officer of a large organization whose n members must communicate securely. Your task is to decide whether to accomplish this goal via symmetric or asymmetric encryption. If symmetric encryption is used, each pair of members must share a unique key, and each member of the pair must store a copy of it on her workstation. If asymmetric encryption is used, each member must have a unique public key stored in a public directory on her workstation. For each type of encryption, how many copies of keys, total for all members, must be stored somewhere? Which type of encryption will you choose for your organization?

Chapter 2

Probability Theory

This chapter introduces some basic ideas from Probability Theory. The notion of probability provides a quantitative measure of our expectation of the likelihood of future events. See Feller [43] for more information about this subject.

2.1 Definitions

The reader likely has some experience with games of chance and through them has acquired an intuitive grasp of the notion of probability. The probability in this book is discrete.

Suppose an experiment has a set $X = \{x_1, \ldots, x_n\}$ of n possible outcomes. Each time the experiment is performed exactly one of the outcomes happens. Let each outcome be assigned a real number between 0 and 1, called the **probability** of that outcome. The sum of the probabilities of all of the outcomes must be 1. Write $p(x_i)$ for the probability of x_i. So, $0 \leq p(x_i) \leq 1$ for each i and $\sum_{i=1}^{n} p(x_i) = 1$.

A subset E of X is called an **event**. The event E "happens" if the outcome of the experiment is in E. The probability of an event E is defined to be the sum of the probabilities of the outcomes in E, that is, $p(E) = \sum_{x \in E} p(x)$. It is easy to see that $0 \leq p(E) \leq 1$ and that the probability that E doesn't happen is $1 - p(E)$.

In the examples in this book it often occurs that all n outcomes x_i of an experiment have equal probability. We say the outcomes are **equally likely**. In this case, we have $p(x_i) = 1/n$ for every i, and if the event E contains exactly k outcomes, then $p(E) = k/n$.

When a coin is tossed, there are two possible outcomes, Heads and Tails. If the coin is evenly balanced and well-tossed, the two outcomes are equally likely and $p(\text{Heads}) = p(\text{Tails}) = 1/2$.

Suppose we are making a known-plaintext attack on a cipher with 1,000,000 possible keys. We are given M and C and must find the key K for which

$M = D_K(C)$ or $C = E_K(M)$ (which are equivalent). If we pick one of the 1,000,000 keys, we can tell whether it is the correct key by testing whether $M = D_K(C)$. If we assume that the keys are equally likely to be chosen, each key has probability 10^{-6} of being the correct one.

The event $E_1 \cup E_2$ is the union of the two sets E_1 and E_2. The event $E_1 \cup E_2$ happens if either of the two events E_1 and E_2 happens, that is, if the outcome is in either set. The event $E_1 \cap E_2$ is the intersection of the two sets E_1 and E_2. The event $E_1 \cap E_2$ happens if both of the two events E_1 and E_2 happen, that is, if the outcome is in both sets.

Events E_1 and E_2 are called **mutually exclusive events** if they are disjoint sets, that is, $E_1 \cap E_2$ is empty. If E_1 and E_2 are disjoint, then the probability that either E_1 or E_2 happens is $p(E_1 \cup E_2) = p(E_1) + p(E_2)$. As a simple example of this principle, suppose the keys for the cipher of the preceding paragraph were 6-digit integers. Let us find the probability that the first digit of the key is either a 2 or a 5. Let E_1 be the event, "the first digit is a 2" and E_2 be the event, "the first digit is a 5." Since there are 100,000 six-digit numbers whose first digit is a 2, $p(E_1) = 100,000/1,000,000 = 0.1$. Likewise, $p(E_2) = 0.1$. Since the first digit cannot be both a 2 and a 5, the events are mutually exclusive and the answer is the sum of these two probabilities, that is, 0.2.

Suppose E_1 and E_2 are two events. Suppose $p(E_2) > 0$. We define the **conditional probability** of E_1 **given** E_2 to be $p(E_1|E_2) = p(E_1 \cap E_2)/p(E_2)$. For example, consider the cipher with six-digit integers for keys. Let E_1 be the event, "the first digit is a 2" and E_2 be the event, "the first digit is even." As above, $p(E_1) = 0.1$. Likewise, $p(E_2) = 1/2$ because half of the first digits are even. However, $E_1 \subset E_2$ because if the first digit is 2, then the first digit is even. Therefore, $E_1 \cap E_2 = E_1$ and $p(E_1 \cap E_2) = p(E_1) = 0.1$. The conditional probability is $p(E_1|E_2) = 0.1/0.5 = 0.2$. The formula defining conditional probability is often used in the form $p(E_1 \cap E_2) = p(E_1|E_2)p(E_2)$. Note that swapping E_1 and E_2 gives $p(E_1 \cap E_2) = p(E_2|E_1)p(E_1)$. Therefore, if both $p(E_1) > 0$ and $p(E_2) > 0$, then $p(E_2|E_1)p(E_1) = p(E_1|E_2)p(E_2)$. We have proved

THEOREM 2.1 Bayes's theorem
If both $p(E_1) > 0$ and $p(E_2) > 0$, then

$$p(E_1|E_2) = \frac{p(E_1)p(E_2|E_1)}{p(E_2)}.$$

Bayes's theorem provides a good way to compute $p(E_1|E_2)$ from $p(E_2|E_1)$.

Finally, we define independence. Two events E_1 and E_2 are called **independent** if $p(E_1|E_2) = p(E_1)$. When both events have positive probability, Bayes's theorem shows that this equation is equivalent to $p(E_2|E_1) = p(E_2)$. Also, from the form $p(E_1 \cap E_2) = p(E_1|E_2)p(E_2)$, we obtain the symmetric condition $p(E_1 \cap E_2) = p(E_1) \cdot p(E_2)$ for E_1 and E_2 to be independent.

The events in the example above are not independent because $E_1 \cap E_2 = E_1$ and $p(E_1 \cap E_2) = p(E_1) = 0.1 \neq p(E_1)p(E_2)$.

Let us use the same cipher with 6-digit integers as keys to give another example. Assume all 10^6 keys are equally likely. What is the probability that the first digit of the key is a 2 and the last digit of the key is a 5? Let E_1 be the event, "the first digit is a 2" and E_2 be the event, "the last digit is a 5." As in the previous paragraph, $p(E_1) = p(E_2) = 0.1$ because 100,000 keys have first digit 2 and the same number have last digit 5. Now 10,000 six-digit keys have *both* first digit 2 and last digit 5 because there are 10,000 ways to choose the other four digits. Therefore, $p(E_1 \cap E_2) = 10^4/10^6 = 0.01 = p(E_1)p(E_2)$, and the events E_1 and E_2 are independent.

2.2 The Birthday Problem

We begin with some simple results from combinatorial analysis needed to count outcomes with equal probability.

THEOREM 2.2 Multiple selections
Suppose there are n_1 distinct elements $a_1, \ldots, a_{n_1}$; n_2 distinct elements $b_1, \ldots, b_{n_2}$, etc.; up to n_s distinct elements $x_1, \ldots, x_{n_s}$. Then one can form $n_1 n_2 \cdots n_s$ ordered s-tuples $(a_{i_1}, b_{i_2}, \ldots, x_{i_s})$ containing one element of each kind.

PROOF   Use induction on s. If $s = 2$, arrange the pairs in a $n_1 \times n_2$ matrix with entry (a_i, b_j) in the i-th row and j-th column. Each pair appears exactly once and there are $n_1 n_2$ pairs.

Let $s > 2$ and suppose the theorem has already been proved for $s - 1$. Then one can form $n_2 \cdots n_s$ ordered $s - 1$-tuples $(b_{i_2}, \ldots, x_{i_s})$ containing one element of each kind other than the first kind. Consider these $s - 1$-tuples to be elements of a new kind. By the case $s = 2$ there are $n_1 \cdot n_2 \cdots n_s$ pairs consisting of an a_i and an element of the new kind. But these pairs are just ordered s-tuples $(a_{i_1}, b_{i_2}, \ldots, x_{i_s})$ containing one element of each kind.   ∎

THEOREM 2.3 Probability of no repetition
If an experiment with n equally likely and independent outcomes is performed k times, where $1 \leq k \leq n$, then the probability that all k outcomes differ is

$$\frac{n(n-1)\cdots(n-k+1)}{n^k}.$$

PROOF   To count the number of possible outcomes when the experiment is performed k times, apply the preceding theorem with $s = k$ and $n_1 = n_2 = \cdots = n_s = n$. The total number of possible outcomes is then n^k.

To count the number of outcomes with all k outcomes different when the experiment is performed k times, apply the preceding theorem with $s = k$ and $n_1 = n$, $n_2 = n - 1$, ..., $n_s = n - k + 1$. This is correct because there are n allowed outcomes for the first performance of the experiment. Then its outcome may not be repeated, so there are $n - 1$ allowed outcomes for the second performance of the experiment, and so forth. The total number of possible outcomes without repetition is then $n(n - 1) \cdots (n - k + 1)$. The probability of all outcomes differing is the quotient in the statement of the theorem. ∎

We now consider variations of the following problem which have important applications in certain attacks on cryptographic functions.

What is the smallest positive integer k so that the probability is $\geq 1/2$ that at least two people in a group of k people have the same birthday?

We begin by making a couple of simplifications. First, we ignore Leap Year's Day and assume every year has 365 days. Then we assume that the birth rate is constant throughout the year so that every one of the 365 days is equally likely to be a birthday. We also assume that the birthdays of different people are independent.

We will first find the probability $Q(k)$ that no two people in a group of k people have the same birthday, that is, all k people have different birthdays. We apply Theorem 2.3. The experiment is finding the person's birthday. There are $n = 365$ possible outcomes. Repeating the experiment k times means finding the birthdays of k different people. Theorem 2.3 tells us that the probability is

$$Q(k) = \frac{365 \times 364 \times \cdots \times (365 - k + 1)}{(365)^k}.$$

Thus, the probability that at least two of the k people have the same birthday is

$$P(k) = 1 - Q(k) = 1 - \frac{365 \times 364 \times \cdots \times (365 - k + 1)}{(365)^k}.$$

The following table shows how $P(k)$ increases.

Table 1. Probability that at least two of k people have the same birthday.

k	$P(k)$	k	$P(k)$	k	$P(k)$
2	0.0027	15	0.2529	28	0.6544
3	0.0082	16	0.2836	29	0.6809
4	0.0163	17	0.3150	30	0.7063
5	0.0271	18	0.3469	31	0.7304
6	0.0404	19	0.3791	32	0.7533
7	0.0562	20	0.4114	33	0.7749
8	0.0743	21	0.4436	34	0.7953
9	0.0946	22	0.4756	35	0.8143
10	0.1169	23	0.5072	36	0.8321
11	0.1411	24	0.5383	37	0.8487
12	0.1670	25	0.5687	38	0.8640
13	0.1944	26	0.5982	39	0.8782
14	0.2231	27	0.6268	40	0.8912

These values (and a few more) are plotted in Figure 2.1.

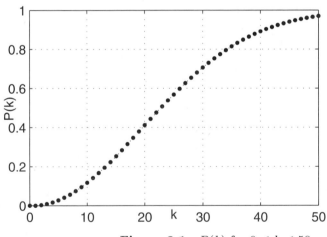

Figure 2.1 $P(k)$ for $0 \le k \le 50$.

The table shows that the answer to the question is that 23 is the smallest size of a group of people so that, with probability $\ge 1/2$, at least two have the same birthday.

Warren Weaver tells a relevant anecdote on page 135 of his book [119]. During World War II, he was explaining the birthday problem to some high-ranking military men at a dinner. They didn't believe that with only 22 or 23 people in a room, there was a 50% chance of two people having the same birthday. One officer noted that there were exactly 22 people at their table and proposed a test of the theory. Each person at the table stated his birthday. They were all different. Weaver was disappointed. Then their waitress piped up, "Excuse me, but I am the twenty-third person in this room and my birthday is May seventeenth, the same as the General over there."

Now we generalize the birthday problem, which is the case $n = 365$ of the following problem.

Suppose $1 \leq k \leq n$ and we choose k integers between 1 and n so that the choices are independent and all n integers are equally likely to be chosen. What is the probability $P(n, k)$ that at least two of the k integers are the same? What value of k makes this probability closest to $1/2$?

Reasoning just as for birthdays, we find

$$P(n, k) = 1 - \frac{n(n-1)\cdots(n-k+1)}{n^k}.$$

Write this as

$$P(n, k) = 1 - \left(1 - \frac{1}{n}\right)\left(1 - \frac{2}{n}\right) \times \cdots \times \left(1 - \frac{k-1}{n}\right).$$

Now we have answered the first question. To answer the second, we must estimate the probability function $P(n, k)$. To do this, note that $1 - x \leq e^{-x}$ for all $x \geq 0$ and $1 - x \approx e^{-x}$ when x is small.

This approximation is illustrated by the graph in Figure 2.2.

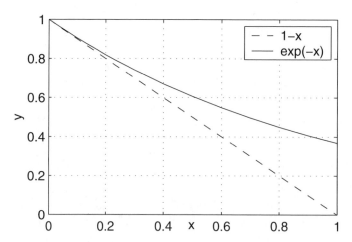

Figure 2.2 Graphs of $y = 1 - x$ and $y = e^{-x}$.

We will use this approximation for each of the factors $1 - (r/n)$ in the product above. We have $1 \leq r \leq k - 1$; so, we are assuming that every $x = r/n$ is small. This will be so provided k is small compared to n.

The approximation $1 - (r/n) \approx e^{-(r/n)}$ gives

$$P(n, k) \approx 1 - e^{-1/n} e^{-2/n} e^{-3/n} \times \cdots \times e^{-(k-1)/n}$$

or

$$P(n, k) \approx 1 - e^{-(1/n + 2/n + 3/n + \cdots + (k-1)/n)}$$

or

$$P(n, k) \approx 1 - e^{-k(k-1)/(2n)}.$$

We will have $P(n, k) = 1/2$ when $1/2 = 1 - e^{-k(k-1)/(2n)}$ or $2 = e^{k(k-1)/(2n)}$, that is, when $\ln 2 = k(k-1)/(2n)$.

We make another approximation. When k is large, the percentage difference between k and $k - 1$ is small, and we may approximate $k - 1 \approx k$. This gives $k^2 \approx 2n \ln 2$ or

$$k \approx \sqrt{2(\ln 2)n} \approx 1.18\sqrt{n}.$$

Note that we have assumed that "k is large" and "k is small compared to n." This means that n must be quite large for the approximations to work.

For $n = 365$, we find $k \approx 1.18\sqrt{365} \approx 22.54$, or $k \approx 23$.

In fact, this is correct, and we see that $n = 365$ and $k = 23$ are large enough.

We state the general result as a theorem. The notation "<<" means "is much less than."

THEOREM 2.4 The birthday paradox
Suppose $1 << k << n$ and we choose k integers between 1 and n so that the choices are independent and all n integers are equally likely to be chosen. The probability $P(n, k)$ that at least two of the k integers are the same is approximately $1 - e^{-k(k-1)/(2n)}$. The value of k that makes this probability closest to $1/2$ is approximately $\sqrt{2(\ln 2)n} \approx 1.18\sqrt{n}$.

We now study the overlap between two sets. Let $1 \leq k \leq n$. Suppose we choose *two* sets of k integers between 1 and n so that all $2k$ choices are independent and all n integers are equally likely to be chosen every time. What is the probability $R(n, k)$ that the two sets overlap, that is, at least one of the n values appears in both sets?

We assume k is small enough ($k < \sqrt{n}$) so that the k integers chosen from each set are probably all different. (A few duplicates won't hurt this analysis.)

The probability that one given element of the first set does not match any element of the second set is $(1 - 1/n)^k$.

The probability that the two sets are disjoint is

$$\left(\left(1 - \frac{1}{n} \right)^k \right)^k = \left(1 - \frac{1}{n} \right)^{k^2}$$

so $R(n, k) = 1 - (1 - 1/n)^{k^2}$.

Using $1 - x \approx e^{-x}$, we get

$$R(n, k) \approx 1 - \left(e^{-1/n}\right)^{k^2} = 1 - e^{-k^2/n}.$$

We will have $R(n, k) = 1/2$ when $\frac{1}{2} = 1 - e^{-k^2/n}$, or $2 = e^{k^2/n}$, or $\ln 2 = k^2/n$
or

$$k = \sqrt{(\ln 2)n} \approx 0.83\sqrt{n}.$$

We state this result as a theorem.

THEOREM 2.5 Probability of overlapping sets

*Suppose $1 << k << n$ and we choose two sets of k integers between 1 and n
so that all $2k$ choices are independent and all n integers are equally likely to be
chosen every time. The probability $R(n, k)$ that the two sets overlap, that is,
at least one of the n values appears in both sets, is approximately $1 - e^{-k^2/n}$.
The value of k that makes this probability closest to 1/2 is approximately
$\sqrt{(\ln 2)n} \approx 0.83\sqrt{n}$.*

2.3 Random Variables

The **sample space** is the set of all possible outcomes E, each of which has
a probability $p(E)$. A **random variable** is a real-valued function r defined
on a sample space. If $x_1, x_2, \ldots$ are all of the possible values of $r(E)$ (in this
book, this set will be finite), then the **probability distribution** of r is the
function f defined by $f(x_i) = p(r(E) = x_i)$, the probability that $r(E) = x_i$.
That is, $f(x_i)$ is the sum of $p(E)$ for all outcomes E for which $r(E) = x_i$.
Several random variables $r_1, \ldots, r_k$ are called **mutually independent** if for
any possible values $y_1, \ldots, y_k$ that they could assume, then the probability
that $r_i(E) = y_i$ for every $1 \leq i \leq k$ equals the product

$$p(r_1(E) = y_1)p(r_2(E) = y_2) \cdots p(r_k(E) = y_k).$$

Example 2.1

Suppose we toss a fair coin n times and observe the sequence of heads and
tails. The sample space has 2^n outcomes E, each an n-tuple of heads and tails,
and each having probability 2^{-n}. Define the random variable $r(E)$ to be the
number of heads in outcome E. Then $r(E)$ is always an integer between 0 and
n. The probability distribution f of r is defined for each integer i between 0
and n, and $f(i)$ is the probability that exactly i heads will appear if a coin is
tossed n times. This number is easily computed to be 2^{-n} times the number
of ways of choosing i tosses from n tosses to show heads. The latter number is
the binomial coefficient, and so

$$f(i) = 2^{-n} \times \binom{n}{i} = \frac{n(n-1)(n-2)\cdots(n-i+1)}{2^n i!}.$$

Statisticians describe a probability distribution of a random variable in a concise way by giving some typical values of it. One such value is the median. The **median** of the probability distribution f of the random variable r is a value x_m assumed by $r(E)$ so that $p(r(E) < x_m) \leq 1/2$ and also $p(r(E) > x_m) \leq 1/2$. That is, the median x_m is chosen so that the probability of $r(E)$ exceeding or falling short of x_m is as close to $1/2$ as possible. The median is the "middle value" of $r(E)$.

Another typical value of a random variable and one with more useful mathematical properties than the median is the mean, or average, or expected value.

DEFINITION 2.1   *The* **mean** *or* **expected value** $\mathbf{E}(r)$ *of a random variable* r *with values* $x_1, x_2, \ldots$ *and probability distribution* f *is*

$$\mu = \mathbf{E}(r) = \sum_i x_i f(x_i).$$

If F is a real-valued function defined on the real numbers and r is a random variable, then $F(r)$ is another random variable, having value $F(r(E))$ on outcome E. Its expected value is

$$\mathbf{E}(F(r)) = \sum_i F(x_i) f(x_i).$$

The t-**th moment** of a random variable r is the expected value of r^t. The **variance** $\mathbf{Var}(r)$ of a random variable r with expected value μ is the second moment of $r - \mu$, that is,

$$\mathbf{Var}(r) = \mathbf{E}((r - \mu)^2) = \mathbf{E}(r^2) - \mu^2.$$

(The last equation is an easy theorem.) The nonnegative square root of the variance of r is called the **standard deviation** of r. It is a measure of how much $r(E)$ varies from the mean μ.

One can prove the following theorem easily from the definitions.

THEOREM 2.6 Mean and variance of linear combinations and sums
If a and b are constants and r and s are random variables, then
 1. $\mathbf{E}(ar + b) = a\mathbf{E}(r) + b$
 2. $\mathbf{Var}(ar + b) = a^2 \mathbf{Var}(r)$
 3. $\mathbf{E}(r + s) = \mathbf{E}(r) + \mathbf{E}(s)$, *and*
 4. *if r and s are mutually independent, then* $\mathbf{Var}(r + s) = \mathbf{Var}(r) + \mathbf{Var}(s)$.

A small variance means that large deviations from the mean are unlikely. The following theorem makes this statement more precise.

THEOREM 2.7 Chebyshev's inequality
Let r be a random variable with mean μ and variance v. For any $t > 0$,

$$p(|r(E) - \mu| \geq t) \leq \frac{v}{t^2}.$$

PROOF   The variance is defined as a sum

$$v = \mathbf{Var}(r) = \mathbf{E}((r - \mu)^2) = \sum_i (x_i - \mu)^2 f(x_i)$$

of nonnegative terms. The sum will not increase if we delete from it all terms for which $|r(E) - \mu| < t$. Hence,

$$v \geq {\sum_i}' (x_i - \mu)^2 f(x_i),$$

where the $\sum'$ indicates that summation extends only over those i for which $|r(E) - \mu| \geq t$. Then it is clear that

$$v \geq {\sum_i}' (x_i - \mu)^2 f(x_i) \geq t^2 {\sum_i}' f(x_i) = t^2 p(|r(E) - \mu| \geq t).$$

∎

There are several theorems in probability theory called "laws of large numbers" that say roughly that if an experiment is performed many times, then large deviations from the expected value are unlikely. For example, if one tosses a true coin a million times, then the number of heads obtained will probably not be far from 500,000. Here is a simple theorem that says this in a precise way.

THEOREM 2.8 A law of large numbers
Let $r_1, r_2, \ldots$ be a sequence of mutually independent random variables with the same probability distribution, and therefore the same mean μ and variance v. Define a new sequence of random variables $s_n = \sum_{i=1}^n r_i$. Then for every $\epsilon > 0$, we have

$$\lim_{n \to \infty} p(|(s_n/n) - \mu| > \epsilon) = 0.$$

PROOF   By Theorem 2.6, $\mathbf{E}(s_n) = n\mu$ and $\mathbf{Var}(s_n) = nv$. By Chebyshev's inequality, we have for every $t > 0$,

$$p(|s_n - n\mu| > t) \leq nv/t^2.$$

When $t > \epsilon n$, the left side is less than $v/(\epsilon^2 n)$, which tends to 0 as $n \to \infty$.

∎

Example 2.2

Let us continue with Example 2.1 and compute the statistics just defined. To determine the median, recall that the binomial coefficients $\binom{n}{i}$ are symmetric about $n/2$: $\binom{n}{i} = \binom{n}{n-i}$. They increase as i increases from 0 to $n/2$ and decrease as i increases from $n/2$ to n. If n is even, there will be one median, namely, $n/2$. If n is odd, there will be two medians, namely, $(n \pm 1)/2$. Thus, the most likely number of heads when a coin is tossed n times is $n/2$ (or the nearest integers to this number if it is not an integer).

The average or mean or expected number of heads h in n tosses is

$$\mu = \mathbf{E}(h) = \sum_{i=0}^{n} i f(i) = \sum_{i=0}^{n} i 2^{-n} \binom{n}{i} = 2^{-n} n 2^{n-1} = \frac{n}{2}.$$

The mean is essentially equal to the median for this particular probability distribution.

The variance of the number of heads is

$$\mathbf{Var}(h) = \sum_{i=0}^{n} (i - \mu)^2 f(i) = \sum_{i=0}^{n} 2^{-n} (i - \mu)^2 \binom{n}{i}$$

$$\mathbf{Var}(h) = \sum_{i=0}^{n} 2^{-n} i^2 \binom{n}{i} - \mu^2 = (n^2 + n)/4 - n^2/4 = n/4;$$

so, the standard deviation is $\sqrt{n}/2$.

We can apply the law of large numbers to this problem if we define random variables r_i. Let r_i be defined on the outcome of the i-th coin toss with value 1 if a head appears and 0 if a tail appears. Then the random variable s_n of Theorem 2.8 is the random variable h above and the theorem says that for every $\epsilon > 0$, we have

$$\lim_{n \to \infty} p(|(h/n) - (n/2)| > \epsilon) = 0.$$

We end with one more example, one which describes a situation similar to one we will see in the chapters on quadratic residues and elliptic curves.

Example 2.3

Suppose we again toss a fair coin n times and observe the sequence of heads and tails. The sample space has 2^n outcomes E, each an n-tuple of heads and tails, and each having probability 2^{-n}. Define the random variable $r(E)$ to be the number of heads in outcome E minus the number of tails. Then $r(E)$ is always an integer between $-n$ and n. The probability distribution f of r is defined for each integer i between $-n$ and n, and $f(i)$ is the probability that exactly i more heads than tails will appear if a coin is tossed n times. (A value $i < 0$ means that there were $|i|$ more tails than heads.) Suppose that h heads and t tails appear in one outcome. Then $h + t = n$ and $i = r(E) = h - t$. We find that $h = (n + i)/2$. (As h is an integer, this shows that n and i are either both odd or both even. In particular, if n is odd, then i can never be 0 because 0 is even.) The probability of having i more heads than tails in n tosses is the same as the probability of having exactly $(n+i)/2$ heads. As in Example 2.1, this probability is seen to be $f(i) = 2^{-n}\binom{n}{(n+i)/2}$ provided i has the same parity (odd or even) as n, and $f(i) = 0$ if i and n have opposite parity.

The median is easily seen to be 0 because of the symmetry property

$$\binom{n}{(n+i)/2} = \binom{n}{n-(n+i)/2} = \binom{n}{(n-i)/2}.$$

If n is even, then 0 will be the only median. If n is odd, both $+1$ and -1 will be medians.

The expected value of r is

$$\mu = \mathbf{E}(r) = \sum_{i=-n}^{n} if(i) = \sum_{\substack{i=-n \\ i+n \text{ even}}}^{n} 2^{-n} i \binom{n}{(n+i)/2} = 0,$$

by the symmetry property. Thus, as common sense suggests, the numbers of heads and tails will balance on average.

We could have derived the mean from the mean of the number h of heads computed in Example 2.2 and Theorem 2.6. Since $r = 2h - n$,

$$\mathbf{E}(r) = \mathbf{E}(2h - n) = 2\mathbf{E}(h) - n = 2(n/2) - n = 0.$$

Likewise, the variance of the number of heads minus the number of tails is

$$\mathbf{Var}(r) = \sum_{i=-n}^{n} (i - \mu)^2 f(i) = \mathbf{Var}(2h - n) = 4\mathbf{Var}(h) = 4\left(\frac{n}{4}\right) = n;$$

so, the standard deviation is $\sqrt{n}$.

We can apply the law of large numbers to this problem if we define r_i on the outcome of the i-th coin toss to have value 1 if a head appears and -1 if a tail appears. Then the random variable s_n of Theorem 2.8 is the random variable r above and the theorem says that for every $\epsilon > 0$, we have

$$\lim_{n \to \infty} p(|(r/n) - 0| > \epsilon) = 0.$$

2.4   *Exercises*

1. Dice are six-sided cubes with the numbers 1 through 6 on the faces. When a die is tossed, each of the six numbers has equal probability of appearing. Suppose two dice are tossed.

   a. What is the probability that a 2 and 5 will show?

   b. What is the probability that the sum of the two numbers will be 9?

   c. If the sum of the two numbers is 5, what is the probability that one of them will be a 1?

   d. What is the probability that the two numbers will be different?

   e. Find the mean, median, variance and standard deviation of the sum of the two numbers.

2. A bag contains 1000 white balls labeled 1, 2, ..., 1000 and another bag contains 1000 black balls labeled 1, 2, ..., 1000.

   a. Suppose 20 balls are removed from each bag. What is the probability that two of the 40 balls have the same label?

   b. Approximately what is the least number k of balls you have to remove from each bag (the same number k of balls from each bag) to make the probability of getting two balls with the same label greater than 1/2?

3. What is the probability that all the students in a class of 35 have different birthdays?

4. A professor posts grades for a class using the last four digits of each student's college identification number. For what size of class is there an even chance that two students have the same four-digit code?

5. Assuming that each month has the same probability of being born in it, what is the probability that two people in a family of five were born in the same month?

6. Assume that the city of Lafayette has 10^5 people, and that all of these people walk past a giant bin and each one drops in a slip of paper having that person's unique identification. The contents of the bin are mixed, and then all the people march by again, each drawing one slip out of the bin.

   a. What is the (approximate) probability that nobody draws his/her own slip?

   b. Same question as part a., but this time every person puts the slip back immediately after drawing and reading it.

7. Prove Theorem 2.6.

Chapter 3

Divisibility and Arithmetic

This chapter concerns the simplest part of number theory, which is the study of integers or whole numbers. We also tell how to perform arithmetic with the very large integers used in cryptography. The reader may consult one of the many excellent number theory texts such as [78], [98] or [51] for more details or alternate proofs. The text by Rosen [99] has the same computational flavor as this book.

3.1   Divisibility

DEFINITION 3.1   When a and b are integers and $a \neq 0$ we say a **divides** b, and write $a|b$, if b/a is a whole number.

This is nearly the same as saying that a divides b if there is a whole number k so that $b = ka$. The only difference is that this definition would allow 0 to divide 0, while 0 does not divide 0 according to Definition 3.1.

THEOREM 3.1 Transitivity of divisibility
Let a, b and c be integers. If $a|b$ and $b|c$, then $a|c$.

PROOF   By hypothesis, the two quotients b/a and c/b are whole numbers. Therefore their product, $(b/a) \times (c/b) = c/a$, is a whole number, which means that $a|c$.   ∎

Theorem 3.1 says that the relation "divides" is transitive.

THEOREM 3.2 Divisibility of linear combinations
Let a, b, c, x and y be integers. If $a|b$ and $a|c$, then $a|bx + cy$.

PROOF   We are given that the two quotients b/a and c/a are whole numbers. Therefore the linear combination $(b/a) \times x + (c/a) \times y = (bx + cy)/a$ is a whole number, which means that $a|(bx + cy)$.   ∎

THEOREM 3.3 The division algorithm
Suppose $a > 0$ and b are two integers. Then there exist two unique integers q and r such that $0 \leq r < a$ and $b = aq + r$.

PROOF   First we show that q and r exist. Let q be the greatest integer $\leq b/a$. Then $b/a = q + \alpha$, where $0 \leq \alpha < 1$. We have $b = a(q + \alpha) = aq + a\alpha$. Now $r = a\alpha$ must be an integer because it is the difference $b - aq$ of two integers. Also, $0 \leq r < a$ because $0 \leq \alpha < 1$.

Now we show that q and r are unique. Suppose that we had $b = aq + r$ with $0 \leq r < a$ and also $b = aq' + r'$ with $0 \leq r' < a$. Subtracting the two equations and dividing by a gives

$$q - q' = (r' - r)/a.$$

Subtracting $a > r \geq 0$ from $0 \leq r' < a$ and dividing by a gives

$$-1 < (r' - r)/a < 1.$$

But $(r' - r)/a = q - q'$ is an integer and the only integer between -1 and 1 is 0. Therefore $q = q'$ and $r = r'$.   ∎

Example 3.1

In Theorem 3.3 let $a = 17$ and $b = 165$. Then $q = 9$ and $r = 12$. We have $165 = 17 \times 9 + 12$.

DEFINITION 3.2   *The integers q and r in Theorem 3.3 are called the* **quotient** *and* **remainder** *when b is divided by a.*

We use the notation $\lfloor x \rfloor$, the **floor** of x, to mean the largest integer $\leq x$, and $\lceil x \rceil$, the **ceiling** of x, to mean the smallest integer $\geq x$. Thus, $\lfloor 5 \rfloor = 5$, $\lfloor 3.14 \rfloor = 3$, $\lfloor -2.7 \rfloor = -3$, $\lceil 5 \rceil = 5$, $\lceil 3.14 \rceil = 4$ and $\lceil -2.7 \rceil = -2$.

With this notation, the quotient q in the definition and in Theorem 3.3 may be written $q = \lfloor b/a \rfloor$. We also use the notation $b \bmod a$ for the remainder r.

We say the integer n is **even** if the remainder is 0 when n is divided by 2, and call n **odd** if this remainder is 1.

3.2 *Arithmetic with Large Integers*

The construction and cryptanalysis of cryptographic algorithms require arithmetic with large integers. These algorithms will run faster if the basic arithmetic operations can be performed swiftly. Computer hardware has a fixed maximum size, such as $2^{31} - 1$, for the integers it can handle directly. Cryptographic algorithms use much larger integers than this hardware maximum value. In this section we explain how computers represent larger integers and how to perform arithmetic with them efficiently. The reader will find different presentations of the material of this section in books by Knuth [56], Rosen [99] and Riesel [96].

Probably because we have ten fingers, we use decimal notation to represent numbers. The character string "6218" represents the integer 6218 with value $6 \times 10^3 + 2 \times 10^2 + 1 \times 10^1 + 8$. Computers usually use binary notation to represent numbers internally. We hope the use of bases 10 and 2 for positional number systems are familiar to the reader. In fact any integer > 1 can be used as a base.

THEOREM 3.4 Positional number systems
Let b be an integer greater than 1. Let n be a positive integer. Then n has a unique representation in the form

$$n = \sum_{i=0}^{k} d_i b^i,$$

where k is a positive integer, the d_i are integers in $0 \le d_i \le b - 1$ and $d_k \ne 0$.

The number b is called the **base** or **radix** of the number system. The numbers d_i for $i = k, k - 1, \ldots, 0$ are called the **digits** in base b of n. The left-most digit d_k is called the first digit or leading digit or most significant digit and the right-most digit d_0 is called the last digit or trailing digit or least significant digit.

PROOF   We use the division algorithm (Theorem 3.3) to construct the representation. First we divide n by b to get $n = bq_0 + d_0$ with $0 \le d_0 \le b-1$. If $q_0 > 0$, divide q_0 by b to get $q_0 = bq_1 + d_1$ with $0 \le d_1 \le b-1$. Continue this process with $q_i = bq_{i+1} + d_{i+1}$ and $0 \le d_{i+1} \le b - 1$ until we get a remainder $q_k = 0$. This condition must occur eventually since

$$n > q_0 > q_1 > \cdots \ge 0,$$

and every decreasing sequence of positive integers must end. Replace q_0 with $bq_1 + d_1$ in $n = bq_0 + d_0$ to obtain

$$n = b(bq_1 + d_1) + d_0 = q_1 b^2 + d_1 b^1 + d_0.$$

When we replace q_1 with $bq_2 + d_2$ and so on to $q_{k-1} = 0b + d_k$, we get the representation in the statement of the theorem.

Now we show the representation is unique. Suppose n had the two representations

$$n = \sum_{i=0}^{k} d_i b^i = \sum_{i=0}^{k} e_i b^i,$$

where k is a positive integer, the d_i and e_i are integers with $0 \le d_i \le b-1$ and $0 \le e_i \le b-1$, and we may have added high-order zero digits to make both sums have $k+1$ digits. If the two representations differ, then there is a least j in $0 \le j \le k$ so that $d_j \ne e_j$. Subtracting the expressions and factoring out a b^j we find

$$0 = \sum_{i=0}^{k}(d_i - e_i)b^i = \sum_{i=j}^{k}(d_i - e_i)b^i = b^j \sum_{i=0}^{k-j}(d_{i+j} - e_{i+j})b^i,$$

and hence

$$\sum_{i=0}^{k-j}(d_{i+j} - e_{i+j})b^i = 0,$$

so that

$$d_j - e_j = -\sum_{i=1}^{k-j}(d_{i+j} - e_{i+j})b^i = -b\sum_{i=1}^{k-j}(d_{i+j} - e_{i+j})b^{i-1}.$$

This shows that b divides $d_j - e_j$. If we subtract $b - 1 \ge e_j \ge 0$ from $0 \le d_j \le b - 1$, we find $-b + 1 \le d_j - e_j \le b - 1$. The only multiple m of b in $-b + 1 \le m \le b - 1$ is $m = 0$. Therefore, $d_j = e_j$. This contradicts our assumption that the two representations differ. It follows that the base b representation is unique. ■

Some special cases of this representation include decimal ($b = 10$), binary ($b = 2$), octal ($b = 8$) and hexadecimal ($b = 16$). The symbols 0, 1, 2, 3, 4, 5, 6, 7, 8, 9, A, B, C, D, E, F are used for the 16 hexadecimal digits. When several bases are being used, the base b in the representation of Theorem 3.4 is indicated as a subscript: $n = (d_k d_{k-1} \ldots d_1 d_0)_b$. Binary digits are called **bits**.

Integers greater than the natural word size are stored in arrays with a fixed number of bits per word. It would be wasteful memory usage to store only one bit per word. On the other hand, it would be difficult to perform arithmetic on large numbers if each word were filled completely with bits of the large integer. A standard compromise often uses all but two bits of each word to store bits of large numbers. For example, many libraries of procedures for arithmetic with large integers pack 30 bits into each 32-bit word.

Sometimes it is necessary to convert a number from one representation to another. Most computers use binary to represent numbers within them. Humans often prefer the decimal form. When a number is input to a computer program by a human, it is often converted from decimal to binary. Likewise, output procedures convert from binary to decimal for human consumption.

It is easy to convert a number from one base to another when both bases are powers of 2 (or both are powers of some other number). In this case the bits just need to be regrouped, and this is easily done with shift operations on a binary computer. For example, to convert a number from binary to hexadecimal, group the bits in blocks of four bits each, starting from the low-order bit, and replace 0000 by 0, 0001 by 1, etc., 1110 by E, 1111 by F. It may be necessary to prepend up to three high-order 0's to form the high-order block of four bits. To convert from octal to binary, start at the low-order octal digit and work to the left replacing 0 by 000, 1 by 001, etc., 7 by 111.

Conversion of a number from base B to base b is more complicated when the two bases are not powers of the same integer. Say the two representations are

$$\sum_{i=0}^{K} D_i B^i = \sum_{i=0}^{k} d_i b^i.$$

We are given the digits D_K, D_{K-1}, ..., D_0 and want to find the digits d_k, d_{k-1}, ..., d_0. We assume we can perform arithmetic in one of the two bases. Humans can do this in base 10, while most computers work in binary. If one knows how to divide using base B arithmetic, then the conversion algorithm is repeated division by b as in the first part of the proof of Theorem 3.4.

Here is the algorithm in pseudocode. It is often used to output a binary number in decimal notation ($b = 10$).

[Conversion from base B to base b using base B arithmetic]
Input: D_K, D_{K-1}, ..., D_0, the base B digits of n.
Output: d_k, d_{k-1}, ..., d_0, the base b digits of n.
Since we can do base B arithmetic, we can work with n as a number
   in that base.

```
i = 0
while (n > 0) {
        d_i = n mod b
        n = ⌊n/b⌋
        i = i + 1
        }
```

Example 3.2

Convert 99_{10} from base 10 to base 8.

   This table shows the progress of the algorithm as a snapshot taken right after the remainder step $d_i = n \bmod 8$.

i	n	d_i
0	99	3
1	12	4
2	1	1

Thus, the digits are 1, 4, 3, and we have $99_{10} = 143_8$.

If one knows how to add and multiply using base b arithmetic, then the conversion algorithm is to use base b arithmetic to evaluate the polynomial $\sum_{i=0}^{K} D_i B^i$ in the form

$$((\cdots(D_K B + D_{K-1})B + \cdots)B + D_1)B + D_0.$$

The pseudocode for this algorithm is quite simple. It is used to input decimal numbers ($B = 10$) to a program.

[Conversion from base B to base b using base b arithmetic]
Input: $D_K, D_{K-1}, \ldots, D_0$, the base B digits of n.
Output: $d_k, d_{k-1}, \ldots, d_0$, the base b digits of n.
Since we can do base b arithmetic, we can just return n as a number.

```
n = 0
for (i = K down to i = 0) {
        n = n * B + Di
        }
return n
```

Example 3.3

Convert 107_8 from base 8 to base 10.
    This table shows the progress of the algorithm as a snapshot taken at the end of each pass of the for loop. Note that $K = 2$ because 107_8 has three digits.

i	n	D_i
2	1	1
1	8	0
0	71	7

Thus, $107_8 = 71_{10}$.

The basic operations of arithmetic are addition, subtraction, multiplication and division. In order to perform these operations on large integers we represent the numbers in a convenient base with their digits stored in arrays. The first three operations use the algorithms you learned in elementary school. The algorithms given here are the "conventional" ones.
    Suppose we use base b and we wish to add $A = \sum_{i=0}^{k} a_i b^i$ to $B = \sum_{i=0}^{m} b_i b^i$. If $k \neq m$, prepend enough leading 0 digits to the shorter number to give the

two numbers the same length. After this has been done, assume the problem is to add $A = \sum_{i=0}^{k} a_i b^i$ to $B = \sum_{i=0}^{k} b_i b^i$. Call the sum $C = \sum_{i=0}^{k+1} c_i b^i$. Note that the sum might have one more digit than the summands. The addition algorithm is to add corresponding digits of A and B to form each digit of C, and carry a 1 if the digit sum is $\geq b$. Here is the algorithm.

[Addition: $C = A + B$ using base b arithmetic]
Input: The base b digits of A and B.
Output: The base b digits of $C = A + B$.

```
carry = 0
for (i = 0 to k) {
        c_i = a_i + b_i+ carry
        if (c_i < b) { carry = 0 }
        else { carry = 1; c_i = c_i - b }
        }
c_{k+1} = carry
```

Note that in the second line of the for loop, we must have $0 \leq c_i < 2b$ because $0 \leq a_i \leq b - 1$, $0 \leq b_i \leq b - 1$ and carry is either 0 or 1. Thus we need to subtract at most one b from c_i (in the else line) to get it into the legal range for digits. The steps of the for loop are executed no more than $k + 1$ times.

Now suppose we wish to subtract $B = \sum_{i=0}^{m} b_i b^i$ from $A = \sum_{i=0}^{k} a_i b^i$. If $k \neq m$, add enough leading to the shorter number to give the two numbers the same length. After this has been done, assume the problem is to subtract $B = \sum_{i=0}^{k} b_i b^i$ from $A = \sum_{i=0}^{k} a_i b^i$. Assume that $A \geq B$. If this is not true, then the sum is negative with absolute value $B - A$. We have not discussed a way to handle signed numbers. If we allow the difference to have a minus sign, then we should allow A and B to have signs as well. We leave the problem of arithmetic with signed numbers to the reader. Call the difference $C = \sum_{i=0}^{k} c_i b^i$. The subtraction algorithm is to subtract corresponding digits of A and B to form each digit of C, and borrow a 1 if the digit difference is negative. Here is the algorithm in pseudocode.

[Subtraction: $C = A - B$ using base b arithmetic]
Input: The base b digits of A and B.
Output: The base b digits of $C = A - B$.

```
borrow = 0
for (i = 0 to k) {
        c_i = a_i - b_i- borrow
        if (c_i < 0) { borrow = 1; c_i = c_i + b }
        else { borrow = 0 }
        }
if (borrow ≠ 0) Error:  A < B
```

Note that in the second line of the `for` loop, we must have $-b \le c_i < b$ because $0 \le a_i \le b - 1$, $0 \le b_i \le b - 1$ and `borrow` is either 0 or 1. Thus we need to add at most one b to c_i (in the `if` line) to get it into the legal range for digits. The steps of the `for` loop are executed no more than $k + 1$ times.

The product of a k-digit integer times an m-digit integer has either $k + m$ or $k + m - 1$ digits (or is zero). Suppose we wish to multiply $A = \sum_{i=0}^{k-1} a_i b^i$ times $B = \sum_{i=0}^{m-1} b_i b^i$. Call the product $C = \sum_{i=0}^{k+m-1} c_i b^i$. Note that the high-order digit might be 0. The elementary school method forms partial products $b_i \times A$, shifts their digits into appropriate columns and adds the shifted partial products. In a computer, it saves space to do the addition concurrently with the multiplication. Here is the algorithm in pseudocode.

[Multiplication: $C = A \times B$ using base b arithmetic]
Input: The base b digits of A and B.
Output: The base b digits of $C = A \times B$.

```
carry = 0
for (i = 0 to k + m − 1) { c_i = 0 }
for (i = 0 to k − 1) {
        carry = 0
        for (j = 0 to m − 1) {
                t = a_i × b_j + c_{i+j}+ carry
                c_{i+j} = t mod b
                carry = ⌊t/b⌋
        c_{m+i+1} = carry }
        }
```

One can show by induction that in the second line of the inner `for` loop, we must have $0 \le t < b^2$ because $0 \le a_i \le b - 1$, $0 \le b_j \le b - 1$, $0 \le c_{i+j} \le b - 1$, and $0 \le$ `carry` $\le b - 1$. Each step of the inner `for` loop is executed km times.

The last operation of arithmetic is division. The elementary school "algorithm" for division is really not an algorithm because one must guess each digit of the quotient, and sometimes the guess is wrong. One way to improve the guess is explained in Knuth [56]. The trick is to "normalize" the divisor by multiplying it by some number d which makes the high-order digit at least $b/2$. The dividend is also multiplied by d. After this normalization, the algorithm proceeds much like the elementary school method, with each quotient digit guessed using the high-order digit(s) of the divisor and current dividend. Knuth shows that the guesses cannot be wrong by more than 1 or 2. At the end, divide the remainder by d. This assumes that there is an algorithm for dividing a multi-digit integer $A = \sum_{i=0}^{k-1} a_i b^i$ by a single-digit integer B. The results are a quotient $Q = \sum_{i=0}^{k-1} q_i b^i$ and a single-digit remainder r. It is easy to design such an algorithm by analogy to the multiplication algorithm. Here is the algorithm in pseudocode:

[Division by a one-digit divisor: $Q = A/B$ using base b arithmetic]
Input: The base b digits of A and B; B has just one digit.
Output: The base b digits of $Q = A/B$; also return a one-digit
   remainder.

```
r = 0
for (i = k − 1 down to 0) {
        t = r × b + a_i
        q_i = ⌊t/b⌋
        r = t mod b
        }
return r
```

The temporary variable t used in the **for** loop must be able to hold a two-digit number in base b. Some computers have a hardware instruction which divides a 64-bit dividend by a 32-bit divisor to produce a 32-bit quotient and a 32-bit remainder. Such an instruction would be ideal for performing the last two lines of the **for** loop together, computing q_i and r in one operation. If the high-order digit q_{k-1} of the quotient is zero, it should be removed. The division algorithm can be modified easily to return only the quotient or only the remainder by not storing the unneeded result.

In order to analyze the complexity of algorithms that use arithmetic we will need to know the time taken by the four arithmetic operations. We do not concern ourselves with the actual time taken, since this time depends on the computer hardware. Rather we will count the number of basic steps. The basic steps we consider are adding, subtracting or multiplying two 1-bit numbers, or dividing a 2-bit number by a 1-bit number. These are called **bit operations**.

Furthermore, we will not worry about the exact count of bit operations. We will use the big-O notation to approximate the growth rate of the number of bit operations as the length of the operands grows.

DEFINITION 3.3   *If f and g are functions defined and positive for all sufficiently large x, then we say f is $O(g)$ if there is a constant $c > 0$ so that $f(x) < cg(x)$ for all sufficiently large x.*

The big-O notation allows us to focus on the general growth rate of a function and ignore the fine details of its growth. For example, $f(x) = 539x^4 + 212027x^3 - 1852x^2 + 178026x - 348561$ is $O(g)$, where $g(x) = x^4$. Suppose this $f(x)$ is the exact number of steps taken by an algorithm when its input is x bits long. Then the running time will be roughly a positive constant times $g(x)$, that is, proportional to the fourth power of the length of the input. This means that if the length of the input doubles, then the number of steps needed will be multiplied by about $2^4 = 16$.

We summarize the complexity of arithmetic operations discussed in this section in this theorem.

THEOREM 3.5   Complexity of arithmetic

One can add or subtract two k-bit integers in $O(k)$ bit operations. One can multiply two k-bit integers in $O(k^2)$ bit operations. One can divide a $2k$-bit dividend by a k-bit divisor to produce a k-bit quotient and a k-bit remainder in $O(k^2)$ bit operations.

PROOF   The statements about addition, subtraction and multiplication are shown by counting the steps in the three algorithms above. The statement about division can be shown the same way, after one writes the division algorithm.   ▌

The time complexities for addition and subtraction stated in the theorem are best possible (except for a constant). But one can multiply and divide faster than the $O(k^2)$ bit operations mentioned in the theorem. One can multiply two k-bit integers, or divide a $2k$-bit dividend by a k bit divisor, in only $O(k \log k \log \log k)$ bit operations, which is not much slower than addition. However, these fast algorithms do not become become useful or practical until k is very large—larger than numbers which occur in cryptography, at least in this book. See Section 4.3.3 of Knuth [56] or Chapter 9 of Crandall and Pomerance [33] for information about these faster arithmetic algorithms.

DEFINITION 3.4   *We say that an algorithm **runs in polynomial time** if there is a k and a constant $c > 0$ so that for every input I of length b bits, the algorithm on input I finishes in no more than cb^k bit operations.*

Base conversion, addition, subtraction, multiplication and division of integers can be done by algorithms that run in polynomial time.

3.3   Greatest Common Divisors and the Euclidean Algorithm

Now that we can perform arithmetic with integers of any size, we return to our study of divisibility.

DEFINITION 3.5   *When a and b are integers and not both zero we define the **greatest common divisor** of a and b, written $\gcd(a, b)$, as the largest integer which divides both a and b. We say that the integers a and b are **relatively prime** if their greatest common divisor is 1.*

It is clear from the definition that $\gcd(a, b) = \gcd(b, a)$. One way to compute the greatest common divisor of two nonzero integers is to list all of their divisors and choose the largest number which appears in both lists. Since d divides a if and only if $-d$ divides a, it is enough to list the positive divisors. For example, to compute $\gcd(6, 9)$, one finds that the positive divisors of 6 are 1, 2, 3 and 6 and that the positive divisors of 9 are 1, 3 and 9. The largest number common to both lists is 3, so $\gcd(6, 9) = 3$. The following theorem will help us compute greatest common divisors quickly, even when we do not know any divisors of the numbers (other than 1). Although the first equation in the theorem might remind one of the division algorithm, there is no requirement here that $0 \leq r < a$.

THEOREM 3.6 GCDs and division
If a is a positive integer and b, q and r are integers with $b = aq + r$, then
$\gcd(b, a) = \gcd(a, r)$.

PROOF Write $d = \gcd(b, a)$ and $e = \gcd(a, r)$. Since d divides both b and a, it must divide $r = b - aq$, by Theorem 3.2. Then d is a common divisor of a and r, so $d \leq e$ since e is the *greatest* common divisor of a and r. Likewise, since e divides both a and r, it must divide b because $b = aq + r$. Then e is a common divisor of b and a, so $e \leq d$. Therefore, $d = e$. ∎

A systematic way of computing $\gcd(a, b)$ has been known for thousands of years. It was published as Proposition 2 in Book VII of Euclid's book *The Elements* more than 2300 years ago and is called the Euclidean algorithm.

THEOREM 3.7 Simple form of the Euclidean algorithm
Let $r_0 = a$ and $r_1 = b$ be integers with $a \geq b > 0$. Apply the division algorithm (Theorem 3.3) iteratively to obtain

$$r_i = r_{i+1} q_{i+1} + r_{i+2} \text{ with } 0 < r_{i+2} < r_{i+1}$$

for $0 \leq i < n - 1$ and $r_{n+1} = 0$. Then $\gcd(a, b) = r_n$, the last nonzero remainder.

PROOF First of all, the algorithm will end because we will eventually get a zero remainder since $a = r_0 \geq r_1 > r_2 > \cdots \geq 0$; so, there cannot be more than a nonzero remainders. Applying Theorem 3.6 n times, we find

$$\gcd(a, b) = \gcd(r_0, r_1) = \gcd(r_1, r_2) = \cdots =$$
$$= \gcd(r_{n-1}, r_n) = \gcd(r_n, 0) = r_n.$$

Hence, $\gcd(a, b) = r_n$. ∎

Example 3.4

Use this theorem to compute the greatest common divisor of 165 and 285.
    We find

$$285 = 1 \times 165 + 120$$
$$165 = 1 \times 120 + 45$$
$$120 = 2 \times 45 + 30$$
$$45 = 1 \times 30 + 15$$
$$30 = 2 \times 15 + 0,$$

so $\gcd(165, 285) = 15$.

This algorithm may be written concisely in pseudocode. We write $a \bmod b$ for the remainder r in $0 \le r < b$ when a is divided by the positive integer b.

[Simple form of the Euclidean Algorithm]
Input: Integers $a \ge b > 0$.
Output: $\gcd(a, b)$.

```
while (b > 0) {
        r = a mod b
        a = b
        b = r
        }
return a
```

THEOREM 3.8 Division by the GCD
  Let $g = \gcd(a, b)$. Then a/g and b/g are relatively prime integers.

PROOF   Suppose d is a positive common divisor of a/g and b/g. Then there are integers m and n such that $a/g = md$ and $b/g = nd$, that is, $a = gdm$ and $b = gdn$. Hence gd is a common divisor of a and b. Since g is the greatest common divisor of a and b, we must have $gd \le g$, or $d \le 1$. Therefore $d = 1$, and a/g and b/g are relatively prime.   ∎

The next theorem tells us that we can solve $ax + by = 1$ for integers x and y whenever a and b are relatively prime.

THEOREM 3.9 GCD is a linear function
  If the integers a and b are not both 0, then there are integers x and y so that $ax + by = \gcd(a, b)$.

PROOF   At least one positive integer, $a^2 + b^2$, has the form $ax + by$. Let g be the smallest positive integer of this form, say $g = ax + by$. Any common divisor of a and b must divide $ax + by = g$ by Theorem 3.2, and so $\gcd(a, b)$ divides g, which implies that $\gcd(a, b) \leq g$. We claim that g divides a. Suppose not. Then $a = gq + r$ with some $0 < r < g$. Note that $r = a - gq = a - q(ax + by) = a(1 - qx) + b(-qy)$, which contradicts the fact that g is the least positive integer of the form $ax + by$. Hence g divides a. Similarly, g divides b. Therefore $g \leq \gcd(a, b)$ and $g = \gcd(a, b)$. ∎

Example 3.5

In Example 3.4, we found that $\gcd(285, 165) = 15$. Now let us find x and y with $285x + 165y = \gcd(285, 165) = 15$.

Beginning with the next to last equation in that example and working backwards, we find

$$15 = 45 - 30 = 45 - (120 - 2 \times 45) = 3 \times 45 - 120$$
$$15 = 3(165 - 120) - 120 = 3 \times 165 - 4 \times 120$$
$$15 = 3 \times 165 - 4(285 - 165) = 7 \times 165 - 4 \times 285.$$

Thus $x = -4$ and $y = 7$.

This method for finding integers x and y with $ax + by = \gcd(a, b)$ is inconvenient because one must work through the Euclidean algorithm, save all the steps and then work backwards to the beginning. The next algorithm finds the same result and requires working through the algorithm only once.

[Extended Euclidean Algorithm]
Input: Integers $a \geq b > 0$.
Output: $g = \gcd(a, b)$ and x and y with $ax + by = \gcd(a, b)$.

```
x = 1;  y = 0;  g = a;  r = 0;  s = 1;  t = b
while (t > 0) {
        q = ⌊g/t⌋
        u = x - qr;  v = y - qs;  w = g - qt
        x = r;  y = s;  g = t
        r = u;  s = v;  t = w
        }
return (g, x, y)
```

To see that the algorithm works, focus first on the variables g, t and w. In the middle of each pass through the `while` loop, w is set to $g \bmod t$. Then t is copied into g and w is copied into g. This is exactly what happens to the variables a, b and r in the simple Euclidean algorithm. Since g and t are initialized to a and b, and the condition for the `while` loop to end is the same in both algorithms, the variable g has the value $\gcd(a, b)$ when the algorithm finishes.

Now prove by induction that at the beginning and end of the `while` loop, these two equations hold:

$$ax + by = g \quad \text{and} \quad ar + bs = t.$$

The induction step is shown by noting that the assignments in the second line of the while loop subtract q times the second equation from the first one, forming the equation

$$a(x - qr) + b(y - qs) = g - qt.$$

If we apply the extended Euclidean algorithm to Example 3.4, the variables take on the values in this table.

x	y	g	r	s	t	q
1	0	285	0	1	165	1
0	1	165	1	-1	120	1
1	-1	120	-1	2	45	2
-1	2	45	3	-5	30	1
3	-5	30	-4	7	15	2
-4	7	15	11	-19	0	—

THEOREM 3.10 Product of numbers relatively prime to m

Let a, b and $m > 1$ be integers. If $\gcd(a, m) = \gcd(b, m) = 1$, then $\gcd(ab, m) = 1$.

PROOF   By Theorem 3.9, there are integers w, x, y, z so that $aw + mx = 1 = by + zm$. Therefore, $(aw)(by) = (1 - mx)(1 - mz) = 1 - mv$, where $v = x + z - mxz$. From $abwy + mv = 1$ and Theorem 3.2 we see that any common divisor of ab and m must also divide 1. Therefore, $\gcd(ab, m) = 1$. ∎

Although it is not easy to determine the average time complexity of the Euclidean algorithm (see Section 4.5.3 of Knuth [56] for the average complexity), it is fairly easy to give an upper bound on the worst-case complexity using Fibonacci numbers.

DEFINITION 3.6   The **Fibonacci numbers** are defined recursively by $u_0 = 0$, $u_1 = 1$, and $u_{n+1} = u_n + u_{n-1}$ for all $n \geq 1$.

The next few Fibonacci numbers after u_1 are $u_2 = 1$, $u_3 = 2$, $u_4 = 3$, $u_5 = 5$, $u_6 = 8$, $u_7 = 13$, $u_8 = 21$ and $u_9 = 34$.

The next lemma shows that the Fibonacci numbers grow exponentially.

LEMMA 3.1

Let $\alpha = (1 + \sqrt{5})/2$. Then $\alpha^{n-2} < u_n < \alpha^{n-1}$ for all $n \geq 3$.

PROOF Use induction on n. The base step is to verify the inequalities for $n = 3$ and $n = 4$, using the fact that α is approximately 1.618. Note that α is a root of $x^2 - x - 1 = 0$, so $\alpha^2 = \alpha + 1$. Multiply by α^{n-4} and α^{n-3} to get $\alpha^{n-2} = \alpha^{n-3} + \alpha^{n-4}$ and $\alpha^{n-1} = \alpha^{n-2} + \alpha^{n-3}$. Assume by induction that the inequalities hold for $n - 2$ and $n - 1$:

$$\alpha^{n-4} < u_{n-2} < \alpha^{n-3} \text{ and } \alpha^{n-3} < u_{n-1} < \alpha^{n-2}.$$

Add these two inequalities and use the equations for the powers of α and the definition of u_n to get $\alpha^{n-2} < u_n < \alpha^{n-1}$. ∎

THEOREM 3.11 GCD of consecutive Fibonacci numbers
For $n \geq 1$, the Euclidean Algorithm takes exactly n steps to compute the greatest common divisor of u_{n+2} and u_{n+1}, which is 1.

PROOF Since $u_{i+1} = u_i + u_{i-1}$, the quotients in the Euclidean algorithm for $\gcd(u_{n+2}, u_{n+1})$ are all 1, and the n steps are:

$$u_{n+2} = 1 \times u_{n+1} + u_n$$
$$u_{n+1} = 1 \times u_n + u_{n-1}$$
$$\vdots$$
$$u_4 = 1 \times u_3 + u_2$$
$$u_3 = 2 \times u_2.$$

∎

We will show in the middle of the next proof that consecutive Fibonacci numbers provide the worst case for the Euclidean algorithm. That is, u_{n+2} and u_{n+1} are the smallest two numbers that make the Euclidean algorithm take n steps.

THEOREM 3.12 Complexity of the Euclidean algorithm, Lamé, 1845
The number of steps (division operations) needed by the Euclidean algorithm to find the greatest common divisor of two positive integers is no more than five times the number of decimal digits in the smaller of the two numbers.

PROOF The Euclidean algorithm takes only one step if the two numbers are equal. Otherwise, apply the Euclidean algorithm to $a = r_0 > b = r_1 > 0$. Suppose the n steps are

$$r_i = r_{i+1} q_{i+1} + r_{i+2} \text{ with } 0 < r_{i+2} < r_{i+1}$$

for $0 \le i < n - 1$ and $r_{n+1} = 0$. Every quotient q_i must be ≥ 1 and the last one, $q_n \ge 2$, because $r_{n-1} > r_n > r_{n+1} = 0$. Hence,

$$r_n \ge 1 = u_2,$$
$$r_{n-1} \ge 2r_n \ge 2u_2 = u_3,$$
$$r_{n-2} \ge r_{n-1} + r_n \ge u_3 + u_2 = u_4,$$
$$\vdots$$
$$r_2 \ge u_{n-1} + u_{n-1},$$
$$b = r_1 \ge u_n + u_{n-1} = u_{n+1},$$
$$a = r_0 \ge u_{n+1} + u_n = u_{n+2}.$$

This shows that u_{n+2} and u_{n+1} are the smallest two numbers that make the Euclidean algorithm take n steps. By Lemma 3.1, we have $u_{n+1} > \alpha^{n-1}$ for $n \ge 2$, where $\alpha = (1 + \sqrt{5})/2$. Hence, $b > \alpha^{n-1}$. Since $\log_{10} \alpha > 0.2$, we have

$$\log_{10} b > (n - 1) \log_{10} \alpha > (n - 1)/5.$$

Thus, $n - 1 < 5\log_{10} b$. Suppose b has d decimal digits. Then $b < 10^k$ and $\log_{10} b < k$. Hence, $n - 1 < 5k$ and, since n and k are integers, we must have $n \le 5k$. ∎

COROLLARY 3.1

The number of bit operations needed by the Euclidean algorithm to find the greatest common divisor of two positive integers is $O((\log_2 a)^3)$, where a is the larger of the two numbers.

PROOF   By Lamé's theorem, it takes $O(\log_2 a)$ division operations to compute the greatest common divisor. The result follows from Theorem 3.5, which says that each division operation takes $O((\log_2 a)^2)$ bit operations. ∎

The corollary shows that the Euclidean algorithm runs in polynomial time.

3.4   Exercises

1. Show that if $a|b$ and $c|d$, then $ac|bd$.

2. Convert 0x3EB7 from hexadecimal to decimal.

3. Convert 6291 from decimal to hexadecimal.

4. Prove that an integer n is even if and only if its last decimal digit is even.

5. Prove that an integer n is divisible by 5 if and only if its last decimal digit is divisible by 5.

6. Prove that an integer n is divisible by 3 if and only if the sum of its decimal digits is divisible by 3.

7. Prove that an integer n is divisible by 9 if and only if the sum of its decimal digits is divisible by 9.

8. Let $m \geq 1$. Prove that an integer n is divisible by 2^m if and only if the integer k consisting of its last m decimal digits is divisible by 2^m. Note that $k = n \bmod 10^m$.

9. Let $m \geq 1$. Prove that an integer n is divisible by 5^m if and only if the integer k consisting of its last m decimal digits is divisible by 5^m. Note that $k = n \bmod 10^m$.

10. Let $n = \sum_{i=0}^{r} d_i 10^i$. Prove that n is divisible by 11 if and only if the alternating sum $d_0 - d_1 + d_2 - d_3 + \cdots$ of its decimal digits is divisible by 11.

11. Modify the algorithm for multiplying integers to make it nearly twice as fast in the special case $A = B$, that is, when the algorithm computes a square $C = A^2$.

12. In this exercise, we show that one can multiply two k-bit binary numbers A and B faster than in $O(k^2)$ steps when k is large. Make k even by prepending a 0 bit, if necessary. We may have to remove two or three leading 0 bits from the product at the end. Write the numbers in base $b = 2^{k/2}$ as $A = A_1 b + A_0$ and $B = B_1 b + B_0$. Prove that

$$AB = (b^2 + b)A_1 B_1 + b(A_1 - A_0)(B_0 - B_1) + (b + 1)A_0 B_0.$$

This formula shows that the product AB of two k-bit numbers can be formed by multiplying the three $k/2$-bit numbers $(A_1 - A_0)(B_0 - B_1)$, $A_1 B_1$ and $A_0 B_0$, together with simple shifting and adding operations. Note that one can multiply a binary number by b or b^2 by shifting the bits by $k/2$ or k positions. This simple trick can be used recursively. Let $T(k)$ denote the time needed to multiply two k-bit binary numbers. The formula shows that $T(k) \leq 3T(k/2) + ck$, for some constant c. Show that this inequality implies that $T(2^i) \leq c(3^i - 2^i)$, for $i \geq 1$. Deduce from this that $T(k) \leq 3c \cdot 3^{\log_2 k} = 3ck^{\log_2 3}$. Since $\log_2 3 \approx 1.585 < 2$, this method, which is called Karatsuba multiplication, is faster theoretically than conventional multiplication when k exceeds a threshold. In practice, when multiplying large numbers with the same length, one uses the formula recursively down to the threshold.

13. Find the greatest common divisor of 4905 and 32445.

14. Find integers x and y so that $4905x + 32445y = \gcd(4905, 32445)$.

15. If u_n is the n-th Fibonacci number, and i and j are two positive integers, prove that $\gcd(u_i, u_j) = u_{\gcd(i,j)}$.

16. When a, b and c are nonzero integers, define $\gcd(a, b, c) = \gcd(\gcd(a, b), c)$.

    a. Prove that $\gcd(a, b, c) = \gcd(a, \gcd(b, c))$.

    b. Extend the extended Euclidean algorithm to one which will find, in polynomial time, integers x, y and z with $ax + by + cz = \gcd(a, b, c)$ for any given nonzero integers a, b, and c.

Chapter 4

Primes

This chapter introduces the prime numbers, which are the building blocks of the integers with respect to multiplication. Many cryptographic algorithms use large prime numbers. To learn more about primes, the reader should consult books by Riesel [96], Robbins [98], Crandall and Pomerance [33] and Niven, Zuckerman and Montgomery [78].

4.1 The Fundamental Theorem of Arithmetic

DEFINITION 4.1   *A **prime number** is an integer greater than 1 which is divisible only by 1 and itself, and by no other positive integer. A **composite number** is an integer greater than 1 which is not prime.*

A composite number n has a positive divisor other than 1 and itself. This factor must be less than n and greater than 1.

Example 4.1

The first few prime numbers are 2, 3, 5, 7, 11, 13, 17, 19, 23, 29, 31 and 37. The first few composite numbers are 4, 6, 8, 9, 10, 12, 14, 15, 16, 18 and 20. The integers $4 = 2 \cdot 2$, $12 = 2 \cdot 2 \cdot 3$ and $63 = 3 \cdot 3 \cdot 7$ are all composite because they each have divisors other than 1 and themselves.

Some old texts and tables consider 1 to be a prime number. We do not do this, nor do we consider 1 to be composite, because then the beautiful Theorem 4.1 would be false.

LEMMA 4.1
Let a, b and c be positive integers. If $a|bc$ and $\gcd(a, b) = 1$, then $a|c$.

PROOF   Since a and b are relatively prime, by Theorem 3.9 there are integers x and y so that $ax + by = \gcd(a,b) = 1$. Multiply by c to get $axc + bcy = c$. Now $a|axc$ and $a|bcy$ by Theorem 3.1 and the hypothesis. Therefore, a divides $axc + bcy = c$ by Theorem 3.2.   ∎

LEMMA 4.2
If a prime p divides a product $a_1 a_2 \cdots a_k$ of positive integers, then it divides at least one of them.

PROOF   We use mathematical induction on the number n of factors. If $n = 1$, there is nothing to prove. Assume the statement is true for n factors. Suppose the prime p divides a product of $n + 1$ positive integers $a_1 a_2 \cdots a_n a_{n+1}$. If $p|a_1$, we are done. Otherwise, p is relatively prime to a_1 because p has only the divisors 1 and p, and p doesn't divide a_1, so $\gcd(p, a_1) = 1$. By Lemma 4.1, p divides the product $a_2 a_3 \cdots a_n a_{n+1}$ of n factors, and so p must divide one of these n numbers by the induction hypothesis.   ∎

THEOREM 4.1 Fundamental theorem of arithmetic
Every integer greater than 1 can be written as a product of primes, perhaps with just one prime, and this product is unique if the primes are written in nondecreasing order.

PROOF   The integer 2 is a prime number and so is the "product" of just one prime. If some integer cannot be expressed as a product of primes, then there must be a smallest one with this property. Let n be the least integer greater than 1 which is not a product of primes. If n were prime, then it would be the "product" of just one prime. So n must be composite, say, $n = ab$, where $1 < a < n$ and $1 < b < n$. Since a and b are smaller than n, it must be possible to write them as the product of primes. But then $n = ab$ is also the product of primes. This shows that every integer greater than 1 can be written as a product of primes.

We now show that this product is unique if the primes are written in nondecreasing order. Suppose to the contrary that some integer could be written in two different ways as a product of primes, say

$$n = p_1 p_2 \cdots p_k = q_1 q_2 \cdots q_l,$$

where all p_i and all q_j are primes and $p_1 \le p_2 \le \cdots \le p_k$ and $q_1 \le q_2 \le \cdots \le q_l$. Cancel any common prime factors to get

$$p_{i_1} p_{i_2} \cdots p_{i_r} = q_{j_1} q_{j_2} \cdots q_{j_s},$$

where no prime appears on *both* sides of the equation. There must be at least one prime factor on each side since we assumed that the two factorizations of n differ. By Lemma 4.2, the prime p_{i_1} must divide one of the numbers on the right side, say, $p_{i_1} | q_{j_m}$. But q_{j_m} is prime, so $p_{i_1} = q_{j_m}$ and the common prime factors were not all canceled. This contradiction shows that the prime factorization of n is unique. ∎

If we allowed 1 to be a prime number, then the fundamental theorem would fail because n could have two factorizations with different numbers of 1's. We don't want 1 to be composite, either, because it has no prime divisor.

Suppose the positive integer n is factored into the product of primes, and the primes are in nondecreasing order. The fundamental theorem of arithmetic says that this representation is unique. If we collect repeated prime factors and write them as the power p^e of a prime, we have the following **standard representation**:

$$n = p_1^{e_1} p_2^{e_2} \cdots p_k^{e_k} = \prod_{i=1}^{k} p_i^{e_i},$$

where $p_1, p_2, \ldots, p_k$ are the primes that actually divide n and $e_i \geq 1$ is the number of factors of p_i dividing n. We make the convention that $n = 1$ has this representation with the empty product.

Sometimes we allow some exponents e_i to be 0 in the representation. We might do this to compare the prime factorizations of two integers. This device is used to find the greatest common divisors of integers whose factorizations are known. We begin with a simple example.

Example 4.2

Find the gcd g of $41184 = 2^5 \cdot 3^2 \cdot 11 \cdot 13$ and $10920 = 2^3 \cdot 3 \cdot 5 \cdot 7 \cdot 13$.

The highest power of 2 that can divide g must divide both numbers; so, it must be the smaller of 2^5 and 2^3, which is 2^3. Likewise, only one 3 can divide g since only one 3 divides the second number. The primes 5, 7 and 11 divide only one of the two numbers, so cannot divide g. A single 13 divides each number, so $13 | g$. Now we know all prime divisors of g and $g = 2^3 \cdot 3 \cdot 13 = 312$.

THEOREM 4.2 GCD of factored numbers

Let $p_1, p_2, \cdots, p_k$ be all the primes that divide either of the positive integers m and n. Write

$$m = p_1^{e_1} p_2^{e_2} \cdots p_k^{e_k} \text{ and } n = p_1^{f_1} p_2^{f_2} \cdots p_k^{f_k},$$

where all exponents e_i and f_i are ≥ 0. Then

$$\gcd(m, n) = p_1^{\min(e_1, f_1)} p_2^{\min(e_2, f_2)} \cdots p_k^{\min(e_k, f_k)}.$$

PROOF The power of each prime which divides the gcd is the smaller of the two powers of the prime which divide the two numbers. ∎

DEFINITION 4.2   *The* **least common multiple** *of* $r > 1$ *positive integers* $n_1, n_2, \ldots, n_r$, *denoted* $\mathrm{lcm}(n_1, n_2, \ldots, n_r)$, *is the smallest positive integer which is divisible by all of the numbers* $n_1, n_2, \ldots, n_r$.

The definition makes sense because $n_1 n_2 \cdots n_r$ is one positive integer which is divisible by all of the numbers $n_1, n_2, \ldots, n_r$, so that $\mathrm{lcm}(n_1, n_2, \ldots, n_r)$ must be some integer between 1 and $n_1 n_2 \cdots n_r$. By analogy to Theorem 4.2 one can prove the following result, which we state only for the least common multiple of two integers.

THEOREM 4.3 LCM of factored numbers
Let $p_1, p_2, \ldots, p_k$ be all the primes that divide either of the positive integers m and n. Write

$$m = p_1^{e_1} p_2^{e_2} \cdots p_k^{e_k} \text{ and } n = p_1^{f_1} p_2^{f_2} \cdots p_k^{f_k},$$

where all exponents e_i and f_i are ≥ 0. Then

$$\mathrm{lcm}(m, n) = p_1^{\max(e_1, f_1)} p_2^{\max(e_2, f_2)} \cdots p_k^{\max(e_k, f_k)}.$$

COROLLARY 4.1
For any two positive integers m and n, $\gcd(m, n)\mathrm{lcm}(m, n) = mn$.

PROOF   The equation follows from the two theorems just stated and the fact that $\min(x, y) + \max(x, y) = x + y$ for any real numbers x and y.   ∎

THEOREM 4.4 LCM of numbers relatively prime in pairs
If $n_1, \ldots, n_r$ are r positive integers which are relatively prime in pairs, that is, $\gcd(n_i, n_j) = 1$ for all $1 \leq i < j \leq r$, then $\mathrm{lcm}(n_1, n_2, \ldots, n_r) = n_1 n_2 \cdots n_r$.

PROOF   Use induction on r. For $r = 2$, the statement is just Corollary 4.1. Suppose the statement is true for $r - 1$. Then we are given that $\mathrm{lcm}(n_1, n_2, \ldots, n_{r-1}) = n_1 n_2 \cdots n_{r-1}$. We must prove the statement for r. Write $L = \mathrm{lcm}(n_1, n_2, \ldots, n_{r-1}) = n_1 n_2 \cdots n_{r-1}$. Note that if a prime p divides L, then it must divide n_i for some $1 \leq i \leq r - 1$, for otherwise L/p would be a smaller common multiple for $n_1, \ldots, n_{r-1}$, and L is the least one. If a prime p divided both L and n_r, then it would divide both n_r and n_i for some $1 \leq i \leq r - 1$. This cannot happen because n_r and n_i are assumed to be relatively prime. Therefore, $\gcd(L, n_r) = 1$. Now by Corollary 4.1, we have $\mathrm{lcm}(n_1, n_2, \ldots, n_r) = \mathrm{lcm}(L, n_r) = L n_r = n_1 n_2 \cdots n_r$.   ∎

The fundamental theorem of arithmetic has many other uses, one more of which is illustrated in the next example.

Example 4.3

Find all the positive divisors of $364 = 2^2 \cdot 7 \cdot 13$.

The only positive divisors of 364 are positive integers whose prime factorizations contain only the primes 2, 7 and 13, raised to nonnegative integer powers no higher than 2, 1 and 1, respectively. These divisors are:

1	7	13	$7 \cdot 13 = 91$
2	$2 \cdot 7 = 14$	$2 \cdot 13 = 26$	$2 \cdot 7 \cdot 13 = 182$
$2^2 = 4$	$2^2 \cdot 7 = 28$	$2^2 \cdot 13 = 52$	$2^2 \cdot 7 \cdot 13 = 364$

4.2 The Distribution of Prime Numbers

Since some cryptographic algorithms require large prime numbers, we must investigate whether there are enough of them. The first theorem, which offers a tiny bit of comfort in this direction, was already known to Euclid more than 2300 years ago. See Euclid's *Elements*, Book IX, Proposition 20.

THEOREM 4.5 Number of primes is infinite
The number of prime numbers is infinite.

PROOF   Suppose $p_1, p_2, \ldots, p_k$ were all of the prime numbers. Let $n = p_1 \cdot p_2 \cdots p_k + 1$ be 1 plus their product. Then n has a prime divisor p, by Theorem 4.1. The prime p cannot be one of the primes p_i because, if it were, then it would divide $n - p_1 \cdot p_2 \cdots p_k = 1$ by Theorem 3.2. Therefore $p_1, p_2, \ldots, p_k$ were not all of the primes. ∎

This theorem is constructive in that it tells us how to find new primes after we think we know all of them. But the construction is not useful because it is difficult to factor large integers. We will see in the next section and in Chapter 11 that there are much easier ways to construct new primes.

The next theorem tells us that there are arbitrarily long gaps between consecutive primes.

THEOREM 4.6 Long gaps between primes
For every positive integer n, there are n (or more) consecutive composite positive integers.

PROOF   We claim the n consecutive positive integers

$$(n + 1)! + 2, (n + 1)! + 3, \ldots, (n + 1)! + n + 1$$

are all composite. For $2 \le i \le n + 1$ we have $i | (n + 1)!$. Theorem 3.2 implies that $i | (n + 1)! + i$, so $(n + 1)! + i$ is composite. ∎

Example 4.4

There are six consecutive composite numbers beginning with $7! + 2 = 5042$. But the first set of six consecutive composite numbers is 90, 91, 92, 93, 94, 95, which are much smaller than 5042.

Two primes whose difference is 2 are called **twin primes**. Some examples are 3 and 5, 17 and 19, 101 and 103, and 3671 and 3673. Much numerical evidence suggests that there are infinitely many twin prime pairs. A famous unsettled conjecture asserts that this is so. Although the gap between consecutive primes can be arbitrarily large, as shown by the theorem just proved, the smallest possible gap which could occur more than once, 2, probably occurs infinitely often.

A prime p for which $2p+1$ is also prime is called a **Sophie Germain prime**. The first few Sophie Germain primes are 2, 3, 5, 11, 23, 29, 41 and 53. Others have been found with hundreds or thousands of digits. It is conjectured that there are infinitely many Sophie Germain primes. Twin primes and Sophie Germain primes are used in a few cryptographic functions.

DEFINITION 4.3   *For positive real numbers x, let $\pi(x)$ be the number of prime numbers less than or equal to x.*

For example, $\pi(1) = 0$, $\pi(10) = 4$ and $\pi(100) = 25$. The function $\pi(x)$ has been computed for selected x up to about 10^{20}. We know from Theorem 4.5 that $\pi(x)$ increases without bound as $x \to \infty$. To use some ciphers, we will have to choose some large primes, say, 100-digit primes. The growth rate of $\pi(x)$ has a strong effect on the difficulty of finding a large prime. For example, if $\pi(x) \approx \sqrt{x}$, at least for x near 10^{100}, it would be quite hard to find even one prime with 100 digits. On one hand, this approximation would say that there are about 10^{50} primes less than 10^{100}. But another way of looking at this (false) estimate is that the probability would be roughly 10^{-50} that a randomly chosen 100-digit integer would be prime. That would mean that we would have to try about 10^{50} random 100-digit integers to get a prime. Fortunately for cryptography, $\pi(x)$ grows nearly as rapidly as x. The next theorem relates this growth to the natural logarithm function $\ln x$.

THEOREM 4.7 The prime number theorem
The ratio of $\pi(x)$ to $x/\ln x$ tends to 1 as x goes to infinity. In symbols,

$$\lim_{x \to \infty} \frac{\pi(x)}{x/\ln x} = 1.$$

The known proofs of the prime number theorem either are very complicated, although "elementary," or else use advanced mathematics. We do not give a proof here. The theorem was conjectured by Gauss more than 200 years ago.

It was first proved in 1896 (independently) by J. Hadamard and Ch. J. de la Vallée-Poussin.

The prime number theorem says that $\pi(x) \approx (x/\ln x)$ and that the percentage error in this approximation goes to 0 as x goes to infinity. We illustrate how good the approximation is at $x = 10^{10}$. It is known that $\pi(10^{10}) = 455052512$ and $(10^{10}/\ln 10^{10}) \approx 434294482$. The ratio of these two numbers is about 1.048, that is, the approximation is about 5% too small. Better analytic approximations to $\pi(x)$ are known, but are not needed for cryptography.

Although $x/\ln x$ is only a rough approximation to $\pi(x)$, it tells us immediately the probability that a random integer n in $1 \le n \le x$ is prime. There are x integers in this range and $x/\ln x$ of them are prime, so the probability is roughly $(x/\ln x)/x = 1/\ln x$. Since the function $\ln x$ changes slowly when x is large, $1/\ln x$ is also the probability that a random integer near x is prime. The probability that a random 100-digit integer is prime is about $1/\ln(10^{100})$, which is about $1/230$. This means that we would have to try about 230 random 100-digit integers to find one prime. We could shorten the search by skipping numbers that have small prime divisors. If we just omit the even numbers, which after 2 cannot be prime, then the probability of each candidate being prime doubles and we would need to try only about 115 random odd 100-digit integers to find one prime.

There are more precise versions of the prime number theorem than Theorem 4.7. They express $\pi(x)$ as a main term (more accurate than $x/\ln x$) plus an error term, and prove an upper bound on the absolute value of the error term. The proofs of those versions of the theorem study the zeros of a function called the Riemann zeta function. The more one knows about these zeros, the smaller the upper bound on the error term that one can prove. It is known that the error term in the estimate for $\pi(x)$ cannot be better than about $\sqrt{x} \ln x$. The **Riemann Hypothesis** is a statement about the zeros of the Riemann zeta function which would imply that the error term in the prime number theorem is as good as it could be. The statement is a famous unsolved problem in number theory. If it were proved, there would be many applications throughout number theory, not just for counting primes. For example, some fast "algorithms" for identifying primes depend on the Riemann Hypothesis for their correctness or speed. They might give the wrong answer or run for a long time if the Riemann Hypothesis were false.

4.3   Identifying and Finding Primes

Now that we know there are plenty of large primes, how do we distinguish them from composite numbers? This section will not answer that question, which is deferred to Chapter 11, but will take the first steps in that direction. The first theorem tells how to tell in $O(\sqrt{n})$ steps whether n is prime or composite.

THEOREM 4.8 Composites have a divisor below their square root
If the integer $n > 1$ is composite, then n has a prime divisor $p \le \sqrt{n}$. In other words, if the integer $n > 1$ has no prime divisor $p \le \sqrt{n}$, then n is prime.

PROOF   Suppose n is composite. Then we can write $n = ab$, where a and b are integers greater than 1. Swap a and b, if necessary, to make $1 < a \le b < n$. Then $a \le \sqrt{n}$, for if $a > \sqrt{n}$, then $b \ge a > \sqrt{n}$ and $n = ab > \sqrt{n}\sqrt{n} = n$, which is impossible. By Theorem 4.1, a must have a prime divisor $p \le a \le \sqrt{n}$. By Theorem 3.1, p divides n.

The second statement has the same meaning as the first one.   ∎

The theorem suggests a simple algorithm for testing a small number for primality and for factoring it if it is composite.

[Factoring and Prime Testing by Trial Division]
Input: A positive integer n to factor or to test for primeness.
Output: Whether n is prime, or one or more prime factors of n.

```
m = n
p = 2
while (p ≤ √m) {
        if (m mod p = 0) {
                Print "n is composite with factor p"
                m = m/p
                }
        else { p = p + 1 }
        }
if (m = n) { Print "n is prime" }
else if (m > 1) { Print "The last prime factor of n is m" }
```

If n is prime, then trial division will take about $O(\sqrt{n})$ steps to prove this fact. If n is composite, then the number of steps required depends on the size of the prime divisors of n. If we merely wish to know whether n is prime or composite, and n is composite, then the algorithm can stop as soon as it finds the first prime divisor, and the number of steps needed is proportional to the smallest divisor of n. In case we wish the complete prime factorization of n, let n_1 be the largest prime factor of n and n_2 be the second largest prime factor of n. The trial division algorithm will have to continue at least until it finds n_2. When this happens, if $n_2 \ge \sqrt{n_1}$, then the `while` loop will terminate on its next iteration; otherwise, trial division of n_1, the last remaining cofactor, will have to continue until it is recognized to be prime when the variable p passes $\sqrt{n_1}$. We have shown that the number of steps the algorithm takes to factor n completely is $O(\max(n_2, \sqrt{n_1}))$.

There are some obvious ways to accelerate the algorithm. It is inefficient to add only 1 to p in the `else` step, because then we try even numbers $p > 3$, which clearly cannot divide the odd remaining cofactor m. In fact, we should really replace p by the next prime after p in the `else` step. However, this would require having a table of primes up to $\sqrt{n}$ and it might be too expensive to precompute such a table. Usually, a compromise is made to avoid many but not all composite trial divisors p. For example, after $p = 5$, one might alternately add 2 and 4 in the `else` step to determine the next p. Then the sequence of trial divisors would be

$$p = 2, 3, 5, 7, 11, 13, 17, 19, 23, 25, 29, 31, 35, 37, 41, 43, 47, 49, \ldots$$

which excludes all multiples of 2 and 3 except for these numbers themselves.

Theorem 4.8 also suggests a simple algorithm for finding all primes up to some limit. The algorithm, called a **sieve**, was known to Eratosthenes more than 2200 years ago.

[Sieve of Eratosthenes]
Input: A limit $n > 2$.
Output: A list of all the primes between 2 and n.

```
Write all the integers between 2 and n in a list.
p = 2
while (p ≤ √n) {
        i = 2p
        while (i ≤ n) {
                Cross out i from the list
                i = i + p
                }
        Let p = the next number after p not yet crossed out
        }
Print the numbers that were not crossed out.
```

Let us estimate the time complexity of this algorithm. The inner `while` loop is performed once for each prime $p \leq \sqrt{n}$, or fewer than $\sqrt{n}$ times. We simplify the analysis by ignoring the restriction that p is prime. This simplification will make the time estimate larger. For each p, the instructions inside the inner `while` loop are performed n/p times because every p-th number is crossed out. We may estimate the total number of steps for the entire algorithm by the following sum, and then approximate the sum by an integral:

$$\sum_{p=2}^{\sqrt{n}} \left\lceil \frac{n}{p} \right\rceil \approx \int_2^{\sqrt{n}} \frac{n}{p} \, dp = n \ln \sqrt{n} - n \ln 2 = O(n \log n).$$

If we had tested each integer i between 2 and n for primality by trial division, it would take $O(\sqrt{i})$ steps to test i, for a total of $O(n\sqrt{n})$ steps. Thus, the sieve of Eratosthenes is much more efficient than trial division for finding *all* primes up to some limit. Furthermore, the fact that the operations of adding and crossing out (setting a byte) are faster than division tips the scale even more in favor of the sieve of Eratosthenes.

A variation of the sieve of Eratosthenes finds all integers in some interval which have no prime divisor less than some limit. For example, let L be a 50-digit integer. One could compute all integers between L and $L + 10000$ free of prime divisors less than 1000 as follows: First, make a list of the primes below 1000, perhaps by the sieve of Eratosthenes. Second, write tokens (bytes in a computer program) representing the 10001 integers $L, L + 1, \ldots, L + 10000$. Third, for each prime $p < 1000$, find the first integer $i \geq L$ divisible by p, and cross out the tokens representing i and each p-th number after i. Finally, scan the list of tokens and output each i whose token was not crossed out. A sieve like this one lies at the heart of several of the fastest known algorithms for factoring large integers.

4.4   The Largest Prime Factor of a Number

For several purposes later in this book we will need to know the approximate size of the largest prime factor of a "typical" integer n. We need it, together with the size of the second largest prime factor of n, to estimate the complexity of the trial division algorithm in the previous section.

DEFINITION 4.4   *A positive integer n is called y-smooth if all of its prime factors are $\leq y$.*

The de Bruijn [35] function, $\psi(x, y)$, is defined to be the number of y-smooth numbers n in $1 \leq n \leq x$.

For $0 \leq t \leq 1$ and $x > 2$, let $p(x, t)$ be the probability that the largest prime factor of an integer $1 \leq n \leq x$ is less than tx. Then $p(x, t) = \psi(x, tx)/x$. We might hope that if $x \neq y$ are two large numbers, then $p(x, t) \approx p(y, t)$ for all $0 < t \leq 1$. If this should happen, then we could define $p(t) = \lim_{x \to \infty} p(x, t)$ and say that $p(t)$ is the probability that the largest prime factor of n is less than tx. It turns out that this is the wrong way to proceed, because $p(t) = 1$ for all $0 < t \leq 1$. This means that if we choose any fixed t in $0 < t < 1$, then almost all integers n in $1 \leq n \leq x$ have no prime factor larger than tx.

It is better to use a logarithmic scale for the size of the largest prime factor. Let $t \geq 0$. Let $p(x, t)$ be the probability that the largest prime factor of an integer $1 \leq n \leq x$ is less than x^t. Then $p(x, t) = \psi(x, x^t)/x$ or, equivalently, $\psi(x, y) = xp(x, (\ln y)/\ln x)$. Dickman [39] gave a heuristic argument that, for each $t \geq 0$,

$$F(t) = \lim_{x \to \infty} p(x, t) = \lim_{x \to \infty} \psi(x, x^t)/x$$

exists, and gave a functional equation for computing $F(t)$. Later, Ramaswami [94] made Dickman's argument rigorous. Thus, $F(t)$ is the probability that the largest prime factor of n in $1 < n \leq x$ is less than x^t. It is clear that

$F(t) = 1$ for $t \geq 1$, because the largest prime factor of any $n \leq x$ must be $\leq x \leq x^t$. For $0 \leq t \leq 1$, Dickman's functional equation for $F(t)$ is

$$F(t) = \int_0^s F\left(\frac{s}{1-s}\right) \frac{ds}{s}. \tag{4.1}$$

His heuristic argument is roughly as follows: Let $0 < s < 1$. The number of integers $n \leq x$ whose largest prime factor is between x^s and x^{s+ds} is $xF'(s)ds$. By the prime number theorem, the number of primes in that interval is

$$\pi(x^{s+ds}) - \pi(x^s) = \pi(x^s + (\ln x)x^s ds) - \pi(x^s)$$
$$\approx \frac{x^s + (\ln x)x^s ds}{\ln x^s} - \frac{x^s}{\ln x^s}$$
$$= \frac{x^s + (\ln x)x^s ds - x^s}{s \ln x} = \frac{x^s ds}{s}.$$

For each prime p in this interval, the number of n such that $np \leq x$ and the largest prime factor of n is $\leq p$ is the same as the number of $n \leq x^{1-s}$ whose greatest prime factor is $\leq x^s = (x^{1-s})^{s/(1-s)}$, that is, $x^{1-s}F(s/(1-s))$. Hence,

$$xF'(s)ds = \left(\frac{x^s ds}{s}\right)\left(x^{1-s}F\left(\frac{s}{1-s}\right)\right),$$

or $F'(s)ds = F(s/(1-s))ds/s$, and we obtain Equation (4.1) by integration.

Equation (4.1) provides an effective way of computing $F(t)$ approximately by numerical integration. Figure 4.1 shows the graph of $F(x)$.

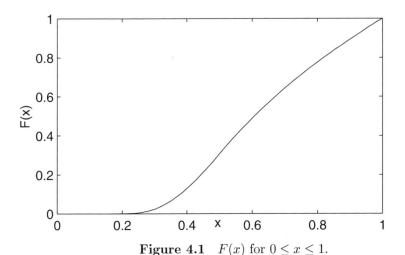

Figure 4.1 $F(x)$ for $0 \leq x \leq 1$.

We will want to use the value of $F(t)$ when t is *very* close to zero. This process becomes easier if we invert the argument. Define $\rho(u) = F(1/u)$ for $u > 0$. Figure 4.2 gives the graph of $\rho(u)$.

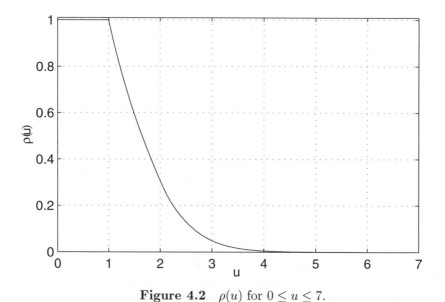

Figure 4.2 $\rho(u)$ for $0 \leq u \leq 7$.

Dickman's heuristic theorem, proved by Ramaswami [94], says this:

THEOREM 4.9 Count of smooth numbers below x
For each fixed real number $u > 0$ there is a real number $\rho(u) > 0$ so that

$$\lim_{x \to \infty} \psi(x, x^{1/u})/x = \rho(u)$$

and $\rho(u)$ is the unique continuous function defined by $\rho(u) = 1$ for $0 \leq u \leq 1$ and the functional equation $\rho'(u) = -\rho(u-1)/u$ for $u > 1$.

The functional equation for $\rho(u)$ is easy to derive from that for $F(t)$. From Equation (4.1) we find $F'(s) = F(s/(1-s))/s$. (In fact that equation was the step before Equation (4.1) in the heuristic argument for Equation (4.1).) Write $u = 1/s$ so that we have $F(s) = \rho(1/s) = \rho(u)$. Differentiating this formula gives $F'(s) = \rho'(1/s)(-1/s^2) = -\rho'(u)u^2$. On the other hand, we have $F(s/(1-s))/s = \rho((1-s)/s)/s = u\rho(u-1)$. Putting it all together, we have

$$-\rho'(u)u^2 = F'(s) = F(s/(1-s))/s = u\rho(u-1),$$

or $\rho'(u) = -\rho(u-1)/u$ for $u > 1$.

One can find a formula for $\rho(u)$ for $1 \le u \le 2$: Since $\rho(u-1) = 1$ for u in this range, the functional equation gives

$$\rho(u) = 1 + \int_1^u \rho'(v)dv = 1 + \int_1^u -\rho(v-1)/v \, dv = 1 - \int_1^u (1/v)dv = 1 - \ln u.$$

However, there is no known closed form for $\rho(u)$ for $u > 2$.

It is possible to compute $\rho(u)$ numerically from the functional equation. It goes to zero rapidly, as shown by the corollary to this theorem.

THEOREM 4.10 Integral for Dickman ρ function

For all $u > 1$, $\int_u^{u+1} \rho(v-1)dv = u\rho(u)$.

PROOF Since $\rho(v-1) = 1$ for $1 \le v \le 2$, we have

$$1 + \int_2^u \rho(v-1)dv = \int_1^u \rho(v-1)dv.$$

By the functional equation, the latter integral is $\int_1^u -v\rho'(v)dv$. Integrating by parts, we obtain

$$1 + \int_2^u \rho(v-1)dv = -v\rho(v)\Big|_1^u + \int_1^u \rho(v)dv = 1 - u\rho(u) + \int_2^{u+1} \rho(v-1)dv.$$

Subtracting the integrals gives $\int_u^{u+1} \rho(v-1)dv = u\rho(u)$. ∎

COROLLARY 4.2

For all positive integers n, $\rho(n) \le 1/n!$.

PROOF The functional equation shows that $\rho(u)$ is strictly decreasing for $u > 1$. This fact and the Theorem imply that $u\rho(u) < \rho(u-1)$ for all $u > 1$. A simple induction gives $\rho(n) < 1/n!$ for all $n > 1$. ∎

In fact, $\rho(u)$ goes to zero about as rapidly as the function u^{-u}, which is a moderately good approximation for it when u is large. The following table illustrates the Corollary and the approximation. See Knuth and Trabb Pardo

[57] for more values of $\rho(u)$.

n	$\rho(n)$	$1/n!$	n^{-n}
1	1.00000000	1.00000000	1.00000000
2	0.30685282	0.50000000	0.25000000
3	0.04860839	0.16666667	0.03703704
4	0.00491093	0.04166667	0.00390625
5	0.00035472	0.00833333	0.00032000
6	0.00001965	0.00138889	0.00002143
7	0.00000087	0.00019841	0.00000121

As a consequence, we have the approximation

$$\psi(x, x^{1/u}) \approx xu^{-u} \tag{4.2}$$

for each fixed u. To estimate the complexity of certain algorithms, we will need a formula like (4.2) even when u is not fixed, but increases with x. Canfield et al. [23] proved that the approximation (4.2) is valid so long as $u < (1 - \epsilon) \ln x / \ln \ln x$ (for any fixed $\epsilon > 0$). This gives a good approximation to $\psi(x, y)$ when $y > \ln^{1+\epsilon} x$ and x is large.

If $y = x^{1/u}$, then $\ln y = \frac{1}{u} \ln x$, so $u = (\ln x) / \ln y$. We can summarize the discussion above by saying that

$$\psi(x, y) \approx x\rho(u) \approx xu^{-u},$$

where $u = (\ln x) / \ln y$. We can also say that the probability that n is y-smooth is $\rho(u) \approx u^{-u}$, where $u = (\ln n) / \ln y$.

Sometimes we will need to estimate the number of integers $\leq x$ whose prime factors are all $\leq z$, *except* for the largest k prime divisors, which must be $\leq y$, where $y > z$. Write the prime factors of n as $n_1 \geq n_2 \geq n_3 \geq \ldots$, so that n_i is the i-th largest prime factor of n. If n doesn't have an i-th largest prime factor, then let $n_i = 1$.

DEFINITION 4.5   An integer n is called k-semismooth with respect to y and z if $n_1 \leq y$ and $n_{k+1} \leq z$. Let $\psi_k(x, y, z)$ denote the number of integers $\leq x$ that are k-semismooth with respect to y and z.

By analogy to Theorem 4.9, one can prove this result. See the doctoral theses of Cavalar [25] and Zhang [131] for a proof.

THEOREM 4.11 Count of semismooth numbers below x
For each integer $k \geq 1$ and fixed real numbers $v > u > 0$ there is a real number $\rho_k(v, u) > 0$ so that

$$\lim_{x \to \infty} \psi_k(x, x^{1/u}, x^{1/v})/x = \rho_k(v, u).$$

There are formulas, like the functional equation for $\rho(u)$, for computing $\rho_k(v, u)$, but their complexity increases rapidly with k.

Example 4.5

In the previous section, we found that the complexity of a simple trial division algorithm is $O(\max(n_2, \sqrt{n_1}))$ to factor n completely. For $0 < t < 0.5$, this complexity will be $\leq n^t$ provided that $n_2 \leq n^t$ and $\sqrt{n_1} \leq n^t$. The second inequality may be written as $n_1 \leq n^{2t}$. This means that the trial division algorithm will factor an integer n in $O(n^t)$ steps provided that n is 1-semismooth with respect to $y = n^{2t}$ and $z = n^t$. By Theorem 4.11, the probability that this will happen is approximately $\rho_1(1/t, 1/2t)$. The following table gives some values of this function.

u	$t = 1/u$	$\rho_1(u, u/2)$
1	1.00000000	1.00000000
2	0.50000000	1.00000000
3	0.33333333	0.44731421
4	0.25000000	0.09639901
5	0.20000000	0.01241348
6	0.16666667	0.00109227
7	0.14285714	0.00007139

See Knuth and Trabb Pardo [57] for more values of $\rho_1(u, u/2)$. This table tells us that trial division will factor n completely in $O(n^{0.25})$ steps with probability 0.0963, and that it will finish in $O(n^{0.2})$ steps for about 1.2% of the numbers n.

4.5   Exercises

1. Factor 10988208 and 17535336 each into the product of primes.

2. Find the greatest common divisor and the least common multiple of $2^6 \cdot 3^2 \cdot 5^2 \cdot 11 \cdot 13$ and $2^3 \cdot 3^5 \cdot 7 \cdot 13$.

3. Prove that there are infinitely many primes of the form $4k + 3$. Model your proof after that of Theorem 4.5. Suppose there were only a finite number of them. Multiply them all and construct a new number which must be divisible by a different prime of this form. To help the last step work, prove that a product of primes of the form $4k + 1$ must have the same form.

4. Use the sieve of Eratosthenes to find all primes between 0 and 200.

5. Use a variation of the sieve of Eratosthenes to find all primes between 2000 and 2100.

6. Estimate the number of 10^4-smooth numbers between $10^{20} - 10^6$ and 10^{20}.

7. Estimate the number of 10^5-smooth numbers between $10^{20} - 10^6$ and 10^{20}.

8. Estimate the number of numbers between $10^{24} - 10^6$ and 10^{24} which will be factored completely by the trial division algorithm by the time the variable p reaches 10^4.

9. Use Dickman's theorem to prove that for any $0 < t \le 1$ we have

$$\lim_{x \to \infty} \psi(x, tx)/x = 1.$$

Chapter 5

Congruences

This chapter introduces the basic facts about congruences. See the number theory texts [99], [78] and [98] for more about congruences. Gauss introduced congruences in 1801 in [45]. A congruence is a statement about divisibility. It is a notation that simplifies reasoning about divisibility. It suggests proofs by its analogy to equations. Congruences are familiar to us as "clock arithmetic." Four hours after 10 AM it will be 2 PM. How do we get the 2 from the 10 and the 4? We add four to ten and then subtract 12. We have used a congruence modulo 12.

5.1 Simple Properties of Congruences

DEFINITION 5.1 *Suppose a and b are integers and m is a positive integer. If m divides $a - b$, then we say a is **congruent to** b **modulo** m and write $a \equiv b \pmod{m}$. If m does not divide $a - b$, we say a is **not congruent to** b **modulo** m and write $a \not\equiv b \pmod{m}$. The formula $a \equiv b \pmod{m}$ is called a **congruence**. The integer m is called the **modulus** (plural **moduli**) of the congruence.*

Usually we will have $m > 1$.

Do not confuse the binary operator "mod" in $a \bmod b$, which means the remainder when a is divided by b, with the "mod" enclosed in parentheses together with the modulus of a congruence. These concepts are related as follows. If m is a positive integer and a and b are integers, then $a \equiv b \pmod{m}$ if and only if $(a \bmod m) = (b \bmod m)$.

We will often use the fact that $a \equiv b \pmod{m}$ if and only if there is an integer k so that $a = b + km$. This fact follows immediately from the definitions of congruence and divide.

The congruence relation has many similarities to equality. The next theorem says that congruence, like equality, is an equivalence relation.

THEOREM 5.1 Congruence is an equivalence relation

Let m be a positive integer. Let a, b and c be integers. Then:

1. $a \equiv a \pmod{m}$.
2. *If* $a \equiv b \pmod{m}$, *then* $b \equiv a \pmod{m}$.
3. *If* $a \equiv b \pmod{m}$ *and* $b \equiv c \pmod{m}$, *then* $a \equiv c \pmod{m}$.

PROOF   Clearly, $m|a - a$. Also, $m|(a - b)$ if and only if $m|(b - a)$. Finally, if $m|(a - b)$ and $m|(b - c)$, then $m|(a - c) = (a - b) + (b - c)$.   ∎

Let $m > 0$ be fixed. For each integer a, the set of all integers $b \equiv a \pmod{m}$ is called the **congruence class** or **residue class** of a modulo m. The congruence class of a modulo m consists of all integers in the arithmetic progression $a + dm$, where d runs through all integers. Each integer in a congruence class is a **representative** of it. If the modulus m is understood and a and b are in the same congruence class, then each is called a **residue** of the other. The smallest nonnegative representative of a congruence class is often used as the standard representative of it. For example, the standard representative of the congruence class of 27 (mod 5) is 2. When we study congruences, we regard all integers in the congruence class of a modulo m as being equivalent. The next theorem says that it makes sense to perform arithmetic on congruence classes without worrying about which representatives we choose.

THEOREM 5.2 Arithmetic with congruences

Let a, b, c and d be integers. Let m be a positive integer. Suppose $a \equiv b \pmod{m}$ and $c \equiv d \pmod{m}$. Then

1. $a + c \equiv b + d \pmod{m}$.
2. $a - c \equiv b - d \pmod{m}$.
3. $ac \equiv bd \pmod{m}$.

PROOF   The first two statements are trivial. For the third, let $m|(a - b)$ and $m|(c - d)$. Then $m|c(a - b) + b(c - d) = ac - bd$.   ∎

COROLLARY 5.1

Let a and b be integers. Let m be a positive integer. Let f be a polynomial with integer coefficients. If $a \equiv b \pmod{m}$, then $f(a) \equiv f(b) \pmod{m}$.

PROOF   Write $f(x) = c_n x^n + \ldots + c_1 x + c_0$, where the c_i are integers. Using $a \equiv b \pmod{m}$ and the last statement of the theorem, we see that $a^2 \equiv b^2 \pmod{m}$, $a^3 \equiv b^3 \pmod{m}$, etc. Using the last statement again, we get $c_i a^i \equiv c_i b^i \pmod{m}$ for each i. Use the first statement n times to add all these congruences and get

$$f(a) = c_n a^n + \ldots + c_1 a + c_0 \equiv c_n b^n + \ldots + c_1 b + c_0 = f(b) \ (\text{mod } m).$$

∎

THEOREM 5.3 Divisor of the modulus
Let a and b be integers. Let m and d be positive integers with $d|m$. If $a \equiv b \ (\text{mod } m)$, then $a \equiv b \ (\text{mod } d)$.

PROOF   We have $d|m$ and $m|(a-b)$. By Theorem 3.1, $d|(a-b)$.   ∎

We saw in Theorem 5.2 that the arithmetic operations of addition, subtraction and multiplication for congruences obey the usual rules for the same operations with integers. However, division does not always work as for integers. For example, $2 \cdot 3 = 6 \equiv 18 = 2 \cdot 9 \ (\text{mod } 12)$, but $3 \not\equiv 9 \ (\text{mod } 12)$.

In general $ac \equiv bc \ (\text{mod } m)$ does not always imply $a \equiv b \ (\text{mod } m)$. We now investigate when this implication will be true.

LEMMA 5.1
If $\gcd(a, m) = 1$ and $0 \le i < j < m$, then $ai \not\equiv aj \ (\text{mod } m)$.

PROOF   If not, then $m|a(i-j)$. Since $\gcd(a, m) = 1$, we have $m|(i-j)$, which contradicts the bounds on i and j.   ∎

THEOREM 5.4 Number relatively prime to modulus has an inverse
If $\gcd(a, m) = 1$, then there is a unique x in $0 < x < m$ such that $ax \equiv 1 \ (\text{mod } m)$.

PROOF   By Lemma 5.1, the function $f(i) = (ai \bmod m)$ for $1 \le i \le m-1$ is one-to-one, and so the set

$$\{ai \bmod m; i = 1, \ldots, m-1\}$$

is a permutation of $\{1, \ldots, m-1\}$. Therefore 1 appears exactly once in the first set, that is, there is exactly one x in $0 < x < m$ such that $ax \equiv 1 \ (\text{mod } m)$.
∎

One can prove Theorem 5.4 from Theorem 3.9 as follows. The latter theorem says that there exist integers x_1 and y so that $ax_1 + my = \gcd(a, m) = 1$. Therefore, m divides $ax_1 - 1$ and we have $ax_1 \equiv 1 \ (\text{mod } m)$. Now use the division algorithm (Theorem 3.3) to find integers q and x with $0 \le x < m$ and $x_1 = mq + x$. Then $ax \equiv 1 \ (\text{mod } m)$. Clearly, x cannot be 0, so $0 < x < m$.

Note that the x in this theorem is like "a^{-1}," the reciprocal of a modulo m. Sometimes we even use the notation "a^{-1} (mod m)" to mean the x of Theorem 5.4.

Now we have enough theory to tell when cancellation is allowed in congruences.

THEOREM 5.5 Division in congruences
If $m > 1$, a, b, c are integers, $(c \neq 0)$, $\gcd(c, m) = 1$, then $ac \equiv bc$ (mod m) implies $a \equiv b$ (mod m).

PROOF   By the previous theorem, there is an x such that $cx \equiv 1$ (mod m). Then $ac \equiv bc$ (mod m) implies $acx \equiv bcx$ (mod m), which implies $a1 \equiv b1$ (mod m), which implies $a \equiv b$ (mod m).   ∎

DEFINITION 5.2   A set of m integers $r_1, \ldots, r_m$ is a **complete set of residues (CSR) modulo** m if every integer is congruent modulo m to exactly one of the r_i's.

The set $\{1, \ldots, m\}$ is called the standard CSR modulo m.

Example 5.1

The set $\{0, \ldots, m-1\}$ is a CSR modulo m, as is the set of all integers between $-m/2$ and $m/2$ (including exactly one of these endpoints if m is even). The set $\{-10, 91, -3, 13, 109\}$ is a CSR modulo 5.

5.2   Linear Congruences

In this section we tell how to solve congruences like $ax \equiv b$ (mod m), where a, b and $m > 1$ are given integers and x is an unknown integer. The solution to an equation $ax = b$, where $a \neq 0$, is the single number $x = a/b$. In contrast, if the congruence $ax \equiv b$ (mod m) has any solution, then infinitely many integers x satisfy it.

For example, the solution to the congruence $2x \equiv 1$ (mod 5) is all integers of the form $x = 5k + 3$, where k may be any integer, that is, x lies in the arithmetic progression

$$\ldots, -12, -7, -2, 3, 8, 13, 18, \ldots.$$

This set of integers may be described compactly as $x \equiv 3$ (mod 5). We could have written this solution as $x \equiv 28$ (mod 5), but we generally use the least nonnegative residue as the standard representative of its congruence class.

Suppose $f(x)$ is a polynomial with integer coefficients and $m > 1$. If x_0 is an integer for which $f(x_0) \equiv 0$ (mod m), then $f(x_0 + km) \equiv 0$ (mod m) for every integer k by Corollary 5.1. Thus $x = x_0 + km$ is a solution to $f(x) \equiv 0 \pmod{m}$ for every integer k. The whole congruence class $(x_0 \pmod{m})$ satisfies $f(x) \equiv 0 \pmod{m}$. However, when considering solutions of congruences we consider these numbers to be just one solution. By the **solution** to the congruence $f(x) \equiv 0 \pmod{m}$ we mean a list of the different congruence classes which satisfy it. These may be described by giving one representative from each class, typically the least nonnegative one.

DEFINITION 5.3   *If $f(x)$ is a polynomial with integer coefficients and m is a positive integer, then the **number of solutions** to $f(x) \equiv 0 \pmod{m}$ is the number of numbers r in a fixed CSR modulo m such that $f(r) \equiv 0 \pmod{m}$.*

By Corollary 5.1, the number of solutions does not depend on which CSR is used. Also, the number of solutions cannot exceed m. When m is small we can solve $f(x) \equiv 0 \pmod{m}$ by testing each x between 0 and $m - 1$.

Example 5.2

Solve the congruence $f(x) = x^2 + x + 1 \equiv 0 \pmod{7}$.
    Evaluate $f(x)$ for $x = 0, 1, \ldots, 6$. We find that $f(2) = 7 \equiv 0 \pmod{7}$ and $f(4) = 21 \equiv 0 \pmod{7}$, but $f(x) \not\equiv 0 \pmod{7}$ for $x = 0, 1, 3, 5, 6$. Therefore, the solution to $f(x) = x^2 + x + 1 \equiv 0 \pmod{7}$ is $x \equiv 2$ or $4 \pmod{7}$.

DEFINITION 5.4   *Let $f(x) = a_n x^n + a_{n-1} x^{n-1} + \ldots + a_0$ be a polynomial with integer coefficients. If m divides every coefficient a_i, then the congruence $f(x) \equiv 0 \pmod{m}$ has no degree. Otherwise, the **degree of the congruence** $f(x) \equiv 0 \pmod{m}$ is the largest integer d in $0 \leq d \leq n$ for which $a_d \not\equiv 0 \pmod{m}$.*

The degree of the congruence $f(x) \equiv 0 \pmod{m}$ is not always the same as the degree of the polynomial $f(x)$. Also, the congruence $f(x) \equiv 0 \pmod{m}$ may have different degrees for different m.

Example 5.3

Let $f(x) = 10x^4 + 8x^3 - 20x + 6$. Then the polynomial $f(x)$ has degree 4. The congruence $f(x) \equiv 0 \pmod{m}$ has degree 4 when $m = 3$ or $m = 7$, degree 3 when $m = 5$ or $m = 10$, and no degree when $m = 2$.

The next two theorems tell how to solve a congruence of degree 1.

THEOREM 5.6 Solvability of a linear congruence
Let $m > 1$, a and b be integers. Then $ax \equiv b \pmod{m}$ has a solution if and only if $\gcd(a, m)$ divides b.

PROOF   The congruence $ax \equiv b \pmod{m}$ is equivalent to saying that there is a whole number y so that $ax - b = my$. Let $g = \gcd(a, m)$. Since $g|a$ and $g|m$, if there is a solution, then $g|b$ by Theorem 3.2. This proves that if g does not divide b, there is no solution.

Conversely, suppose $g|b$. Suppose $b = gt$, where t is an integer. By Theorem 3.9, there are integers u and v so that $au + mv = g$. Multiply by t to get $a(ut) + m(vt) = gt = b$, and $x = ut$ is a solution to $ax \equiv b \pmod{m}$. If we divide $au + mv = g$ by g, we get $(a/g)u + (m/g)v = 1$, which shows that u is a solution to $(a/g)u \equiv 1 \pmod{(m/g)}$.   ∎

THEOREM 5.7 Solutions of a linear congruence

Let $m > 1$, a and b be integers. Let $g = \gcd(a, m)$. If $g|b$, then $ax \equiv b \pmod{m}$ has g solutions. They are

$$x \equiv \frac{b}{g}x_0 + t\frac{m}{g} \pmod{m}, \quad t = 0, 1, \ldots, g-1,$$

where x_0 is any solution of $\frac{a}{g}x_0 \equiv 1 \pmod{\frac{m}{g}}$. This means that

$$x = \frac{b}{g}x_0 + t\frac{m}{g}, \quad t = 0, 1, \ldots,$$

are all integer solutions x.

PROOF   By Theorem 5.6 we know the congruence has a solution. The last two sentences of the proof of Theorem 5.6 show that one solution is $x = (b/g)x_0$, where x_0 is any solution of $(a/g)x_0 \equiv 1 \pmod{(m/g)}$.

Now suppose $x = y$ and $x = z$ are two solutions to $ax \equiv b \pmod{m}$ and assume that y and z are the least nonnegative representatives of their congruence classes. Then $ay \equiv az \equiv b \pmod{m}$. This means that $ay - az = mk$ for some integer k. Divide by g to get $(a/g)y - (a/g)z = (m/g)k$, which shows that $(a/g)y \equiv (a/g)z \pmod{(m/g)}$. Now a/g and m/g are relatively prime by Theorem 3.8; so, by Theorem 5.5 we can cancel them and get $y \equiv z \pmod{(m/g)}$. This means that $y = z + j(m/g)$ for some integer j. Use the division algorithm (Theorem 3.3) to write $j = sg + t$, where s and t are integers and $0 \leq t \leq g-1$. Then $y = z + (sg+t)(m/g) = z + sm + t(m/g)$ and we have $y \equiv z + t(m/g) \pmod{m}$. As we let t run through the g different values $t = 0, 1, \ldots, g-1$ we get all solutions to $ax \equiv b \pmod{m}$, one of which must be the solution $x = (b/g)x_0$ we found above. To show that these solutions are distinct, that is, incongruent modulo m, suppose to the contrary that $z + t(m/g) \equiv z + u(m/g) \pmod{m}$, where $0 \leq t < u \leq g-1$. Then m divides $t(m/g) - u(m/g) = (t-u)(m/g)$. Hence g divides $(t-u)$, which is impossible since $0 < u-t < g$ because $0 \leq t < u \leq g-1$. Therefore there are exactly g distinct solutions, those in the statement of the theorem.   ∎

Example 5.4

Solve $7x \equiv 3 \pmod{12}$.

We find $g = \gcd(7, 12) = 1$ and $g \mid b$ since $1 \mid 3$, so there is one solution. Since $7 \cdot 7 \equiv 1 \pmod{12}$, we have $x_0 = 7$, and the solution to $7x \equiv 3 \pmod{12}$ is $x = 3 \cdot 7 + t \cdot 12 = 21 + 12t \equiv 9 \pmod{12}$.

Example 5.5

Solve $165x \equiv 100 \pmod{285}$.

In Example 3.4, we calculated that $\gcd(165, 285) = 15$. Since 15 does not divide 100, the congruence has no solution.

Example 5.6

Solve $165x \equiv 105 \pmod{285}$.

As in the preceding example, we have $\gcd(165, 285) = 15$. Since $15 \mid 105$, the congruence has fifteen solutions modulo 285. To find them, we first solve the congruence $(165/15)x_0 \equiv 1 \pmod{(285/15)}$, or $11x_0 \equiv 1 \pmod{19}$. The extended Euclidean algorithm gives $11(7) + 19(-4) = 1$, so $x_0 = 7$. Then $x = (105/15)(7) + t(19) = 7 \cdot 7 + 19t$. The fifteen solutions are

$$x = 7, 7 + 19, 7 + 2 \cdot 19, \ldots, 7 + 14 \cdot 19,$$

that is, $x \equiv 7, 26, 45, 64, \ldots$, or $273 \pmod{285}$. These solutions are the same numbers as $x \equiv 7 \pmod{19}$.

We can solve a system of two linear congruences, with the same modulus, in two unknowns by a method much like that used to solve two equations in two unknowns. The difference is that we must be more careful when dividing. Division by a number relatively prime to the common modulus is simple. It is performed by multiplying by the multiplicative inverse of the divisor modulo the common modulus. When the divisor has a common factor > 1 with the modulus, we must use Theorem 5.7. The following examples illustrate what possibilities exist.

Example 5.7

Solve the system of two linear congruences

$$5x + 4y \equiv 1 \pmod{11}$$
$$x + 2y \equiv 5 \pmod{11}.$$

Subtract 2 times the second congruence from the first to get $3x \equiv 1 - 2 \cdot 5 \equiv 2 \pmod{11}$. When we apply Theorem 5.7 to $3x \equiv 2 \pmod{11}$, we find $g = \gcd(3, 11) = 1 \mid 2$, so there is one solution. We find $x_0 \equiv 4 \pmod{11}$, $x \equiv 8 \pmod{11}$ and $y \equiv 4 \pmod{11}$.

Example 5.8

Solve the system of two linear congruences

$$5x + 4y \equiv 2 \pmod{15}$$
$$x + 2y \equiv 5 \pmod{15}.$$

Subtract 2 times the second congruence from the first to get $3x \equiv 2 - 2 \cdot 5 \equiv 7 \pmod{15}$. When we apply Theorem 5.7 to $3x \equiv 7 \pmod{15}$, we find $g = \gcd(3, 15) = 3$ does not divide 7, so there is no solution. Hence the system has no solution either.

Example 5.9

Solve the system of two linear congruences

$$5x + 4y \equiv 1 \pmod{12}$$
$$x + 2y \equiv 5 \pmod{12}.$$

Subtract 2 times the second congruence from the first to get $3x \equiv 1 - 2 \cdot 5 \equiv 3 \pmod{12}$. When we apply Theorem 5.7 to $3x \equiv 3 \pmod{12}$, we find $g = \gcd(3, 12) = 3|3$, so there are 3 solutions. Here x_0 is the solution to $(3/3)x_0 \equiv 1 \pmod{(12/3)}$ or $x_0 \equiv 1 \pmod 4$. Then

$$x \equiv (b/g)x_0 + t(m/g) \equiv 1 + t \cdot 4 \ (t = 0, 1, 2) \ = 1, 5 \text{ or } 9.$$

Consider first the case $x \equiv 1 \pmod{12}$. With 1 for x the two congruences become

$$4y \equiv 1 - 5 \cdot 1 \equiv 8 \pmod{12}$$
$$2y \equiv 5 - 1 \equiv 4 \pmod{12}.$$

The first of these is implied by the second (multiply by 2), but the reverse implication is false. Apply Theorem 5.7 to the second congruence. We find $g = \gcd(2, 12) = 2|4$, so there are two solutions. Next we compute $x_0 \equiv 1 \pmod 6$ and $y \equiv 2$ or $8 \pmod{12}$. If we had applied Theorem 5.7 to the first congruence, we would find $g = \gcd(4, 12) = 4|8$, so there are four solutions. We get $x_0 \equiv 1 \pmod 3$ and $y \equiv 2, 5, 8$ or $11 \pmod{12}$. However, 5 and 11 must be discarded because they do not satisfy the second congruence. In summary, two solution pairs to the original system of congruences are $(x, y) = (1, 2)$ and $(x, y) = (1, 8)$. In a similar fashion, we find that when $x \equiv 5 \pmod{12}$ we have $y \equiv 0$ or $6 \pmod{12}$ and when $x \equiv 9 \pmod{12}$ we have $y \equiv 4$ or $10 \pmod{12}$. The system of congruences has six solution pairs modulo 12.

We close this section by showing that a congruence modulo a prime may have no more solutions than its degree. The hypothesis that the modulus be prime is essential, as is shown by the example, $f(x) = x^2 - 1 \equiv 0 \pmod 8$, which has the four solutions $x \equiv 1, 3, 5, 7 \pmod 8$. In other words, the square of every odd number is $\equiv 1 \pmod 8$.

THEOREM 5.8 No more solutions than degree
Let p be prime. Let $f(x)$ be a polynomial with integer coefficients and degree d modulo p. Then the congruence $f(x) \equiv 0 \pmod{p}$ has at most d solutions.

PROOF   Use induction on d. If $d = 0$, then $f(x) = a_0 \not\equiv 0 \pmod{p}$, and this congruence has no solution. If $d = 1$, then the congruence has exactly one solution by Theorem 5.6 and the fact that the modulus is prime. Assume the theorem is true for all congruences modulo p with degree $< d$. Suppose that the congruence $f(x) \equiv 0 \pmod{p}$ with degree d has more than d solutions. Let $r_1, \ldots, r_d, r_{d+1}$ be $d + 1$ incongruent solutions modulo p. Let the leading term of $f(x)$ be $a_d x^d$, where $a_d \not\equiv 0 \pmod{m}$. Define a new polynomial

$$g(x) = f(x) - a_d(x - r_1) \cdots (x - r_d).$$

Note that the degree, if any, of the congruence $g(x) \equiv 0 \pmod{p}$ must be less than d because the terms $a_d x^d$ cancel in $f(x)$ and $a_d(x - r_1) \cdots (x - r_d)$. Thus, if the congruence $g(x) \equiv 0 \pmod{p}$ has a degree, then it must have fewer than d solutions by the induction hypothesis. But it is easy to see that this congruence has at least the d solutions $r_1, \ldots, r_d$. Therefore, the congruence $g(x) \equiv 0 \pmod{p}$ has no degree. This means that p divides every coefficient of $g(x)$, and so $g(x) \equiv 0 \pmod{p}$ for every x. In particular, $g(r_{d+1}) \equiv 0 \pmod{p}$. But also, $f(r_{d+1}) \equiv 0 \pmod{p}$. Therefore, $x = r_{d+1}$ is a solution to the congruence

$$a_d(x - r_1) \cdots (x - r_d) \equiv 0 \pmod{p},$$

and this contradicts Lemma 4.2.   ∎

5.3   *The Chinese Remainder Theorem*

The Chinese remainder theorem is a clever, useful and old idea which gets its name because it first appeared in a book, *The Art of War*, by Sun Tsu (or Sun Che, depending on dialect) more than 1500 years ago. It allows one to deduce an integer from its approximate size and its least nonnegative remainder modulo m for a few small m. One early application was to count soldiers by ordering them to, "Count off by sevens," then, "Count off by tens," etc., and remember the numbers shouted by the last soldier. Suppose an army of a few hundred men was assembled and the "Count off" orders showed that the remainders of the number x of soldiers modulo 7, 10 and 13 were 1, 3 and 8, respectively. Then x satisfies the system of congruences

$$\begin{aligned} x &\equiv 1 \pmod{7} \\ x &\equiv 3 \pmod{10} \\ x &\equiv 8 \pmod{13}. \end{aligned} \tag{5.1}$$

THEOREM 5.9 Chinese remainder theorem

Let $n_1, \ldots, n_r$ be r positive integers relatively prime in pairs. (That is, $\gcd(n_i, n_j) = 1$ whenever $1 \le i < j \le r$.) Let $a_1, \ldots, a_r$ be any r integers. Then the r congruences

$$x \equiv a_i \pmod{n_i}$$

for $i = 1, \ldots, r$ have common solutions. Any two common solutions are congruent modulo $n = n_1 \cdots n_r$.

Note that the hypotheses are satisfied in the example in Congruences (5.1) above because any two of the moduli 7, 10, 13 are relatively prime. The proof of the Chinese remainder theorem gives a reasonably efficient algorithm for computing the common solution.

PROOF   For $j = 1, \ldots, r$, the number n/n_j is an integer. We claim that $\gcd(n/n_j, n_j) = 1$. If not, then some prime d would have to divide both n/n_j and n_j. But n/n_j is the product of the n_i for $i \ne j$, so d would have to divide some n_i for $i \ne j$ by Lemma 4.2. Then n_i and n_j would not be relatively prime, contrary to hypothesis. Since $\gcd(n/n_j, n_j) = 1$, by Theorem 5.4, there is an integer b_j such that $(n/n_j)b_j \equiv 1 \pmod{n_j}$. Clearly, $(n/n_j)b_j \equiv 0 \pmod{n_i}$ if $i \ne j$ because n_i divides (n/n_j). Let $x_0 = \sum_{j=1}^{r}(n/n_j)b_j a_j$. Let $\delta_{ij} = 1$ if $i = j$ and $= 0$ if $i \ne j$. Then $x_0 = \sum_{j=1}^{r}(n/n_j b_j)a_j \equiv \sum_{j=1}^{r} \delta_{ij} a_j \equiv a_i \pmod{n_i}$. Thus there is a common solution x_0.

If x_1 is another common solution, then $x_0 \equiv a_i \equiv x_1 \pmod{n_i}$ and thus $n_i | (x_0 - x_1)$ for each i. Now use the division algorithm (Theorem 3.3) to divide $(x_0 - x_1)$ by n: $(x_0 - x_1) = qn + r$, where $0 \le r < n$. Suppose $r > 0$. Note that n is the least common multiple of $n_1, \ldots, n_r$, by Theorem 4.4. But r is a smaller nonnegative common multiple of these numbers, by Theorem 3.2. Therefore, $r = 0$ and $x_0 \equiv x_1 \pmod{n}$.   ∎

Example 5.10

Solve the problem of counting the soldiers in Congruences (5.1).

We have $n_1 = 7$, $n_2 = 10$, $n_3 = 13$, $a_1 = 1$, $a_2 = 3$, $a_3 = 8$ and $n = 910$. Then $n/n_1 = 10 \cdot 13 \equiv 4 \pmod 7$. The extended Euclidean algorithm gives $b_1 \equiv 4^{-1} \equiv 2 \pmod 7$. Likewise, $b_2 \equiv 1^{-1} \equiv 1 \pmod{10}$ and $b_3 \equiv 5^{-1} \equiv 8 \pmod{13}$. Then

$$x \equiv 130 \cdot 2 \cdot 1 + 91 \cdot 1 \cdot 3 + 70 \cdot 8 \cdot 8 = 5013 \equiv 463 \pmod{910}.$$

Since there are a few hundred soldiers, there must be exactly 463 of them, as the next integer solution would be $463 + 910 = 1373$.

Here is a simple algorithm based on the proof of the Chinese remainder theorem.

[Solve simultaneous congruences via the Chinese remainder theorem]
Input: Integers $r > 1$, $n_1, \ldots, n_r$, all > 1, with $\gcd(n_i, n_j) = 1$
    for $1 \leq i < j \leq r$, and integers $a_1, \ldots, a_r$.
Output: The solution x of $x \equiv a_i \pmod{n_i}$ for $1 \leq i \leq r$.

```
n = ∏ⁱ₌₁ʳ nᵢ
for  (i = 1 to r)  {
        mᵢ = n/nᵢ
        bᵢ = (mᵢ mod nᵢ)⁻¹  (mod  nᵢ)
        cᵢ = mᵢ · bᵢ
        }
x = 0
for  (i = 1 to r)  {
        x = x + cᵢ · aᵢ
        }
return x
```

Note that if we have many systems of simultaneous congruences to solve
with the *same set of moduli* $n_1, \ldots, n_r$, and different $a_1, \ldots, a_r$, then the first
`for` loop needs to be performed only once. Of course, the first `for` loop takes
most of the time because it uses the extended Euclidean algorithm to find the
modular inverses.

THEOREM 5.10 Complexity of Chinese remainder algorithm
*The first `for` loop in the Chinese remainder algorithm takes $O(r(\log n)^3)$ bit
operations. The second `for` loop in the Chinese remainder algorithm takes
$O(r(\log n)^2)$ bit operations.*

PROOF   The calculations are performed modulo n or modulo some $n_i \leq n$,
so by Corollary 3.5, each arithmetic operation can be done in $O((\log n)^2)$ bit
operations. Each `for` loop runs from 1 to r. If we ignore the time for the
modular inverses, each loop requires $O(r(\log n)^2)$ bit operations. A total of r
modular inverses are performed in the first `for` loop. By Corollary 3.1 each
takes $O((\log n)^3)$ bit operations; so, the total number of bit operations for the
first `for` loop is $O(r(\log n)^3)$.   ∎

Let the positive integers $n_1, \ldots, n_r$ be relatively prime in pairs and let
$n = n_1 \cdots n_r$, as in the hypothesis of the Chinese remainder theorem. Let $\mathcal{S}$
be the set of integers between 0 and $n - 1$ and let $\mathcal{T}$ be the set of all r-tuples
$(a_1, \ldots, a_r)$ of integers, where $0 \leq a_i \leq n_i$ for $1 \leq i \leq r$. Both sets $\mathcal{S}$ and $\mathcal{T}$
contain n elements. The function $f(x)$ from $\mathcal{S}$ into $\mathcal{T}$ defined by

$$f(x) = (x \bmod n_1, \ldots, x \bmod n_r)$$

gives a one-to-one correspondence between the two sets. It is one-to-one because if

$$(x \bmod n_1, \ldots, x \bmod n_r) = (y \bmod n_1, \ldots, y \bmod n_r),$$

then $x \equiv y \pmod{n}$ by the Chinese remainder theorem, so that $x = y$. The proof of the Chinese remainder theorem and the algorithm above show how to compute the inverse function to $f(x)$.

One use of the Chinese remainder theorem is to solve congruences modulo composite numbers when we already know how to solve them modulo prime powers. For example, in Chapter 7, we will learn how to to find modular square roots, that is, how to solve congruences like $x^2 \equiv a \pmod{q}$, where q is a power of a prime. Suppose we wish to solve $x^2 \equiv a \pmod{m}$, where m has the standard factorization $m = p_1^{e_1} \cdots p_r^{e_r}$. Suppose we can solve $x^2 \equiv a \pmod{p_i^{e_i}}$ for each i, and that $x = a_i$ is a solution to this congruence. Since the prime powers $p_i^{e_i}$ are relatively prime in pairs, we can use the Chinese remainder theorem to solve the system $x \equiv a_i \pmod{p_i^{e_i}}$ for $1 \le i \le r$ and get a solution x to $x^2 \equiv a \pmod{m}$. Furthermore, different r-tuples of solutions $(a_1, \ldots, a_r)$ will give different solutions x.

Another use of the Chinese remainder theorem is to perform complex calculations with large integers. Suppose we wish to compute the determinant of a square matrix with integer entries. The determinant is an integer. Suppose we have an upper bound B on its absolute value. (For example, Hadamard [50] proved that the absolute value of the determinant of a $k \times k$ matrix whose entries are $\le M$ cannot exceed $B = k^{k/2} M^k$.) Choose moduli $n_1, \ldots, n_r$ of convenient size and relatively prime in pairs. For example, one might choose the moduli to be distinct primes slightly less than 2^{30} whose product exceeds $2B$. For each i, compute the determinant modulo n_i, for example, by using Gaussian elimination to transform the matrix into a diagonal matrix. Remember that when the algorithm requires dividing by some integer d, one must instead multiply by $d^{-1} \pmod{n_i}$. Let a_i be the value of the determinant modulo n_i. Then use the Chinese remainder theorem to compute the determinant modulo $n = n_1 \cdots n_r$. Call the value x, where $0 \le x < n$. Since $n > 2B > 2|\text{determinant}|$, we know that if $x \le n/2$, then the determinant is x and if $x > n/2$, then the determinant is $x - n < 0$. Note that there are no congruences in the question or the answer.

5.4   Exercises

1. A computer job starts at 9 PM and runs for 100 hours. At what time of day will it end?

2. Let $m > 1$. Prove that $a \equiv b \pmod{m}$ if and only if $a \bmod m = b \bmod m$.

3. Write a congruence to say that $x = 12k + 5$ for some integer k.

4. Show that the unit's digit of a square written in decimal must be one of 0, 1, 4, 5, 6 or 9.

5. Which decimal digits can occur as the last digit of the fourth power of an integer?

6. Find a complete residue system modulo 13 consisting only of multiples of 5.

7. Write a single congruence equivalent to the pair of congruences $x \equiv 3 \pmod 5$ and $x \equiv 4 \pmod 6$.

8. Solve the congruence $15x \equiv 9 \pmod{36}$.

9. Solve the congruence $9x \equiv 15 \pmod{36}$.

10. Solve the congruence $36x \equiv 9 \pmod{15}$.

11. Solve the system of two congruences $3x + 4y \equiv 2 \pmod 7$ and $1x + 2y \equiv 5 \pmod 7$.

12. What are the degrees of the congruence $12x^3 + 2x - 3 \equiv 0 \pmod m$ when $m = 2$, when $m = 3$ and when $m = 5$?

13. Solve the system of simultaneous congruences $x \equiv 3 \pmod 5$, $x \equiv 2 \pmod 7$, $x \equiv 1 \pmod 8$.

14. Find an integer that leaves a remainder of 3 when divided by 6 or 7, and which is a multiple of 5.

15. For which positive integers m is it true that

$$1 + 2 + 3 + \cdots + (m - 1) \equiv 0 \pmod m?$$

16. Let u_n denote the n-th Fibonacci number. Prove that a multiplicative inverse of u_n modulo u_{n+1} is $(-1)^{n+1}u_n$.

17. Show that the system of congruences $x \equiv a_i \pmod{n_i}$, for $i = 1, \ldots, r$, has a common solution if and only if $\gcd(n_i, n_j)$ divides $a_i - a_j$ for every pair (i, j) with $1 \leq i < j \leq r$. Prove that if a common solution exists, then it is unique modulo $\mathrm{lcm}(n_1, \ldots, n_r)$.

Chapter 6

Euler's Theorem and Its Consequences

Fermat's little theorem is a statement about primes that nearly characterizes them. Euler generalized Fermat's theorem to a statement about any positive integer. Many interesting properties about congruences follow from Euler's theorem. See the number theory texts [99], [78] and [98] for more about this subject.

6.1 Fermat's Little Theorem

This exceedingly useful theorem was proved more than 350 years ago by Pierre de Fermat.

THEOREM 6.1 Fermat's "little" theorem
Let p be prime and a be an integer that is not a multiple of p. Then $a^{p-1} \equiv 1 \pmod{p}$.

Fermat proved this theorem before congruences were invented. He expressed the conclusion by saying that p divides $a^{p-1} - 1$ (in French, of course).

PROOF Since $\gcd(a, p) = 1$, as we saw in the proof of Theorem 5.4, the set $\{ai \bmod p; i = 1, \ldots, p - 1\}$ is a permutation of the set $\{1, \ldots, p - 1\}$. Therefore,

$$a^{p-1} \prod_{i=1}^{p-1} i = \prod_{i=1}^{p-1} (ai) \equiv \left(\prod_{i=1}^{p-1} i \right) \cdot 1 \pmod{p}.$$

Now $\gcd \left(\prod_{i=1}^{p-1} i, p \right) = 1$ because p, being prime, has no divisor between 2 and $p - 1$. Thus, by Theorem 5.5, we can cancel the product $(p - 1)!$ and get

$a^{p-1} \equiv 1 \pmod{p}$.   ∎

COROLLARY 6.1
If p is prime and a is an integer, then $a^p \equiv a \pmod{p}$.

PROOF   If p does not divide a, then $a^{p-1} \equiv 1 \pmod{p}$ by Theorem 6.1, and the corollary follows when we multiply both sides by a. If p does divide a, then $a^p \equiv 0 \equiv a \pmod{p}$.   ∎

Fermat's little theorem has many uses. It provides an alternate way to compute the multiplicative inverse a^{-1} of a modulo a prime p: Recall that a^{-1} is the residue class modulo p such that $a^{-1}a \equiv aa^{-1} \equiv 1 \pmod{p}$. It is defined only when p does not divide a. In that situation we have $a^{p-1} \equiv 1 \pmod{p}$ by Fermat's little theorem. Factoring out one a gives $a \cdot a^{p-2} \equiv 1 \pmod{p}$, whence $a^{-1} \equiv a^{p-2} \pmod{p}$.

For large p, computing $a^{-1} \bmod p$ by this formula requires roughly the same number of bit operations as computing $a^{-1} \bmod p$ by the extended Euclidean algorithm, provided one uses the following fast exponentiation algorithm, which we state first for integers.

[Fast Exponentiation]
Input: An integer $n \geq 0$ and a number a.
Output: The value a^n.

```
e = n
y = 1
z = a
while (e > 0) {
        if (e is odd) y = y · z
        z = z · z
        e = ⌊e/2⌋
        }
return y
```

Note that we did not require a to be an integer. We said a is a "number." In fact, the algorithm works when a is anything that can be multiplied associatively, such as a real number or even a matrix. When a is a congruence class modulo m, we can use the algorithm to compute $a^n \bmod m$ while keeping the numbers small (smaller than m, that is), by reducing modulo m after each multiplication. The modulus m need not be prime in this application.

THEOREM 6.2 Complexity of fast exponentiation
The fast exponentiation algorithm computes a^n in $O(\log n)$ multiplications.

PROOF   The algorithm works by computing $z = a^{2^i}$ while simultaneously computing the bits of the binary representation of the exponent n (the statement if (e is odd) means, "if the next bit is a 1") from low order to high order, and multiplying the powers a^{2^i} selected by the 1 bits into a partial product y. If we write $n = \sum_{i=0}^{k} b_i 2^i$, where each bit b_i is 0 or 1, then $k = \lceil \log_2 n \rceil$ and the algorithm computes

$$a^n = a^{\left(\sum_{i=0}^{k} b_i 2^i\right)} = \prod_{i=0}^{k} a^{b_i 2^i}.$$

The while loop computes $z = a^{2^i}$ for $0 \le i \le k$ and multiplies a^{2^i} into the partial product y whenever $b_i = 1$. The instructions inside the while loop are performed $k = \lceil \log_2 n \rceil$ times. No more than $2 \log_2 n$ multiplications are performed.   ∎

If we use this theorem to estimate the number of bit operations needed to compute a modular inverse modulo p, the number is $O((\log p)^3)$, the same complexity as the extended Euclidean algorithm would take to find the same inverse. Often, the architecture of the machine will determine which algorithm is faster.

Fermat's little theorem shows that powers modulo a prime p are periodic with period dividing $p - 1$.

THEOREM 6.3 Powers are periodic modulo a prime
Let p be prime and a, e and f be positive integers such that $e \equiv f \pmod{p-1}$ and p does not divide a. Then $a^e \equiv a^f \pmod p$.

PROOF   Since $e \equiv f \pmod{p-1}$, we can write $e = f + m(p-1)$ for some integer m. Then, by Fermat's little theorem, $a^{p-1} \equiv 1 \pmod p$, and we have

$$a^e \equiv a^{f+m(p-1)} \equiv a^f \left(a^{p-1}\right)^m \equiv a^f 1^m \equiv a^f \pmod p.$$

∎

Theorem 6.3 allows us to find the last digits of powers of integers.

Example 6.1

Find the low-order decimal digit of 3^{1234}.

Note that $10 = 2 \cdot 5$. We will compute $3^{1234} \bmod 2$ and $3^{1234} \bmod 5$, and then use the Chinese remainder theorem to combine the residues and find the digit $3^{1234} \bmod 10$. First, since 3 is odd, any power of it will be odd because the product of odd numbers is always odd. Hence, $3^{1234} \bmod 2 = 1$. Now, $1234 \bmod (5-1) = 1234 \bmod 4 = 2$. Since $1234 \equiv 2 \pmod 4$, Theorem 6.3 tells us that $3^{1234} \equiv 3^2 = 9 \equiv 4 \pmod 5$. Therefore, the low-order decimal digit of 3^{1234} is the least nonnegative solution x to the system

$$x \equiv 1 \pmod 2$$
$$x \equiv 4 \pmod 5,$$

which is easily seen to be $x = 9$.

Fermat's little theorem can almost be used to find large primes. The theorem says that if p is prime and p does not divide a, then $a^{p-1} \equiv 1 \pmod p$. Thus, this theorem gives a test for *compositeness*: If p is odd and p does not divide a, and $a^{p-1} \not\equiv 1 \pmod p$, then p is not prime.

If the converse of Fermat's theorem were true, it would give a fast test for *primality*. The converse would say, if p is odd and p does not divide a, and $a^{p-1} \equiv 1 \pmod p$, then p is prime. This converse is not a true statement, although it is true for most p and most a. If p is a large random odd integer and a is a random integer in $2 \le a \le p - 2$, then the congruence $a^{p-1} \equiv 1 \pmod p$ almost certainly implies that p is prime. However, later we will see that there are more reliable tests for primality having the same complexity. In Theorem 6.10, we will prove a true statement quite similar to the false converse.

DEFINITION 6.1   *An odd integer $p > 2$ is called a **probable prime to base** a if $a^{p-1} \equiv 1 \pmod p$. A composite probable prime to base a is called a **pseudoprime to base** a.*

If one had a list of all base a pseudoprimes $< L$, then the following instructions would form a correct primality test for odd integers $p < L$:

1. Compute $r = a^{p-1} \bmod p$.
2. If $r \ne 1$, then p is composite.
3. If p appears on the list of pseudoprimes $< L$, then p is composite.
4. Otherwise, p is prime.

Although this algorithm has occasionally been used, there are much better tests, some having the same complexity.

There are only three pseudoprimes to base 2 below 1000. The first one is $p = 341 = 11 \cdot 31$. By fast exponentiation or otherwise, one finds $2^{340} \equiv 1 \pmod{341}$.

The second pseudoprime p to base 2 has a remarkable property. This number is $p = 561 = 3 \cdot 11 \cdot 17$. Not only is $2^{560} \equiv 1 \pmod{561}$, but also $a^{560} \equiv 1 \pmod{561}$ for *every* integer a with $\gcd(a, 561) = 1$. It is a Carmichael number.

DEFINITION 6.2   *A **Carmichael number** is an odd composite positive integer which is a pseudoprime to every base relatively prime to it.*

The bad news is that there are infinitely many Carmichael numbers [2]. The good news is that Carmichael numbers are so rare that if you choose a random large odd number, it almost certainly will not be a Carmichael number.

6.2 Euler's Theorem

Euler's theorem generalizes Fermat's little theorem to composite moduli and is even more useful for cryptography than Fermat's little theorem. For a composite modulus, the analogue of the exponent $p-1$ is the size of a reduced set of residues.

DEFINITION 6.3 *A* **reduced set of residues (RSR) modulo** *m is a set of integers R, each relatively prime to m, so that every integer relatively prime to m is congruent to exactly one integer in R.*

THEOREM 6.4 The GCD depends only on the residue class
If $a \equiv b$ (mod m), then $\gcd(a, m) = \gcd(b, m)$.

PROOF The congruence means that $a = b + mq$ for some integer q. The theorem then follows from Theorem 3.6. ∎

In view of Theorem 6.4, one may construct a reduced set of residues by starting with a complete set of residues and deleting the members which are not relatively prime to the modulus. If one begins with the standard CSR $\{1, \ldots, m\}$ and deletes the numbers not relatively prime to m, one gets the standard RSR, namely the set of all i in $1 \leq i \leq m$ with $\gcd(i, m) = 1$. All RSR's modulo m have the same size because the definition gives a one-to-one correspondence between the elements of two RSR's.

DEFINITION 6.4 *The common size of all RSR's modulo m is called the* **Euler phi function,** *$\phi(m)$,* **of** *m.*

The Euler phi function is sometimes called the **totient function**.
If we consider the size of the standard RSR, we get this alternate definition: $\phi(m)$ is the number of i in $1 \leq i \leq m$ with $\gcd(i, m) = 1$.

THEOREM 6.5 A multiple of an RSR is an RSR
Let a be relatively prime to m. Let $\{r_1, \ldots, r_{\phi(m)}\}$ be an RSR modulo m. Then $\{ar_1, \ldots, ar_{\phi(m)}\}$ is also an RSR modulo m.

PROOF We first show that each integer ar_i is relatively prime to m.

Suppose not. Then $\gcd(ar_i, m) > 1$ for some i. Hence there is a prime p which divides the gcd, so p divides both m and ar_i. Since $\gcd(a, m) = 1$ and $p|m$, p cannot divide a. Thus $\gcd(p, a) = 1$ and by Lemma 4.1, $p|r_i$. Hence p divides both m and r_i, so $p|\gcd(m, r_i)$ and $\gcd(m, r_i) > 1$, contrary to the hypothesis that $\gcd(r_i, m) = 1$ because r_i is a member of an RSR modulo m.

So far we know that the elements of $\{ar_1, \ldots, ar_{\phi(m)}\}$ are relatively prime to m and that the set has the correct size $\phi(m)$ to be an RSR modulo m. We need only show that no two members of the set are congruent to each other. Suppose to the contrary that $ar_i \equiv ar_j \pmod{m}$. Then Theorem 5.5 shows that $r_i \equiv r_j \pmod{m}$, and so $i = j$ because $\{r_1, \ldots, r_{\phi(m)}\}$ is an RSR modulo m. ∎

Now we can prove the main theorem of this chapter.

THEOREM 6.6 Euler's theorem
Let $m > 1$ and $\gcd(a, m) = 1$. Then $a^{\phi(m)} \equiv 1 \pmod{m}$.

PROOF   Let $\{r_1, \ldots, r_{\phi(m)}\}$ be an RSR modulo m. By Theorem 6.5, the set $\{ar_1, \ldots, ar_{\phi(m)}\}$ is an RSR modulo m, too. Therefore, for all i, there is a unique j so that $r_i \equiv ar_j \pmod{m}$. Then

$$a^{\phi(m)} \prod_{i=1}^{\phi(m)} r_i = \prod_{i=1}^{\phi(m)} (ar_i) \equiv \prod_{i=1}^{\phi(m)} r_i \pmod{m}.$$

Since $\gcd\left(\prod_{i=1}^{\phi(m)} r_i, m\right) = 1$, we can cancel, by Theorem 5.5, and get $a^{\phi(m)} \equiv 1 \pmod{m}$. ∎

Euler's theorem has many corollaries. First, we can derive Fermat's little theorem (Theorem 6.1) as a corollary: If $m = p$ is prime, then $\phi(p) = p - 1$ because the numbers $1, 2, \ldots, p - 1$ are all relatively prime to p, and so they form an RSR modulo p. Also since p is prime, the statement "a is relatively prime to p" has the same meaning as the statement "p does not divide a." Thus, Euler's theorem for a prime modulus says that if p does not divide a, then $a^{p-1} \equiv 1 \pmod{p}$, which is just Fermat's theorem.

Another corollary gives an alternative to the extended Euclidean algorithm for computing modular inverses.

COROLLARY 6.2
Suppose $m > 1$ and $\gcd(a, m) = 1$. Then $a^{\phi(m)-1} \bmod m$ is a multiplicative inverse of a modulo m.

PROOF   We have $a\left(a^{\phi(m)-1} \bmod m\right) \equiv a^{\phi(m)} \equiv 1 \pmod{m}$. ∎

For large m, computing $a^{-1} \bmod m$ by this formula requires roughly the same number of bit operations as computing $a^{-1} \bmod m$ by the extended Euclidean algorithm. However, the latter must be used if $\phi(m)$ is unknown.

Here is another corollary of Euler's theorem, useful in cryptography.

COROLLARY 6.3

Let $m > 1$, x, y and g be positive integers with $\gcd(g, m) = 1$. If $x \equiv y \pmod{\phi(m)}$, then $g^x \equiv g^y \pmod{m}$.

PROOF   We have $x = y + k\phi(m)$ for some integer k, so

$$g^x = g^{y+k\phi(m)} = g^y \left(g^{\phi(m)}\right)^k \equiv g^y \pmod{m},$$

since $g^{\phi(m)} \equiv 1 \pmod{m}$, by Euler's theorem.   ∎

This corollary allows us to give a different solution to an earlier example, this time without using the Chinese remainder theorem.

Example 6.2

Find the low-order decimal digit of 3^{1234}.

We need to find the residue $3^{1234} \bmod 10$. It is easy to see that an RSR modulo 10 is $\{1, 3, 7, 9\}$ because these four integers are relatively prime to 10, while the other integers between 1 and 10 are not. Thus, $\phi(10) = 4$. Now $1234 \bmod 4 = 2$, so that $1234 \equiv 2 \pmod 4$. By Corollary 6.3,

$$3^{1234} \equiv 3^2 = 9 \pmod{10}.$$

The answer 9 is the same answer we obtained in the previous section.

In order to use Euler's theorem, we must be able to compute $\phi(m)$. The next two theorems provide a way to do this when we know the factorization of m.

DEFINITION 6.5   *A real-valued function $f(x)$ defined for positive integers is called **multiplicative** if $f(ab) = f(a)f(b)$ whenever $\gcd(a, b) = 1$.*

DEFINITION 6.6   *The **Cartesian product** of two sets $\mathcal{S}$ and $\mathcal{T}$ is the set $\mathcal{S} \times \mathcal{T}$ of all ordered pairs (s, t) with $s \in \mathcal{S}$ and $t \in \mathcal{T}$.*

The set $\mathcal{S} \times \mathcal{T}$ is called the Cartesian *product* because when $\mathcal{S}$ and $\mathcal{T}$ are finite sets, the number of elements in $\mathcal{S} \times \mathcal{T}$ equals the product of the number of elements in $\mathcal{S}$ times the number of elements in $\mathcal{T}$. It is called *Cartesian* because Descartes invented it.

THEOREM 6.7 The Euler phi function is multiplicative

The Euler phi function, $\phi(x)$, is multiplicative, that is, $\phi(mn) = \phi(m)\phi(n)$ whenever m and n are relatively prime.

PROOF   The statement is trivial if $m = 1$ or $n = 1$. We assume now that $m > 1$ and $n > 1$.

For positive integers k, let $R(k)$ be the standard RSR modulo k, that is, the set of all i in $1 \leq i \leq k$ with $\gcd(i, k) = 1$. Note that $R(1) = \{1\}$, but k is not an element of $R(k)$ for $k > 1$. The size of $R(k)$ is $\phi(k)$.

We will show that $\phi(mn) = \phi(m)\phi(n)$ by constructing a one-to-one correspondence between $R(mn)$ and $R(m) \times R(n)$. Define a function f from $R(mn)$ into $R(m) \times R(n)$ by

$$f(x) = (x \bmod m, x \bmod n).$$

We must show that the function is well defined. Suppose x is relatively prime to mn. Then $\gcd(x, m) = 1$ and $\gcd(x \bmod m, m) = \gcd(x, m) = 1$. Also, $1 \leq x \bmod m \leq m - 1$ because $m > 1$. Therefore, $x \bmod m$ is in $R(m)$. Similarly, $x \bmod n$ is in $R(n)$. Therefore f is well defined.

Now define a function g from $R(m) \times R(n)$ to $R(mn)$ as follows. Given a pair (a, b) in $R(m) \times R(n)$, use the Chinese remainder theorem to find the unique solution $x = g((a, b))$ to the congruences

$$x \equiv a \pmod{m}$$
$$x \equiv b \pmod{n}$$

with $0 \leq x < mn$. We now show that x really is an element of $R(mn)$. Note first that $\gcd(x, m) = \gcd(a, m) = 1$ because $x \equiv a \pmod{m}$ and a is in $R(m)$. (We have used Theorem 6.4 here.) Similarly, $\gcd(x, n) = 1$. If x were not in $R(mn)$, then $\gcd(x, mn) > 1$. Suppose a prime p divides $\gcd(x, mn)$. Then $p|mn$ and hence $p|m$ or $p|n$ by Lemma 4.2. If $p|m$, then $p| \gcd(x, m)$, which cannot happen. Likewise, p cannot divide n either. Therefore, x is in $R(mn)$ and the function g is well defined.

It is clear from their definitions that f and g are inverse functions to each other. Therefore, $R(mn)$ and $R(m) \times R(n)$ have the same size and $\phi(mn) = \phi(m)\phi(n)$. ∎

THEOREM 6.8 Formulas for the Euler ϕ function

Let p be prime, and e and m be positive integers. Then:

1. $\phi(p) = p - 1$,
2. $\phi(p^e) = p^e - p^{e-1}$,
3. if $m = \prod_{i=1}^{k} p_i^{e_i}$, then $\phi(m) = \prod_{i=1}^{k} \left(p_i^{e_i} - p_i^{e_i-1}\right)$, and
4. $\phi(m) = m \prod_{q|m,\ q \text{ prime}} \left(1 - \frac{1}{q}\right)$.

PROOF   1. The $p-1$ numbers $1, \ldots, p-1$ are all relatively prime to p.

2. The p^e numbers $1, \ldots, p^e$ are all relatively prime to p except for the p^{e-1} multiples of p: $p, 2p, \ldots, p^{e-1}p$.

3. Using Theorem 6.7 and the fact that powers of different primes are relatively prime, we have that if $m = \prod_{i=1}^{k} p_i^{e_i}$, then

$$\phi(m) = \phi\left(\prod_{i=1}^{k} p_i^{e_i}\right) = \phi(p_1^{e_1})\phi\left(\prod_{i=2}^{k} p_i^{e_i}\right) = \cdots = \prod_{i=1}^{k} \phi(p_i^{e_i}).$$

Now use Part 2.

4. Factor out $p_i^{e_i}$ from the i-th factor in the product in Part 3. These factors multiply to produce m and the remaining factors give $\prod_{i=1}^{k}\left(1 - \frac{1}{p_i}\right)$.
∎

The following corollary is just a special case of Part 3 of the theorem, but it is important for the RSA cipher.

COROLLARY 6.4
If p and q are distinct primes, then $\phi(pq) = (p-1)(q-1)$.

The next corollary leads to a true converse of Fermat's little theorem.

COROLLARY 6.5
If $m > 1$ is composite, then $\phi(m) \leq m - 2$.

PROOF   If m is a prime power p^e with $e > 1$, then $\phi(m) = p^e - p^{e-1} \leq p^e - 2$ because $p \geq 2$.

If m has at least two different prime factors p and q, then these two primes, at least, are between 1 and $m - 1$ and not relatively prime to m, so $\phi(m) \leq (m-1) - 2$.   ∎

DEFINITION 6.7   If $m > 1$ and $\gcd(a, m) = 1$, then the **order of a modulo m** is the smallest positive integer e for which $a^e \equiv 1 \pmod{m}$.

The order e of a modulo m is well defined because, by Euler's theorem, $a^{\phi(m)} \equiv 1 \pmod{m}$ when $\gcd(a, m) = 1$, so that $1 \leq e \leq \phi(m)$.

Classical number theory books write "a belongs to the exponent e" for "a has order e."

THEOREM 6.9   Multiples of the order
Let $m > 1$, $\gcd(a, m) = 1$ and e be the order of a modulo m. Then the positive integer x is a solution of $a^x \equiv 1 \pmod{m}$ if and only if $e | x$.

PROOF   If $e|x$, say $x = ek$, then $a^x = (a^e)^k \equiv 1^k \equiv 1 \pmod{m}$ because $a^e \equiv 1 \pmod{m}$. Now suppose $a^x \equiv 1 \pmod{m}$. Use the division algorithm, Theorem 3.3, to write $x = eq + r$ with $0 \le r < e$. Then $a^x \equiv a^{eq+r} \equiv (a^e)^q a^r \equiv a^r \pmod{m}$. But $a^x \equiv 1 \pmod{m}$, so $a^r \equiv 1 \pmod{m}$. Since e is the smallest positive integer for which $a^e \equiv 1 \pmod{m}$ and $0 \le r < e$, we must have $r = 0$. Therefore $e|x$.   ∎

COROLLARY 6.6
If $m > 1$ and $\gcd(a, m) = 1$, then the order of a modulo m divides $\phi(m)$.

PROOF   Euler's theorem tells us that $a^{\phi(m)} \equiv 1 \pmod{m}$; so, the order of a modulo m divides $\phi(m)$ by Theorem 6.9.   ∎

Now we can give our first prime-proving theorem. It is a true converse to Fermat's little theorem.

THEOREM 6.10 Lucas-Lehmer $m - 1$ primality test
Let $m > 1$ and a be integers such that $a^{m-1} \equiv 1 \pmod{m}$, but $a^{(m-1)/p} \not\equiv 1 \pmod{m}$ for every prime p dividing $m - 1$. Then m is prime.

PROOF   The congruence $a^{m-1} \equiv 1 \pmod{m}$ and Theorem 6.9 imply that $\gcd(a, m) = 1$ and the order e of a modulo m divides $m - 1$. The second condition, $a^{(m-1)/p} \not\equiv 1 \pmod{m}$ for every prime p dividing $m - 1$, shows that e is not a proper divisor of $m - 1$. Therefore, e must equal $m - 1$. But by Corollary 6.6, e divides $\phi(m)$. Hence, $m - 1 \le \phi(m)$. But by Corollary 6.5, if $m > 1$ is composite, then $\phi(m) \le m - 2$. Thus, m cannot be composite.   ∎

This theorem can be used to prove primeness of almost any prime m for which we know the factorization of $m - 1$. If m is an odd prime, then usually a small prime a can be found quickly which will satisfy all the conditions. The principal difficulty in using the theorem to prove that a prime m is prime is not the search for a, but rather finding the factorization of $m - 1$. If $m - 1$ has been factored, then one can use this simple algorithm to try to prove it is prime.

[Lucas-Lehmer $m - 1$ primality test]
  1. Choose $a = 2$ or choose a random a in $2 \le a \le m - 1$.
  2. Compute $r = a^{m-1} \bmod m$.
  3. If $r \ne 1$, then m is composite.
  4. Check that $a^{(m-1)/p} \not\equiv 1 \pmod{m}$ for each prime p dividing $m - 1$.
  5. If all these incongruences are true, then m has been proved prime.

6. If they are not satisfied, then either choose another a (either the next small prime or a new random $2 \leq a \leq m - 1$) and go back to Step 2, or else give up if many a have already been tried.

If m is a large prime, then the expected number of a this algorithm must try before finding one that works is known to be $< 2 \ln \ln m$. See Theorem 6.18. If m is composite, but not a Carmichael number (Definition 6.2), then the algorithm will almost certainly stop in Step 3. If m is a Carmichael number, then the algorithm will probably stop when you give up in Step 6.

THEOREM 6.11 Complexity of Lucas-Lehmer $m - 1$ primality test
If the input m of the Lucas-Lehmer $m - 1$ primality test is prime and the complete prime factorization of $m - 1$ is given, then the average time complexity of the algorithm is $O(\log^4 m(\log \log m))$ bit operations.

PROOF   We have already mentioned that the expected number of a which must be tried is $< 2 \ln \ln m$ and will not prove this here. See Theorem 6.18. This estimate gives the factor $\log \log m$ in the theorem statement.

It is easy to see that for each a, most of the work is the calculation of $a^{(m-1)/p} \bmod m$ in Step 4. Each of these exponentiations takes $O(\log m)$ multiplications by Theorem 6.2, and each multiplication takes $O(\log^2 m)$ bit operations by Theorem 3.5. For each a, the exponentiation must be done for each prime p dividing $m - 1$. No integer n can have more than $\log n$ prime divisors because each prime is ≥ 2 and n is the product of its prime divisors, some of which may be repeated. The total complexity is

$$O((\log m - 1)(\log^2 m)(\log m)(\log \log m)) = O(\log^4 m(\log \log m))$$

bit operations.   ∎

We finish this section by stating a generalization of the prime number theorem (Theorem 4.7). Sometimes we need a prime that lies in a specific congruence class modulo d, that is, it lies in a specific arithmetic progression $a + dn$. If the first term a and common difference d have a common factor > 1, then every number in the arithmetic progression is divisible by this common factor, so there cannot be more than one prime.

Example 6.3

In the congruence class 9 mod 12, which is the same as the arithmetic progression $9 + 12n$, every number is divisible by $\gcd(9, 12) = 3$, and there are no primes. The congruence class 3 mod 12 contains only the prime 3 since every number of the form $3 + 12n$ is divisible by 3.

But if the first term a and common difference d are relatively prime, then the arithmetic progression $a + dn$ contains infinitely many primes. In fact, for fixed d every arithmetic progression $a + dn$ which could have infinitely many primes has asymptotically the same number of primes $\leq x$. There are $\phi(d)$ congruence classes relatively prime to d, one for each element of an RSR modulo m, and each class has about $1/\phi(d)$ of the primes.

THEOREM 6.12 Dirichlet's theorem, the prime number theorem for arithmetic progressions
Suppose a and $d > 1$ are integers with $\gcd(a, d) = 1$. Let $\pi_{a,d}(x)$ the number of primes $\equiv a$ (mod d) which are $\leq x$, that is, the number of primes of the form $a + dn \leq x$. Then

$$\lim_{x \to \infty} \frac{\pi_{a,d}(x)}{x/(\phi(d)\ln x)} = 1.$$

The theorem says that, roughly speaking, half of the primes are $\equiv 1$ (mod 4) and half are $\equiv 3$ (mod 4).

There is a more precise version of this theorem that expresses $\pi_{a,d}(x)$ as a main term plus an error term. The **extended Riemann Hypothesis**, ERH, is a statement about the zeros of certain functions, which would imply that the error term in the theorem is as good as possible. The ERH is a famous unsolved problem in number theory. If proved, the ERH would have many consequences throughout number theory.

6.3   *Primitive Roots*

In the previous section, when $\gcd(a, m) = 1$, we defined the order of a modulo m to be the smallest positive integer e for which $a^e \equiv 1$ (mod m). We showed that the order divides $\phi(m)$ and so cannot be larger than $\phi(m)$. Numbers a whose order modulo m equals $\phi(m)$ have important uses in cryptography.

DEFINITION 6.8   *An integer g whose order modulo m is $\phi(m)$ is called a primitive root modulo m.*

If g is a primitive root modulo m, then $\gcd(g, m) = 1$ because the order of g would be undefined if $\gcd(g, m) > 1$.
Some positive integers m have primitive roots and some do not.

THEOREM 6.13 Which integers have primitive roots
A positive integer m has a primitive root if and only if $m = 2$, or 4, or p^e or $2p^e$, where p is an odd prime and e is a positive integer. If m has at least one primitive root, then it has exactly $\phi(\phi(m))$ of them.

Some versions of this theorem in number theory texts state that 1 has order

1 modulo 1, so 1 is a primitive root modulo 1, and 1 is added to the list of the integers m having primitive roots. We omit the proof of this theorem, which has many steps. We focus instead on how to find primitive roots.

Consider first the powers of 2. It is easy to see that 1 is a primitive root modulo 2 and 3 is a primitive root modulo 4. There is no primitive root modulo 8 because $\phi(8) = 4$ and the possible candidates 3, 5 and 7, for primitive root modulo 8, all have order 2. Although there is no primitive root modulo 2^e for $e \geq 3$, the number 5 has order 2^{e-2} modulo 2^e for every $e \geq 3$, and this order is $\phi(2^e)/2$ and is as large as possible.

Theorem 6.13 says that every prime p has primitive roots, in fact $\phi(\phi(p)) = \phi(p-1)$ of them. If one proves p to be prime via the Lucas-Lehmer $m-1$ test, Theorem 6.10, then the number a satisfying all the hypotheses is a primitive root modulo p. The Lucas-Lehmer $m - 1$ primality test provides an efficient way of finding primitive roots for large primes. The complexity of this method of finding primitive roots is given in Theorem 6.11.

If p is an odd prime and g is a primitive root modulo p, then either g or $g + p$, whichever one is odd, is a primitive root modulo p^e for every $e \geq 1$. (However, some even numbers are primitive roots modulo p^e.) If p is an odd prime and g is a primitive root modulo p^2, then g is a primitive root modulo p^e for every $e \geq 1$. If p is an odd prime and g is a primitive root modulo p^e, then either g or $g + p^e$, whichever one is odd, is a primitive root modulo $2p^e$ for every $e \geq 1$. (Of course, no even number can be a primitive root modulo $2p^e$.)

Example 6.4

Let $p = 3$. Then $\phi(\phi(3)) = \phi(2) = 1$; so, there is only one primitive root modulo 3, namely $g = 2$. Since $2 + 3 = 5$ is odd, 5 is a primitive root modulo 3^e for all $e \geq 1$. Actually, 2 is a primitive root modulo 9 because $\phi(9) = 6$ and the powers of 2 modulo 9 are: 2, 4, 8, 7, 5, 1. Therefore, 2 is a primitive root modulo 3^e for every $e \geq 1$. Since, $\phi(\phi(9)) = \phi(6) = 2$, we have found all of the primitive roots modulo 9. They are 2 and 5.

Example 6.5

Let $p = 5$. Then $\phi(\phi(5)) = \phi(4) = 2$, so there are two primitive roots modulo 5, namely $g = 2$ and 3. Since $2 + 5 = 7$ is odd, 7 is a primitive root modulo 5^e for all $e \geq 1$, as is 3.

The following theorems give useful properties of primitive roots. In many applications of the first three theorems, a will be a primitive root modulo m and its order $h = \phi(m)$.

THEOREM 6.14 Order of a power
If $\gcd(a, m) = 1$ and a has order e modulo m, then a^k has order $e/\gcd(e, k)$ modulo m.

PROOF   Let j be the order of a^k modulo m. Let $d = \gcd(e, k)$, so we may write $e = bd$ and $k = cd$ for some integers b and c with $\gcd(b, c) = 1$. We have

$$(a^k)^b \equiv a^{cdb} \equiv (a^{bd})^c \equiv (a^e)^c \equiv 1^c \equiv 1 \pmod{m}.$$

Therefore, by Theorem 6.9, $j|b$. On the other hand, $a^{kj} \equiv (a^k)^j \equiv 1 \pmod{m}$ since j is the order of a^k. Using Theorem 6.9 once more, we see that $e|kj$. This may be written as $bd|cdj$, so $b|cj$. But $\gcd(b, c) = 1$, so $b|j$ by Lemma 4.1. Therefore, $j = b = e/d = e/\gcd(e, k)$. ∎

The next theorem generalizes Theorem 6.3 and Corollary 6.3. It shows that if we wish to compute powers of a modulo m, then we should work modulo the order of a in the exponent.

THEOREM 6.15 Order is the modulus for the exponent
 If $\gcd(a, m) = 1$ and a has order e modulo m, then $a^i \equiv a^j \pmod{m}$ if and only if $i \equiv j \pmod{e}$.

PROOF   Suppose first that $i \equiv j \pmod{e}$. Then $i = j + en$ for some integer n and we have

$$a^i \equiv a^{j+en} \equiv a^j(a^e)^n \equiv a^j 1^n \equiv a^j \pmod{m}.$$

Conversely, suppose $a^i \equiv a^j \pmod{m}$. Interchanging i and j if necessary, we may assume that $i \geq j$, so that $i - j \geq 0$. Now $\gcd(a, m) = 1$ by hypothesis. Therefore, we may cancel a^j from each side and obtain $a^{i-j} \equiv 1 \pmod{m}$. Therefore, by Theorem 6.9, $e|(i - j)$, so $i \equiv j \pmod{e}$. ∎

THEOREM 6.16 The power residues are distinct
 Suppose $\gcd(a, m) = 1$ and a has order e modulo m. Then the powers $1, a, a^2, \ldots, a^{e-1}$ are all different modulo m.

PROOF   Suppose $a^i \equiv a^j \pmod{m}$ where $0 \leq i \leq j \leq e - 1$. Then $i \equiv j \pmod{e}$ by Theorem 6.15. Hence, $e|(j - i)$, and this cannot happen for $0 \leq i \leq j \leq e - 1$ unless $i = j$. Therefore, the e powers of a must be distinct. ∎

THEOREM 6.17 The powers of a primitive root form an RSR
 If g is a primitive root modulo m and $\gcd(b, m) = 1$, then there is exactly one exponent k in $0 \leq k < \phi(m)$ with $g^k \equiv b \pmod{m}$.

PROOF   There cannot be more than one such k by Theorem 6.16. Since $\gcd(g, m) = 1$, every power g^i is relatively prime to m. Therefore, the $\phi(m)$

numbers $g^i \bmod m$ for $i = 0, 1, \ldots, \phi(m) - 1$ are all relatively prime to m. Thus, they must be contained in an RSR modulo m. They are all different by Theorem 6.16. There are enough of the powers to form an RSR modulo m. Hence they must be an RSR modulo m, and so every integer b relatively prime to m must be congruent to one of them. ∎

Using analytic number theory to prove a lower bound on $\phi(p-1)$, one can obtain the following result.

THEOREM 6.18 A lower bound on the number of primitive roots
If $p > 10^{12}$ is prime, then there are at least $(p-1)/(2 \ln \ln p)$ primitive roots modulo p.

See Exercise 4.1 of [33] for an outline of the proof. We used Theorem 6.18 in the proof of Theorem 6.11.

6.4 Discrete Logarithms

If a modulus m and a primitive root g are fixed, then the exponents on the powers of g have properties similar to those of logarithms. Number theorists and cryptographers give different names to these exponents.

DEFINITION 6.9 *Let g be a primitive root modulo m. If the integer b is relatively prime to m, then by Theorem 6.17 there is a unique integer k such that $g^k \equiv b \pmod{m}$ and $0 \le k < \phi(m)$. This integer k is called (by number theorists) the* **index** *of b to base g modulo m and (by cryptographers) the* **discrete logarithm** *of b to base g modulo m.*

The notations $k = \operatorname{ind}_g b$ and $k = \operatorname{Log}_g b$ are used for the index or discrete logarithm of b to base g modulo m (which are the same thing). Both notations suppress the modulus m, which is assumed to be fixed. We will call k a discrete logarithm rather than an index. We write "Log" to emphasize that it is different from ordinary logarithms, which are denoted by "log." In this notation, we have $b \equiv g^{\operatorname{Log}_g b} \pmod{m}$.

Remember that by Theorem 6.13 only the moduli $m = 2$, 4, p^e and $2p^e$ have primitive roots. Thus, m must be one of these numbers in order for discrete logarithms modulo m to be defined.

Many useful properties of discrete logarithms are given in the following theorem. Note their similarity to properties of ordinary logarithms.

THEOREM 6.19 Properties of discrete logarithms
Let g be a primitive root modulo m. Let a and b be integers relatively prime

to m. Then

1. $\text{Log}_g 1 = 0$ and $\text{Log}_g g = 1$,
2. $a \equiv b \pmod{m}$ if and only if $\text{Log}_g a = \text{Log}_g b$,
3. $\text{Log}_g(ab) \equiv \text{Log}_g a + \text{Log}_g b \pmod{\phi(m)}$,
4. $\text{Log}_g(g^k) \equiv k \pmod{\phi(m)}$, and
5. $\text{Log}_g(a^k) \equiv k(\text{Log}_g a) \pmod{\phi(m)}$.

PROOF 1. This is immediate from $g^0 = 1$ and $g^1 = g$.

2. Let $i = \text{Log}_g a$ and $j = \text{Log}_g b$. Then $0 \le i, j < \phi(m)$, $a \equiv g^i \pmod{m}$ and $b \equiv g^j \pmod{m}$. By Theorem 6.15, $g^i \equiv g^j \pmod{m}$ if and only if $i \equiv j \pmod{\phi(m)}$. Then Part 2 follows from the definition.

3. By the definition of discrete logarithm,

$$g^{\text{Log}_g(ab)} \equiv ab \pmod{m}$$

and

$$g^{\text{Log}_g a + \text{Log}_g b} \equiv g^{\text{Log}_g a} \cdot g^{\text{Log}_g b} \equiv ab \pmod{m}.$$

Therefore,

$$g^{\text{Log}_g(ab)} \equiv g^{\text{Log}_g a + \text{Log}_g b} \pmod{m},$$

and Part 3 follows from Theorem 6.15.

4. Let $i = \text{Log}_g(g^k)$. We have $g^i \equiv g^k \pmod{m}$ by the definition of discrete logarithm. Hence, $i = \text{Log}_g(g^k) \equiv k \pmod{\phi(m)}$ by Theorem 6.15.

5. Note first that $g^{\text{Log}_g(a^k)} \equiv a^k \pmod{m}$ and

$$g^{k \cdot \text{Log}_g a} \equiv (g^{\text{Log}_g a})^k \equiv a^k \pmod{m}$$

by definition. Hence, $g^{\text{Log}_g(a^k)} \equiv g^{k \cdot \text{Log}_g a} \pmod{m}$, and Part 5 follows from Theorem 6.15. ∎

For example, 2 is a primitive root modulo 13. The powers of 2 modulo 13 are given in this table.

k:	0	1	2	3	4	5	6	7	8	9	10	11
$2^k \bmod 13$:	1	2	4	8	3	6	12	11	9	5	10	7

Therefore, the discrete logarithms modulo 13 are given in this table.

b:	1	2	3	4	5	6	7	8	9	10	11	12
$\text{Log}_2 b$:	0	1	4	2	9	5	11	3	8	10	7	6

The second table may be formed from the first by sorting the columns in the first table in increasing order of the numbers in the second row and then swapping the two rows.

6.5 Exercises

1. Find the last digit of the base 13 expansion of 7^{200}. (The exponent 200 is in decimal.)

2. The fast exponentiation algorithm processes the bits of the exponent n from right to left as it computes a^n. The following algorithm uses the same bits from left to right to compute a^n. Show that it is correct for $n \geq 2$, and compare its complexity to that of fast exponentiation.

   [Left to Right Fast Exponentiation]
   Input: An integer $n \geq 2$ and a number a.
   Output: The value a^n.

   ```
   write n in binary as n = ∑ᵏᵢ₌₀ bᵢ2ⁱ,
           with bᵢ = 0 or 1, and bₖ = 1
   y = a
   for (i = k − 1 down to 0) {
           y = y²
           if (bᵢ = 1) y = y · x
           }
   return y
   ```

3. Find a reduced residue system modulo 12 consisting entirely of multiples of 5.

4. Prove that if $m > 2$, then the sum of the numbers in any reduced residue system modulo m is a multiple of m.

5. Show that if p and q are distinct primes, then $p^{q-1} + q^{p-1} \equiv 1 \pmod{pq}$.

6. Find the last hexadecimal digit of 7^{1234}. (The exponent 1234 is in decimal.)

7. Find the last two decimal digits of 7^{1234}.

8. Find $\phi(m)$ for each integer m between 20 and 30.

9. For which positive integers m is $\phi(m)$ odd?

10. Solve a quadratic equation to find the primes p and q, given that $n = pq = 4386607$ and $\phi(n) = 4382136$.

11. Show that every odd composite integer is a pseudoprime to base 1 and to base -1.

12. Find a primitive root modulo 19. How many primitive roots modulo 19 are there?

13. Use the Lucas-Lehmer $m - 1$ primality test to prove that 17 is prime.

14. Use the Lucas-Lehmer $m - 1$ primality test to prove that 23 is prime.

15. Multiply (7×9) mod 13 by adding the discrete logarithms of 7 and 9, using the tables at the end of Section 6.4.

16. Find $\mathrm{Log}_g(m - 1)$ when $m > 2$ and g is a primitive root modulo m.

Chapter 7

Second Degree Congruences

In Section 5.2, we learned how to solve linear congruences. This chapter introduces quadratic congruences. Some integer factoring algorithms and some protocols require the rapid solution of quadratic congruences. Certain primality testing methods become improved by the ability to tell whether some second degree congruences have solutions, although one need not find them. See the number theory texts [99], [78] and [98] for more about second degree congruences.

The most general quadratic congruence is

$$ax^2 + bx + c \equiv 0 \pmod{m} \tag{7.1}$$

where a, b and c are integers. If instead we had to solve a quadratic equation $ax^2 + bx + c = 0$, we could use the quadratic formula

$$x = \frac{-b \pm \sqrt{b^2 - 4ac}}{2a}. \tag{7.2}$$

Suppose a, b and c are integers and we wish to try to use Formula (7.2) to solve Congruence (7.1). We could perform the addition, subtraction and multiplication modulo m in Formula (7.2) without difficulty. We could perform the division by $2a$ modulo m provided $\gcd(2a, m) = 1$. (Note that problems arise here when m is even.) The part of the problem that is new in this chapter is taking a square root modulo m. We must solve $y^2 \equiv r \pmod{m}$, where $r = b^2 - 4ac$. This congruence may be solved by first solving it modulo each prime power divisor of m and then combining the solutions via the Chinese remainder theorem. We will see that the solutions modulo a prime power p^k are obtained by first solving $z^2 \equiv r \pmod{p}$ and then "lifting" those solutions to solutions modulo p^k. We begin by studying $z^2 \equiv r \pmod{p}$, where p is prime.

7.1   The Legendre Symbol

The solution to the congruence $x^2 \equiv r \pmod 2$ is simple. There is always one solution modulo 2. If $r = 0$, then $x \equiv 0 \pmod 2$. If $r = 1$, then $x \equiv 1 \pmod 2$.

When p is an odd prime, the congruence $x^2 \equiv r \pmod p$ has solutions for some r and no solution for other r. Consider this table of squares modulo 11:

$x:$	0	1	2	3	4	5	6	7	8	9	10
$x^2 \bmod 11:$	0	1	4	9	5	3	3	5	9	4	1

Note that the congruence classes 0, 1, 3, 4, 5, 9 are squares and the classes 2, 6, 7, 8, 10 are not squares modulo 11. The class 0 has only one square root, but the other classes which are squares each have two square roots. For example, the solutions to $x^2 \equiv 3 \pmod{11}$ are $x \equiv 5$ and $6 \pmod{11}$.

The congruence $x^2 \equiv 0 \pmod p$ always has the unique solution $x \equiv 0 \pmod p$ when p is prime. We exclude this special case from the next definition.

DEFINITION 7.1   *If m is a positive integer and r is relatively prime to m, then we say r is a **quadratic residue (QR) modulo** m if the congruence $x^2 \equiv r \pmod m$ has a solution, and we say r is a **quadratic nonresidue (QNR) modulo** m if the congruence $x^2 \equiv r \pmod m$ has no solution.*

Thus, 1, 3, 4, 5 and 9 are the quadratic residues modulo 11 and 2, 6, 7, 8 and 10 are the quadratic nonresidues modulo 11.

In fact, 1 is always a quadratic residue modulo m. It always has 1 and -1 as square roots. When m is prime, there are no other square roots.

THEOREM 7.1 Square roots of 1 modulo p
If p is prime, then $x^2 \equiv 1 \pmod p$ if and only if $x \equiv \pm 1 \pmod p$.

PROOF   We may write the quadratic congruence as $(x - 1)(x + 1) = x^2 - 1 \equiv 0 \pmod p$. It holds if and only if $p|(x-1)(x+1)$. By Lemma 4.2, this means either $p|(x - 1)$ or $p|(x+1)$, that is, $x \equiv 1 \pmod p$ or $x \equiv -1 \pmod p$.
∎

An alternate proof of Theorem 7.1 uses Theorem 5.8. That theorem says that the congruence $x^2 - 1 \equiv 0 \pmod p$ has no more solutions than its degree. Clearly, 1 and -1 are solutions. There can be no more solutions.

COROLLARY 7.1
If g is a primitive root modulo an odd prime p, then $g^{(p-1)/2} \equiv -1 \pmod p$.

PROOF   Let $x = g^{(p-1)/2}$. Then $x^2 \equiv g^{p-1} \equiv 1 \pmod{p}$, so by Theorem 7.1, either $x \equiv 1 \pmod{p}$ or $x \equiv -1 \pmod{p}$. But if $x \equiv 1 \pmod{p}$, then g would not be a primitive root. Therefore, $x \equiv -1 \pmod{p}$.   ∎

THEOREM 7.2 There are either 0 or 2 square roots of r modulo p

  If p is an odd prime and r is not a multiple of p, then the congruence $x^2 \equiv r \pmod{p}$ has either no solution or exactly two incongruent solutions modulo p.

PROOF   Suppose $x = a$ is a solution to $x^2 \equiv r \pmod{p}$. Then $x = -a$ is also a solution because $(-a)^2 = a^2 \equiv r \pmod{p}$ and $-a \not\equiv a \pmod{p}$ since the odd prime p does not divide $a - (-a) = 2a$. (If $p|a$, then $p|a^2$ and so $p|r$.)

  Suppose b were a third solution to $x^2 \equiv r \pmod{p}$. Then $b^2 \equiv r \equiv a^2 \pmod{p}$, so p divides $b^2 - a^2 = (b - a)(b + a)$. By Lemma 4.2, this means either $p|(b - a)$ or $p|(b + a)$, that is, $b \equiv a \pmod{p}$ or $b \equiv -a \pmod{p}$.

  We have shown that if the congruence $x^2 \equiv r \pmod{p}$ has a solution, then it has exactly two of them. Therefore it has either no solution or exactly two solutions.   ∎

THEOREM 7.3 Equal numbers of quadratic residues and nonresidues

  If p is an odd prime, then there are exactly $(p - 1)/2$ quadratic residues among $1, 2, \ldots, p - 1$, and the same number of quadratic nonresidues.

PROOF   Every one of the $p - 1$ numbers $x = 1, 2, \ldots, p - 1$ satisfies one of the congruences $x^2 \equiv r \pmod{p}$, namely, the one with $r = x^2 \bmod p$. But by Theorem 7.2, each congruence $x^2 \equiv r \pmod{p}$ has either zero or two solutions. Therefore, as r goes from 1 to $p - 1$, half of the congruences $x^2 \equiv r \pmod{p}$ have two solutions x and the other half have no solution. The $(p - 1)/2$ values of $r \pmod{p}$ for which $x^2 \equiv r \pmod{p}$ has two solutions are the quadratic residues modulo p and the other $(p - 1)/2$ values of r are the quadratic nonresidues.   ∎

DEFINITION 7.2   *Let p be an odd prime and r be an integer. The* **Legendre symbol** (r/p) *is defined to be $+1$ if r is a quadratic residue modulo p, -1 if r is a quadratic nonresidue modulo p and 0 if p divides r.*

  This notation was introduced by the French mathematician A.-M. Legendre more than 200 years ago.

THEOREM 7.4 Euler's criterion for r being a quadratic residue

  Let p be an odd prime and r an integer not divisible by p. Then $r^{(p-1)/2} \bmod$

$p = 1$ or $p - 1$. If it is 1, then r is a quadratic residue modulo p, and if it is $p - 1$, then r is a quadratic nonresidue modulo p. In terms of the Legendre symbol,

$$\left(\frac{r}{p}\right) \equiv r^{(p-1)/2} \pmod{p}.$$

PROOF   By Fermat's little theorem, Theorem 6.1, p divides $r^{p-1} - 1 = (r^{(p-1)/2} - 1)(r^{(p-1)/2} + 1)$. By Lemma 4.2 p divides either $(r^{(p-1)/2} - 1)$ or $(r^{(p-1)/2} + 1)$. However, p cannot divide both of these numbers because in that case, by Theorem 3.2, it would divide their difference, which is 2, and p is odd. Thus, $r^{(p-1)/2} \equiv \pm 1 \pmod{p}$.

If r is a quadratic residue modulo p, then by definition there is an a so that $a^2 \equiv r \pmod{p}$, and we have

$$r^{(p-1)/2} \equiv (a^2)^{(p-1)/2} = a^{p-1} \equiv +1 \pmod{p},$$

by Fermat's little theorem. Thus the $(p-1)/2$ quadratic residues are solutions to $r^{(p-1)/2} \equiv +1 \pmod{p}$. As this congruence has degree $(p-1)/2$, it can have no more than $(p - 1)/2$ solutions, by Theorem 5.8. Therefore, the $(p - 1)/2$ quadratic nonresidues must be solutions to $r^{(p-1)/2} \equiv -1 \pmod{p}$.   ∎

If one merely wishes to know whether the congruence $x^2 \equiv a \pmod{p}$ has a solution x, then Euler's criterion, with fast exponentiation to evaluate the power, provides an ideal solution. For many purposes in cryptography, this algorithm is sufficient.

COROLLARY 7.2

Let g be a primitive root modulo an odd prime p. Then the quadratic residues modulo p are the powers of g with even exponents and the quadratic nonresidues modulo p are the powers of g with odd exponents.

PROOF   By Corollary 7.1, $g^{(p-1)/2} \equiv -1 \pmod{p}$. So if i is even, then $(g^i)^{(p-1)/2} \equiv (-1)^i \equiv +1$, while if i is odd, then $(g^i)^{(p-1)/2} \equiv (-1)^i \equiv -1$. Now apply Euler's criterion, Theorem 7.4.   ∎

COROLLARY 7.3

Every primitive root modulo an odd prime p is a quadratic nonresidue modulo p.

The following theorem lists some useful properties of the Legendre symbol.

THEOREM 7.5 Properties of the Legendre symbol

Let p be an odd prime and a and b be integers. Then

1. *the number of solutions to the congruence* $x^2 \equiv a \pmod{p}$ *is* $1 + (a/p)$,
2. $(a/p) \equiv a^{(p-1)/2} \pmod{p}$,
3. $(ab/p) = (a/p)(b/p)$,
4. *if* $a \equiv b \pmod{p}$, *then* $(a/p) = (b/p)$,
5. $(1/p) = +1$ *and* $(-1/p) = (-1)^{(p-1)/2}$, *and*
6. *if* p *does not divide* a, *then* $(a^2/p) = +1$ *and* $(a^2 b/p) = (b/p)$.

PROOF   Part 1 is clear from the definition of quadratic residue and quadratic nonresidue, and from Theorem 7.2. Part 2 follows from Euler's criterion if $\gcd(a, p) = 1$. If p divides a, then Part 2 is true because both sides are $\equiv 0 \pmod{p}$. The other parts follow easily from Part 2.   ∎

One can prove Part 3 directly when $(a/p) = (b/p) = +1$ as follows: Let $x^2 \equiv a \pmod{p}$ and $y^2 \equiv b \pmod{p}$. Then $z = xy$ satisfies $z^2 \equiv ab \pmod{p}$. Similar easy proofs can be given when $(a/p) = -(b/p)$, but not when $(a/p) = (b/p) = -1$.

In Part 5, observe that $(-1)^{(p-1)/2} = +1$ when $p \equiv 1 \pmod{4}$ and $= -1$ when $p \equiv 3 \pmod{4}$, because if $p = 4k + 1$, then $(p-1)/2 = 2k$ is even, and if $p = 4k + 3$, then $(p-1)/2 = 2k + 1$ is odd.

By Part 5 of Theorem 7.5, the number 1 is the smallest positive quadratic residue modulo any prime p. Some number theoretic algorithms require us to find a quadratic nonresidue modulo p. Suppose we try consecutive positive integers looking for one. How far must we search to find the first quadratic nonresidue?

THEOREM 7.6 A bound on the least quadratic nonresidue
If p *is an odd prime and* n *is the smallest positive quadratic nonresidue modulo* p, *then* $n < 1 + \sqrt{p}$.

PROOF   Let $m = \lceil p/n \rceil$ so that $(m-1)n \le p < mn$. Since $n \ge 2$ and p is an odd prime, we actually must have $(m-1)n < p < mn$. Therefore, $0 < mn - p < n$. Since n is the smallest positive quadratic nonresidue, $mn - p$ must be a quadratic residue modulo p. Thus,

$$1 = ((mn - p)/p) = (mn/p) = (m/p)(n/p) = (m/p)(-1),$$

and $(m/p) = -1$ by Parts 4 and 3 of Theorem 7.5. This shows that m is a quadratic nonresidue modulo p, and so $m \ge n$. Then

$$(n - 1)^2 < (n - 1)n \le (m - 1)n < p,$$

and $n - 1 < \sqrt{p}$, as required.   ∎

Theorem 7.6 is not the best upper bound on the size of the first quadratic residue. Burgess [20] has shown that for every $\epsilon > 0$ there is a $p_0(\epsilon)$ so that the least positive quadratic nonresidue modulo p is $< p^{c+\epsilon}$ for all primes $p > p_0(\epsilon)$, where $c = 1/(4\sqrt{e}) \approx 0.1516$. A similar bound holds for the number of consecutive integers we must try, beginning at any number, until we find the first quadratic nonresidue. Even these results seem far from the truth. Vinogradov conjectured that for every $\epsilon > 0$ there is a $p_0(\epsilon)$ so that the least positive quadratic nonresidue modulo p is $< p^\epsilon$ for all primes $p > p_0(\epsilon)$. Although one cannot prove a good upper bound for the least positive quadratic nonresidue, the *average* number of positive integers which must be tried to find a quadratic nonresidue modulo p is 2, by Theorem 7.3. Indeed, the quadratic residues and nonresidues seem to be very evenly distributed in the interval from 1 to $p - 1$.

7.2   The Law of Quadratic Reciprocity

The Law of Quadratic Reciprocity is a beautiful theorem proved by Gauss [45] more than 200 years ago. We do not prove it here because all known proofs are long and complicated.

THEOREM 7.7 Law of Quadratic Reciprocity
If p and q are distinct odd primes, then $(p/q) = (q/p)$ if at least one of p, q is $\equiv 1 \pmod 4$, but $(p/q) = -(q/p)$ if $p \equiv q \equiv 3 \pmod 4$.

The theorem is often stated in the concise form

$$\left(\frac{p}{q}\right)\left(\frac{q}{p}\right) = (-1)^{\frac{p-1}{2} \cdot \frac{q-1}{2}}.$$

It is easy to see that the power of (-1) on the right side is $+1$ if either p or q is $\equiv 1 \pmod 4$, and -1 if $p \equiv q \equiv 3 \pmod 4$; so, this formula is equivalent to the statement above. Another way of stating the theorem is that when p and q are different odd primes, the two congruences

$$x^2 \equiv p \pmod q$$
$$y^2 \equiv q \pmod p$$

are either both solvable or neither is solvable in case either p or q is $\equiv 1 \pmod 4$, and exactly one of the two congruences is solvable if $p \equiv q \equiv 3 \pmod 4$.

Most proofs of Theorem 7.7 prove the following theorem on the way.

THEOREM 7.8 Supplement to the Law of Quadratic Reciprocity
If p is an odd prime, then the congruence $x^2 \equiv 2 \pmod p$ is solvable if $p \equiv 1$ or $7 \pmod 8$, but not solvable if $p \equiv 3$ or $5 \pmod 8$.

Theorem 7.8 has the concise form $(2/p) = (-1)^{(p^2-1)/8}$, which one may verify by evaluating the exponent modulo 2 for p in the four odd congruence classes modulo 8. If p is an odd prime number written in binary with the three low-order bits $b_2 b_1 b_0$ (where $b_0 = 1$), then 2 is a quadratic residue modulo p if and only if $b_2 = b_1$.

Recall that Part 5 of Theorem 7.5 says that $(-1/p) = (-1)^{(p-1)/2}$, that is, the congruence $x^2 \equiv -1$ (mod p) is solvable when $p \equiv 1$ (mod 4), but not when $p \equiv 3$ (mod 4).

One application of the results just stated is in evaluating Legendre symbols by hand with numbers of modest size.

Example 7.1

Is the congruence $x^2 \equiv -22$ (mod 59) solvable?

If we used Euler's criterion, we would have to compute $(-22)^{(59-1)/2}$ mod 59. This is a simple matter, even without fast exponentiation, on a small computer, but it is tedious to perform correctly by hand.

By Part 3 of Theorem 7.5, we can write

$$\left(\frac{-22}{59}\right) = \left(\frac{-1}{59}\right)\left(\frac{2}{59}\right)\left(\frac{11}{59}\right).$$

Part 5 of Theorem 7.5 shows that $(-1/59) = -1$. Theorem 7.8 tells us that $(2/59) = -1$. Theorem 7.7 shows that $(11/59) = -(59/11)$, and then $(59/11) = (4/11) = +1$ by Parts 3 and 6 of Theorem 7.5.

Finally, $(-22/59) = (-1)(-1)(-(+1)) = -1$, so the original congruence has no solution. Note that we performed no exponentiation at all in this solution.

A more important application of the Law of Quadratic Reciprocity is in determining which primes q are quadratic residues modulo a given odd prime p. Theorem 7.8 tells us that the odd primes of the forms $8k + 1$ and $8k + 7$ are the ones which have 2 as a quadratic residue.

Example 7.2

Which odd primes p have 3 as a quadratic residue?

By the Law of Quadratic Reciprocity, $(3/p) = (p/3)(-1)^{(p-1)/2}$. Now $(p/3) = (1/3) = +1$ when $p \equiv 1$ (mod 3) and $(p/3) = (2/3) = -1$ when $p \equiv 2$ (mod 3). Also, $(-1)^{(p-1)/2} = +1$ when $p \equiv 1$ (mod 4) and $(-1)^{(p-1)/2} = -1$ when $p \equiv 3$ (mod 4). This means that $(3/p) = +1$ if and only if $p \equiv 1$ (mod 3) and $p \equiv 1$ (mod 4), or $p \equiv 2$ (mod 3) and $p \equiv 3$ (mod 4), that is, $p \equiv 1$ or 11 (mod 12) by the Chinese remainder theorem.

Example 7.3

Which odd primes p have 5 as a quadratic residue?

The Law of Quadratic Reciprocity says that $(5/p) = (p/5)$. The quadratic residues modulo 5 are 1 and 4; so, the answer is all primes $p \equiv 1$ or 4 (mod 5). Since p must be odd, this condition is the same as $p \equiv 1$ or 9 (mod 10), that is, all primes whose last decimal digit is 1 or 9.

Analogous questions for larger primes may be answered in the same way. Suppose p is fixed and we ask, which odd primes q have p as a quadratic residue? We need to evaluate the Legendre symbol (p/q). If $p \equiv 1 \pmod 4$, then $(p/q) = (q/p)$ and the answer is the primes in $(p-1)/2$ residues classes modulo p, namely the ones which are quadratic residues. Since p and q are odd, this set of residue classes is equivalent to a set of $(p-1)/2$ residues classes modulo $2p$. If $p \equiv 3 \pmod 4$, then $(p/q) = (-1)^{(q-1)/2}(q/p)$, and the answer is the primes in $p-1$ congruence classes modulo $4p$, just as in Example 7.2.

Another use of the Law of Quadratic Reciprocity is to evaluate the Jacobi symbol, which leads to an algorithm for computing the Legendre symbol without factoring the "numerator" and also to improved primality tests.

7.3   *The Jacobi Symbol*

DEFINITION 7.3   *Let m be an odd positive integer with prime factorization $m = \prod_{i=1}^{k} p_i^{e_e}$ and let a be an integer. The **Jacobi symbol** (a/m) is defined by*

$$\left(\frac{a}{m}\right) = \prod_{i=1}^{k} \left(\frac{a}{p_i}\right)^{e_i},$$

where the symbols on the right side are Legendre symbols. We allow $m = 1$, and define $(a/1) = 1$ for every a.

If $\gcd(a, m) > 1$, then some prime factor p of m will also divide a and the Legendre symbol (a/p) in the definition of (a/m) will be 0. Thus, the Jacobi symbol $(a/m) = 0$ when a is not relatively prime to m.

The Jacobi symbol shares many properties with the Legendre symbol. Compare the next theorem with Theorem 7.5.

THEOREM 7.9 Properties of the Jacobi symbol
Let m and n be odd positive integers and a and b be integers. Then
  1. if $a \equiv b \pmod m$, then $(a/m) = (b/m)$,
  2. $(ab/m) = (a/m)(b/m)$,
  3. $(a/mn) = (a/m)(a/n)$, and
  4. if $\gcd(a, m) = 1$, then $(a^2/m) = (a/m^2) = +1$, $(a^2 b/m) = (b/m)$ and $(a/(m^2 n)) = (a/n)$.

PROOF   1. If $a \equiv b \pmod m$, then $a \equiv b \pmod p$ for every prime p dividing m. Hence, $(a/p) = (b/p)$ for every p dividing m by Part 4 of Theorem 7.5. Then $(a/m) = (b/m)$ by Definition 7.3.

2. This formula follows from Definition 7.3 and Part 3 of Theorem 7.5.

3. This formula is immediate from Definition 7.3.

4. These formulas follow from Parts 2 and 3 and $(-1)^2 = 1$. ∎

The next theorem shows that the Jacobi symbol enjoys the same Law of Quadratic Reciprocity as the Legendre symbol, and leads to an efficient algorithm for computing Legendre symbols.

THEOREM 7.10 Law of Quadratic Reciprocity for Jacobi symbols
1. If m is an odd positive integer, then $(-1/m) = +1$ if $p \equiv 1$ (mod 4), and $(-1/m) = -1$ if $p \equiv -1$ (mod 4).
2. If m is an odd positive integer, then $(2/m) = +1$ if $p \equiv 1$ or 7 (mod 8), and $(2/m) = -1$ if $p \equiv 3$ or 5 (mod 8).
3. If m and n are relatively prime positive integers, then $(m/n) = (n/m)$ if at least one of m, n is $\equiv 1$ (mod 4), but $(m/n) = -(n/m)$ if $m \equiv n \equiv 3$ (mod 4).

PROOF Part 5 of Theorem 7.5 says that if p is an odd prime, then the Legendre symbol $(-1/p) = (-1)^{(p-1)/2}$. We will show that $(-1/m) = (-1)^{(m-1)/2}$, which is equivalent to the statement in Part 1 above.

Let m have the prime factorization $m = \prod_{i=1}^{k} p_i^{e_i}$. From Definition 7.3, we have

$$\left(\frac{-1}{m} \right) = \prod_{i=1}^{k} \left(\frac{-1}{p_i} \right)^{e_i} = (-1)^s,$$

where $s = \sum_{i=1}^{k} e_i(p_i-1)/2$ by Euler's criterion. Now $m = \prod_{i=1}^{k}(1+(p_i-1))^{e_i}$. Because each $p_i - 1$ is even, we have

$$(1 + (p_i - 1))^{e_i} \equiv 1 + e_i(p_i - 1) \pmod 4$$

and

$$(1 + e_i(p_i - 1))(1 + e_j(p_j - 1)) \equiv 1 + e_i(p_i - 1) + e_j(p_j - 1) \pmod 4.$$

Using these congruences repeatedly, we find

$$m \equiv 1 + \sum_{i=1}^{k} e_i(p_i - 1) \pmod 4.$$

Hence, $(m - 1)/2 \equiv s$ (mod 2) and we have $(-1/m) = (-1)^{(m-1)/2}$.

Parts 2 and 3 are proved the same way, except that the modulus 64 replaces modulus 4 in the proof of Part 2. ∎

The following algorithm uses Part 2 of Theorem 7.9 and Parts 2 and 3 of Theorem 7.10 to evaluate a Jacobi symbol. It is a recursive algorithm, which means that function calls itself with smaller values of its parameters. The first few instructions check the input data and return special values. The while loop removes factors of 4 from a. The last few lines use the Law of Quadratic Reciprocity to reduce the evaluation of (a/m) to that of a Jacobi symbol with smaller parameters.

[Compute the Jacobi symbol]
Input: An integer a and an odd positive integer m.
Output: The value of the Jacobi symbol (a/m).

```
Recursive function Jac(a, m)
if (m is even or m ≤ 0) Error:  bad input to Jac.
if (m = 1) return 1
if (a ≥ m or a < 0) a = a mod m
if (a = 0) return 0
if (a = 1) return 1
while (4 divides a) { a = a/4 }
if (2 divides a) {
        if (m ≡ 1 or 7 (mod 8)) { return Jac(a/2, m) }
              else { return −Jac(a/2, m) }
      }
if (a ≡ 1 (mod 4) or m ≡ 1 (mod 4)) { return Jac(m mod a, a) }
      else { return −Jac(m mod a, a) }
```

If the algorithm did not remove powers of 2, then the sequence of recursive calls would take the variables a and m through the same sequences of values as in the Euclidean algorithm for computing $\gcd(a, m)$. With a slightly more careful analysis (see Theorem 5.9.3 of Bach and Shallit [8]), one can prove that it is no harder to evaluate the Jacobi symbol (a/m) than to compute $\gcd(a, m)$. Compare this estimate for the complexity with Theorem 3.12.

THEOREM 7.11 Complexity of evaluating Jacobi symbol

Let a and m be relatively prime integers with $0 < a < m$. Then the Jacobi symbol (a/m) can be evaluated, using $O(\log^3 m)$ bit operations, by the algorithm above.

Recall that we can compute a modular inverse in about the same time using either Euler's theorem or the extended Euclidean algorithm. Similarly, it takes about as long to evaluate a Legendre symbol by the Euler criterion as by the algorithm above. Of course, you cannot use the Euler criterion to evaluate a Jacobi symbol (a/m) when m is composite; you must use the algorithm above.

Example 7.4

Let us use the algorithm to solve the problem of Example 7.1, namely, compute the Legendre symbol $(-22/59)$. The algorithm does it this way:

$$\left(\frac{-22}{59}\right) = \left(\frac{37}{59}\right) = \left(\frac{59}{37}\right) = \left(\frac{22}{37}\right) = \left(\frac{2}{37}\right)\left(\frac{11}{37}\right) =$$
$$= -\left(\frac{11}{37}\right) = -\left(\frac{37}{11}\right) = -\left(\frac{4}{11}\right) = -1.$$

In this example, all of the Jacobi symbols just happened to be Legendre symbols because the "denominators" are all prime.

Example 7.5

Evaluate the Legendre symbol $(133/401)$.

Here 401 is prime, but 133 is composite. We find

$$\left(\frac{133}{401}\right) = \left(\frac{401}{133}\right) = \left(\frac{2}{133}\right) = -1.$$

If we knew only about Legendre symbols, then we would have had to factor 133 as the first step, or else use Euler's criterion.

Consider what we have done so far in this chapter. We defined the Legendre symbol (a/p), which tells when we can solve the congruence $x^2 \equiv a \pmod{p}$. Then we defined the Jacobi symbol (a/m), which is easier to compute than the Legendre symbol. In case m is prime, the Jacobi symbol (a/m) is the same as the Legendre symbol (a/m). Jacobi symbols provide a convenient way of computing Legendre symbols. But when m is composite, the fact that the Jacobi symbol (a/m) is $+1$ does not mean that one can solve the congruence $x^2 \equiv a \pmod{m}$. If the congruence is solvable and $\gcd(a, m) = 1$, then $(a/m) = +1$ because the Legendre symbols $(a/p) = +1$ for every prime divisor p of m. This means that if the Jacobi symbol $(a/m) = -1$ and $\gcd(a, m) = 1$, then a is a quadratic nonresidue modulo m.

The reader may wonder why we didn't define the Jacobi symbol (a/m) to be $+1$ or -1 according as the congruence $x^2 \equiv a \pmod{m}$ has a solution or not. If we had made that definition, then the Jacobi symbol would not satisfy the Law of Quadratic Reciprocity, and it would be difficult to compute for large m. For example, $(5/9) = (9/5) = +1$. The congruence $x^2 \equiv 9 \pmod 5$ has the solutions $x \equiv 2$ or $3 \pmod 5$, but the congruence $x^2 \equiv 5 \pmod 9$ has no solution.

7.4 Euler Pseudoprimes

By Fermat's little theorem, if the prime p does not divide a, then $a^{p-1} \equiv 1 \pmod{p}$. We mentioned that if m is odd and $\gcd(a, m) = 1$ and $a^{m-1} \equiv 1 \pmod{m}$, then m is probably prime.

We can devise an analogous probable prime test using Euler's criterion. It says that if the prime p does not divide a, then $a^{(p-1)/2} \equiv (a/p) \pmod{p}$. It turns out that if m is odd and $\gcd(a, m) = 1$ and $a^{(m-1)/2} \equiv (a/m) \pmod{m}$, then m is probably prime.

DEFINITION 7.4   *An odd integer $m > 2$ is called an* **Euler probable prime to base** *a if $a^{(m-1)/2} \equiv (a/m) \pmod{m}$ and $\gcd(a, m) = 1$. (The symbol (a/m) here is Jacobi.) A composite probable prime to base a is called an* **Euler pseudoprime to base** *a.*

Every prime $p > 2$ is an Euler probable prime to every base a which is not a multiple of p.

THEOREM 7.12 Euler probable primes are probable primes

If m is an Euler probable prime to base a, then m is a probable prime to base a.

PROOF   We have $a^{(m-1)/2} \equiv (a/m) \pmod{m}$ and $\gcd(a, m) = 1$ by hypothesis. The Jacobi symbol $(a/m) = \pm 1$ since $\gcd(a, m) = 1$. Square both sides of the congruence to get $a^{m-1} \equiv 1 \pmod{p}$, so m is a probable prime.   ∎

Thus, every Euler pseudoprime is a pseudoprime (to the same base). However, some pseudoprimes are not Euler pseudoprimes. For example, $341 = 11 \cdot 31$ is a pseudoprime to base 2, but it is not an Euler pseudoprime to base 2 because $2^{(341-1)/2} = 2^{170} \equiv +1 \pmod{341}$ while $(2/341) = -1$ by Theorem 7.10. Therefore the Euler probable primality test is more discriminating than the simple probable prime test. The two tests have essentially the same complexity.

7.5   *Solving Quadratic Congruences Modulo m*

We now return to the task of solving quadratic congruences modulo m, which we considered at the beginning of this chapter. So far, we have found an efficient way, actually two of them, for deciding whether $x^2 \equiv a \pmod{p}$ has a solution where p is prime. Let us find the solutions x when there are any. The answer is easy when p is a Blum prime. A **Blum prime** is a prime $\equiv 3 \pmod{4}$. The name arises because M. Blum used these primes in the oblivious transfer protocol and in a random number generator.

THEOREM 7.13 Square roots modulo a Blum prime

If $p \equiv 3 \pmod{4}$ is prime and a is a quadratic residue modulo p, then the solutions to $x^2 \equiv a \pmod{p}$ are $x \equiv \pm \left(a^{(p+1)/4}\right) \pmod{p}$.

PROOF   Note that

$$x^2 \equiv a^{(p+1)/2} \equiv a \cdot a^{(p-1)/2} \equiv a(a/p) \equiv a \pmod{p}$$

by Euler's criterion and the fact that a is a quadratic residue modulo p. ∎

What happens if we apply the formula in the theorem when a is a quadratic nonresidue modulo p? Obviously, we won't get a solution to the congruence, for it has none. When $p \equiv 3 \pmod 4$, $(-1/p) = -1$, so $(-a/p) = (-1/p)(a/p) = -(a/p)$. If $(a/p) = -1$, then $(-a/p) = +1$, so $-a$ is a quadratic residue. The numbers x computed by Theorem 7.13 are solutions to the congruence $x^2 \equiv -a \pmod{p}$, as one can see from the proof of the Theorem.

Example 7.6

Solve $x^2 \equiv 6 \pmod{47}$.
    We first compute

$$(6/47) = (2/47)(3/47) = (+1)(-1)(47/3) = -(2/3) = -(-1) = +1,$$

so 6 is a quadratic residue modulo 47. The solutions are

$$x \equiv \pm \left(6^{(47+1)/4}\right) \equiv \pm \left(6^{12}\right) \equiv \pm 37 \pmod{47}.$$

One checks that $37^2 \equiv 6 \pmod{47}$.

It is slightly harder to find square roots modulo primes $p \equiv 5 \pmod 8$.

THEOREM 7.14 Square roots modulo a prime $p \equiv 5 \pmod 8$
    If $p \equiv 5 \pmod 8$ and a is a quadratic residue modulo p, then the solutions to $x^2 \equiv a \pmod{p}$ are $\pm x$, where x is computed by this algorithm.

$x = a^{(p+3)/8} \bmod p$
if $(x^2 \not\equiv a \pmod{p})$ $x = x2^{(p-1)/4} \bmod p$

PROOF   Note first that with $x = a^{(p+3)/8} \bmod p$, we have

$$x^4 \equiv a^{(p+3)/2} \equiv a^2 a^{(p-1)/2} \equiv a^2(a/p) \equiv a^2 \pmod{p}$$

by Euler's criterion and the fact that a is a quadratic residue modulo p. Therefore, $x^2 \equiv \pm a \pmod{p}$. If $x^2 \equiv a \pmod{p}$, then the algorithm returns x. Otherwise, the algorithm multiplies x by $2^{(p-1)/4} \bmod p$. Now 2 is a quadratic nonresidue modulo p by Theorem 7.8, so $2^{(p-1)/2} \equiv -1 \pmod{p}$. In this case, we have

$$x^2 \equiv \left(a^{(p+3)/8} \cdot 2^{(p-1)/4}\right)^2 \equiv (-a)(-1) \equiv a \pmod{p}.$$

▮

Example 7.7

Solve $x^2 \equiv 3 \pmod{37}$.

Let $x \equiv 3^{(37+3)/8} \equiv 3^5 \equiv 21 \pmod{37}$. Then $x^2 \equiv 21^2 \equiv 34 \not\equiv 3 \pmod{37}$; so, we multiply x by $2^{(37-1)/4} \equiv 2^9 \equiv 31$ to obtain a new $x \equiv 21 \cdot 31 \equiv 22 \pmod{37}$. This x works because $22^2 \equiv 3 \pmod{37}$.

Next we present an algorithm which will find square roots of quadratic residues modulo any odd prime p. When $p \not\equiv 1 \pmod 8$ it reduces to the algorithm in one of the last two theorems. The algorithm begins by choosing a random quadratic nonresidue n modulo p. There is no known deterministic polynomial-time algorithm for finding a quadratic nonresidue n modulo p. Just try random n and use Euler's criterion to determine (n/p). This procedure makes the algorithm probabilistic. However, usually it is easy to find n quickly because half of the integers between 1 and $p-1$ are quadratic nonresidues; so, the expected number of n's that must be tried is 2. The algorithm uses the quadratic nonresidue n to construct an integer N whose order is 2^e, where $p-1 = 2^e f$, with f odd. The `for` loop determines the correct power of N to multiply times the first guess $a^{(f+1)/2}$ for x to get the true solution x.

The algorithm returns just one solution x. The other one is $-x$ or $p-x$. Of course, one must be sure that a is a quadratic residue modulo p before using the algorithm.

The average complexity of the algorithm is $O(\log^3 p)$ bit operations, that is, averaged over many random primes, but the worst case (when e is large) is $O(\log^4 p)$ bit operations plus the time needed to find n.

[Square root of a modulo p]
Input: An odd prime p and an integer a with $(a/p) = +1$.
Output: A solution x to $x^2 \equiv a \pmod p$.

```
Find a (random) quadratic nonresidue n modulo p
Compute e ≥ 0 and odd f so that p − 1 = 2^e f
A = a^f mod p
N = n^f mod p
j = 0
for (1 ≤ i < e) {
         if ((AN^j)^{2^{e-i-1}} ≡ −1 (mod p)) { j = j + 2^i }
         }
x = a^{(f+1)/2} N^{j/2} mod p
```

See Theorem 7.1.3 of Bach and Shallit [8] for a proof that the algorithm works. The algorithm is similar to Algorithm 2.3.8 of Crandall and Pomerance [33].

Example 7.8

Solve $x^2 \equiv 2 \pmod{17}$.

In the algorithm we have $p = 17$ and $a = 2$. Note that $(2/17) = +1$ by Theorem 7.8. We have $p - 1 = 16 = 2^4$, so $e = 4$ and $f = 1$. Trying small n, we find that the first quadratic nonresidue is $n = 3$, since $(3/17) = (17/3) = (2/3) = -1$. We find $A = a^f = 2^1 = 2$ and $N = n^f = 3^1 = 3$. We set $j = 0$ and begin the for loop.

When $i = 1$, the test is whether $(2)^{2^2} \equiv -1 \pmod{17}$. This is true and j becomes 2.

When $i = 2$, the test is whether $(2 \cdot 3^2)^{2^1} \equiv -1 \pmod{17}$. The left side is $18^2 \equiv 1 \pmod{17}$, the test fails and j remains 2.

When $i = 3$, the test is whether $(2 \cdot 3^2)^{2^0} \equiv -1 \pmod{17}$. The left side is $18^1 \equiv 1 \pmod{17}$, the test fails and j remains 2.

Finally, $x = 2^{(1+1)/2} 3^{2/2} \equiv 2^1 3^1 \equiv 6 \pmod{17}$. One verifies that $6^2 \equiv 2 \pmod{17}$, so the solutions are $x \equiv \pm 6 \pmod{17}$.

Now we know how to solve $x^2 \equiv a \pmod{m}$ when m is prime. Next we will solve this congruence when m is a prime power. The first theorem applies to a congruence with a general polynomial. It "lifts" zeros of the polynomial modulo p^i to zeros modulo p^{i+1}. The procedure is just like using Newton's method to refine an approximate zero to a polynomial.

THEOREM 7.15 Hensel's lemma

Let p be prime and $f(x)$ be a polynomial with integer coefficients. If $f(a) \equiv 0 \pmod{p^i}$ and $f'(a) \not\equiv 0 \pmod{p}$, then there is a unique t so that $f(a+tp^i) \equiv 0 \pmod{p^{i+1}}$.

PROOF Let $f(x)$ have degree d modulo p^{i+1}. Expand $f(a + tp^i)$ in a Taylor series

$$f(a + tp^i) = f(a) + tp^i f'(a) + \cdots + t^d p^{di} f^{(d)}(a)/d!$$

Derivatives after the d-th are zero polynomials. We claim that this expansion reduced modulo p^{i+1} is

$$f(a + tp^i) \equiv f(a) + tp^i f'(a) \pmod{p^{i+1}}. \tag{7.3}$$

If cx^e is a typical term in $f(x)$, then the corresponding term in $f^{(k)}(a)$ is

$$ce(e - 1)(e - 2) \cdots (e - k + 1)a^{e-k}$$

But $e(e - 1)(e - 2) \cdots (e - k + 1)/k! = \binom{e}{k}$ is an integer, so $k!$ divides $e(e - 1)(e - 2) \cdots (e - k + 1)$. Therefore, in the Taylor expansion above, the term $t^k p^{kj} f^{(k)}(a)$ with $2 \le k \le d$ is divisible by p^{kj} and so by p^{i+1}, which proves congruence 7.3.

Congruence 7.3 shows that if $f(a + tp^i) \equiv 0 \pmod{p^{i+1}}$, then $f(a) + tp^j f'(a) \equiv 0 \pmod{p^{i+1}}$. Since $f(a) \equiv 0 \pmod{p^i}$, this is equivalent to $f'(a)t \equiv -f(a)/p^i \pmod{p}$, which is a linear congruence in t. By Theorem 5.7, it may have zero, one or p solutions. But when $f'(a) \not\equiv 0 \pmod{p}$, it has exactly one solution. ∎

If $f(a) \equiv f'(a) \equiv 0 \pmod{p}$, then the root a is called singular. We do not discuss how to "lift" singular solutions here. See Section 2.6 of [78] to learn how to do it.

THEOREM 7.16 Solution of $x^2 \equiv a \pmod{p^i}$
Let a be a quadratic residue modulo an odd prime p. Then for every $n \geq 1$ the congruence $x^2 \equiv a \pmod{p^n}$ has exactly two solutions, $x \equiv \pm a_n \pmod{p^n}$. Also, $\gcd(a_n, p) = 1$.

PROOF   Use induction on n. The base step $n = 1$ holds since a is a quadratic residue modulo p. The induction hypothesis says $x^2 \equiv a \pmod{p^{n-1}}$ has only the two solutions $x \equiv \pm a_{n-1} \pmod{p^{n-1}}$, and $\gcd(a_{n-1}, p) = 1$. If x is a solution to $x^2 \equiv a \pmod{p^n}$, then it must be a solution to $x^2 \equiv a \pmod{p^{n-1}}$, and so $x \equiv \pm a_{n-1} \pmod{p^{n-1}}$. Thus, $x \equiv \pm(a_{n-1} + tp^{n-1}) \pmod{p^n}$. Write $f(x) = x^2 - a$. By Theorem 7.15 if $f'(a_{n-1}) \not\equiv 0 \pmod{p}$, there is a unique t so that $f(a_{n-1} + tp^{n-1}) \equiv 0 \pmod{p^n}$. Now $f'(a_{n-1}) = 2a_{n-1}$ is not a multiple of p because p is odd and, by induction, $\gcd(a_{n-1}, p) = 1$. Thus, a_{n-1} lifts to a unique solution $a_n = a_{n-1} + tp^{n-1}$ of $x^2 \equiv a \pmod{p^n}$. The same argument shows that $-a_{n-1}$ lifts to a unique solution, which must be $-a_n$ because $(-x)^2 = x^2$ and there are no other solutions. Finally, $\gcd(a_n, p) = 1$ because $a_n \equiv a_{n-1} \pmod{p}$ and $\gcd(a_{n-1}, p) = 1$. ∎

Example 7.9

Solve $x^2 \equiv 2 \pmod{17^2}$.
    We saw in Example 7.8 that the solutions of $x^2 \equiv 2 \pmod{17}$ are $x \equiv \pm 6 \pmod{17}$. Let us lift the solution 6 (mod 17). Write $f(x) = x^2 - 2$. Then the solution $a_1 = 6$ lifts to a unique solution $a_2 = 6 + 17t$ where t satisfies

$$f'(a_1)t \equiv -\frac{f(a_1)}{17} \pmod{17},$$

or $2 \cdot 6t \equiv -(6^2 - 2)/17 \pmod{17}$. That is, $12t \equiv -2 \pmod{17}$, or $t \equiv 14 \pmod{17}$. Finally, $a_2 = 6 + 14 \cdot 17 = 244 = 17^2 - 45$, and the solutions to $x^2 \equiv 2 \pmod{17^2}$ are $x \equiv \pm 45 \pmod{289}$.

Now we can solve $x^2 \equiv a \pmod{m}$ when m is an odd prime power. What about modulo a power of 2? We cannot use Hensel's lemma to lift a solution of $x^2 \equiv a \pmod{2^i}$ to a solution modulo 2^{i+1} because every root b of $f(x) = x^2 - a$

is singular since $f'(x) = 2x \equiv 0 \pmod 2$. Nevertheless, it is easy to lift a solution directly.

THEOREM 7.17 Solution of $x^2 \equiv a \pmod{2^i}$

The solutions to $x^2 \equiv a \pmod{2^i}$, where a is odd, are as follows:

1. If $i = 1$, then $a = 1$ gives one solution $x \equiv 1 \pmod 2$.

2. If $i = 2$, then $a = 1$ gives two solutions $x \equiv \pm 1 \pmod 4$ and $a = 3$ gives no solution.

3. If $i \geq 3$, then there are four solutions $x \pmod{2^i}$ if $a \equiv 1 \pmod 8$ and no solution $x \pmod{2^i}$ if $a \not\equiv 1 \pmod 8$. If x is one solution, then the other three are $-x$, $x \pm 2^{i-1}$. Solutions may be lifted as follows: If y is a solution to $y^2 \equiv a \pmod{2^i}$, then either $x = y$ or $x = y + 2^{i-1}$ is a solution to $x^2 \equiv a \pmod{2^{i+1}}$.

Example 7.10

Solve $x^2 \equiv 9 \pmod{32}$.

The solutions to $x^2 \equiv 9 \pmod 8$ are $x \equiv 1, 3, 5, 7 \pmod 8$. The solution 3 clearly lifts to a solution modulo 16. The solution 1 does not lift, but $1 + 4 = 5$ is a solution modulo 16. The other solutions modulo 16 are $16 - 3 = 13$ and $16 - 5 = 11$.

Of the solutions 3, 5, 11, 13, modulo 16, one find that 3 and 13 lift to solutions modulo 32, but 5 and 11 do not. However, $5 + 8 = 13$ and $11 + 8 = 19$ are solutions. We already knew that 13 was a solution. The fourth solution modulo 32 is $32 - 3 = 29$.

Of course, a congruence $x^2 \equiv a \pmod{p^i}$ may have solutions when $p | a$, so that a is not a quadratic residue modulo p. For example, $x^2 \equiv 4 \pmod 8$ has the obvious solutions $x \equiv 2, 6 \pmod 8$ and $x^2 \equiv 0 \pmod 8$ has the solutions $x \equiv 0, 4 \pmod 8$. Solutions like this exist when $a \equiv 0 \pmod{p^i}$ or when a power of p with an even exponent exactly divides a. A congruence of the latter type may be reduced to one of lower degree. For example, solving $x^2 \equiv p^2 a \pmod{p^5}$, where $\gcd(a, p) = 1$, is equivalent to solving $y^2 \equiv a \pmod{p^3}$ and letting $x = py$.

Finally, to solve $x^2 \equiv a \pmod m$, solve it first modulo p^i for each prime power dividing m, and combine the solutions with the Chinese remainder theorem. One special case is so important in cryptography that we record it here as a theorem.

THEOREM 7.18 Four square roots modulo pq

Let p and q be distinct odd primes and let a be a quadratic residue modulo pq. Then there are exactly four solutions to $x^2 \equiv a \pmod{pq}$.

PROOF The hypothesis implies that a is a quadratic residue modulo each of p and q, so the two congruences $x^2 \equiv a \pmod{p}$ and $x^2 \equiv a \pmod{q}$ each have two solutions by Theorem 7.2. By the Chinese remainder theorem, each of the four pairs of solutions gives rise to a solution modulo pq. ∎

7.6 *Exercises*

1. Evaluate the Legendre symbols $(r/103)$ for $1 \le r \le 10$. Use Theorem 7.5 to simplify your work.

2. Evaluate the Legendre symbols $(10/79)$, $(11/43)$ and $(6/23)$.

3. Find the smallest positive quadratic nonresidue modulo 71.

4. Prove that if p is an odd prime, then $\sum_{r=0}^{p-1}(r/p) = 0$.

5. Find the odd primes that have -2 as a quadratic residue. Express your answer as a set of congruence classes modulo 8.

6. Find the odd primes that have 7 as a quadratic residue. Express your answer as a set of congruence classes modulo 28.

7. If a is a quadratic nonresidue modulo each of the odd primes p and q, what is the Jacobi symbol (a/pq)? How many solutions does $x^2 \equiv a \pmod{pq}$ have?

8. Show that $x^8 \equiv 16 \pmod{p}$ has a solution for every prime p. (Hint: Factor $x^8 - 16$ into the product of *four* quadratic polynomials.)

9. Solve $x^2 \equiv 3 \pmod{11}$.

10. Solve $x^2 \equiv 3 \pmod{13}$.

11. Solve $x^2 \equiv 3 \pmod{11^2}$.

12. Solve $x^2 \equiv 3 \pmod{143}$.

13. Solve $x^2 \equiv 41 \pmod{64}$.

14. Find all the square roots of 58 modulo 77.

15. Find a quadratic nonresidue modulo the composite integer 4009 without factoring this modulus.

16. Prove that if p and $p+2$ are twin primes, then $(p/(p+2)) = ((p+2)/p)$.

17. Prove that if p is a Sophie Germain prime, then $(p/(2p+1)) = (-1/p)$.

18. Prove that $1+\sum_{i=0}^{2000}\left(\sum_{j=1}^{i+1} j^{2002}\right) 2003^i$ is not the square of an integer.

Chapter 8

Information Theory

This chapter introduces information theory and its use in analyzing simple ciphers. See Denning [36] for another view of much of the material in this chapter. This subject was created by Shannon [106] to give a theoretical foundation for communication and, in particular, for cryptography. He measured the secrecy of a cipher by the uncertainty in the plaintext given the ciphertext. The most secret ciphers are the ones for which an eavesdropper learns nothing at all about the plaintext by seeing the ciphertext. Most ciphers leave some information about the plaintext in the ciphertext. If an eavesdropper has enough ciphertext, he may obtain enough information to break the cipher, at least theoretically. Many ciphers can be broken from just a hundred or so bits of ciphertext. These ciphers are not necessarily insecure, because an enormous computation might be required to break them, and the cryptanalyst might not have enough resources to do it.

Shannon applied his information theory also to "noisy channels," in which Alice sends a redundant message to Bob over a communication channel, which may change the message randomly through imperfections. Bob tries to recover the original message from its redundancy. Ordinary English is redundant. One may regard encryption as a kind of "noise" added to a message before an eavesdropper receives it. The eavesdropper tries to recover the plaintext from the ciphertext. The same theory of information that predicts how much noise must be added to a message before Bob can no longer recover it from its redundancy also predicts how well a cipher protects a message from an eavesdropper.

8.1 Entropy

The amount of information contained in a message is measured by its entropy. In other words, entropy measures the uncertainty about a message before it is received or deciphered. Suppose there are n possible messages $x_1, \ldots, x_n$ which could be sent. Let p_i be the probability that x_i is the message sent, so that $p_1 + \cdots + p_n = 1$. The entropy of the message should depend only on these probabilities and not on the particular set of messages because if $y_1, \ldots, y_n$ were another set of possible messages with the same probability distribution, then it would have the same uncertainty. Therefore, we may write $H(p_1, \ldots, p_n)$ for the entropy of the set of messages.

In defining entropy, Shannon [106] required that it satisfy three properties. First, it should be a continuous function of the variables $p_1, \ldots, p_n$, subject to $p_1 + \cdots + p_n = 1$. Second, when the messages are equally likely, that is, every $p_i = 1/n$, H should be an increasing function of n. He required this property because there is more choice, or uncertainty, when there are more equally likely messages. The third property said that if the choice of one among n messages is replaced by two successive choices, first of a subset of the messages and then a message in the chosen subset, then the entropy of the set of messages should be a weighted sum of the entropies of the two choices. For example, if there are four equally likely messages, we may choose one of them as follows: (1) Choose a subset of the messages, either the first one or the second one or the last two. (2) If the subset was the last two, choose one of them. Then the third property would say

$$H\left(\tfrac{1}{4}, \tfrac{1}{4}, \tfrac{1}{4}, \tfrac{1}{4}\right) = H\left(\tfrac{1}{4}, \tfrac{1}{4}, \tfrac{1}{2}\right) + \tfrac{1}{2}H\left(\tfrac{1}{2}, \tfrac{1}{2}\right).$$

The coefficient of the last term is $\tfrac{1}{2}$ because the second choice is made half of the time.

From these three properties, Shannon [106] proved that the entropy must be

$$H(p_1, \ldots, p_n) = -K \sum_{i=1}^{n} p_i \log p_i,$$

where K is a positive constant. The constant K may be regarded as a choice of units for entropy. Choosing $K = 1/\log 2$ makes the binary digit the unit of entropy. His theorem motivates the definition of entropy.

DEFINITION 8.1   *If X is a message that takes on the value x_i with probability p_i, for $i = 1, \ldots, n$, then the **entropy** of X in bits is*

$$H(X) = -\sum_{i=1}^{n} p_i \log_2 p_i.$$

We either exclude terms with $p_i = 0$ from the sum or else we define $0 \log_2 0$ to be $\lim_{p \to 0+} p \log_2 p = 0$.

The entropy is always nonnegative. It equals 0 if and only if one outcome is certain.

Example 8.1

Suppose we toss a coin having probability p of showing heads and $1 - p$ of showing tails. Let the message X be the outcome of the coin toss: heads or tails. Then

$$H(X) = -p \log_2 p - (1 - p) \log_2 (1 - p).$$

This function of p has a maximum of 1 at $p = 0.5$ and a minimum of 0 when $p = 0$ or $p = 1$. If the coin is true ($p = 0.5$), then there is one bit of uncertainty in the outcome. A one-bit message could tell the outcome of the coin toss. But there is less uncertainty in the outcome as the coin becomes more unbalanced, with no uncertainty at all if the coin always shows heads.

Note that $H(X)$ is the expected value of the random variable

$$-\log_2 p(X) = \log_2(1/p(X)).$$

Example 8.2

Suppose X is a random n-bit integer, with all 2^n possible integers being equally likely. Then each message has probability 2^{-n} and the entropy is

$$H(X) = -\sum_{i=1}^{2^n} 2^{-n} \log_2 2^{-n} = -(2^n)(2^{-n}(-n)) = n.$$

This example shows that $H(X)$ measures the number of bits of information we learn when we are told the message X. We learn n bits when we are told an n-bit number.

The entropy $H(X)$ is the average number of bits needed to encode all possible messages in an optimal encoding, called a **Huffman code**.

Example 8.3

Suppose there are four messages, x_1, x_2, x_3, x_4, with probabilities $1/2$, $1/4$, $1/8$ and $1/8$, respectively. The entropy is

$$H(X) = -\frac{1}{2} \log_2 \frac{1}{2} - \frac{1}{4} \log_2 \frac{1}{4} - \frac{1}{8} \log_2 \frac{1}{2} - \frac{1}{8} \log_2 \frac{1}{8} =$$
$$= \frac{1}{2}(1) + \frac{1}{4}(2) + (2)\frac{1}{8}(3) = \frac{1}{2} + \frac{2}{4} + \frac{6}{8} = \frac{7}{4}.$$

A Huffman code for the four messages would use one bit for the first message, two bits for the second, and three bits each for the third and fourth. For example, code the messages by the bit strings, 0, 10, 110, 111. Then the average length of the bit string to reveal which x_i was sent is

$$\frac{1}{2}(1) + \frac{1}{4}(2) + (2)\frac{1}{8}(3) = \frac{1}{2} + \frac{2}{4} + \frac{6}{8} = 1.75,$$

the same as $H(X)$.

In cryptography, we measure the entropy of ciphertext and keys, as well as of plaintext. We can define the **conditional entropy** of one of these items given another one. For example, the conditional entropy of the key K given the ciphertext C is

$$H(K|C) = -\sum_c \sum_k p(C = c \text{ and } K = k) \log_2 p(K = k | C = c).$$

The **joint entropy** $H(X,Y)$ is the entropy of the pair (X,Y).

With these definitions, one can prove the following facts about entropy:

1. $H(X,Y) = H(X) + H(Y|X)$. This formula says that joint uncertainty of the pair (X,Y) equals the uncertainty of X plus the uncertainty of Y, given that X is known.

2. $H(Y|X) \le H(Y)$, with equality if and only if X and Y are independent. This inequality tells us that the uncertainty about Y, given that X is known, is no greater than the uncertainty about Y. But if X and Y are independent events, then the uncertainty about Y, given that X is known, is the same as the uncertainty about Y. This means that X can only tell us information about Y; learning X cannot make us more uncertain about Y.

3. $H(X,Y) \le H(X) + H(Y)$. This says that the uncertainty in the pair (X,Y) is no more than the sum of the uncertainties in X and Y separately.

4. $H(X) \le \log_2 n$, where n is the number of possible X's. We have equality if and only if the n X's are equally likely.

Example 8.4

Suppose X and Y each can be one of four equally likely messages, and each Y message limits X to one of two equally likely messages. (For instance, Y_2 might say, "X is X_1 or X_4.") Then each $p(X|Y)$ is $1/2$ or 0, so

$$H(X|Y) = 4[(1/4) \cdot 2(1/2) \log_2 2] = 1.$$

8.2   Perfect Secrecy

Let M, C and K represent plaintext, ciphertext and keys, respectively.

DEFINITION 8.2   *A cipher has **perfect secrecy** if $H(M|C) = H(M)$.*

This definition, taken from Shannon [107], says that if a cipher has perfect secrecy, then an eavesdropper is just as uncertain about the plaintext after seeing the ciphertext as he was before seeing the ciphertext. He learns nothing at all about the plaintext from the ciphertext.

Perfect secrecy is clearly a desirable property for a cipher. Few ciphers enjoy perfect secrecy. However, one fairly simple cipher does have the property.

DEFINITION 8.3   *A* **one-time pad** *is a synchronous stream cipher with a truly random key stream.*

One-time pads are so called because, in early versions of this cipher, the sender and receiver would have identical pads of paper with random key characters printed on them. After using each sheet to encipher or decipher a message, the cryptographer would destroy the sheet. It is important that each page of key characters be used only once, because if one were reused, a cryptanalyst could gain information about the plaintexts by comparing the two ciphertexts.

The **Vernam cipher** is a one-time pad that uses the exclusive-or operation $\oplus$ to encipher (and decipher). If m_i, k_i and c_i are the i-th characters of plaintext, key and ciphertext, respectively, then $c_i = m_i \oplus k_i$ and $m_i = c_i \oplus k_i$. If two plaintexts, $m_1, m_2, \ldots$ and $m'_1, m'_2, \ldots$, were both enciphered using the same random key stream, and a cryptanalyst obtained the two ciphertexts $c_1, c_2, \ldots$ and $c'_1, c'_2, \ldots$, then he could compute $c_i \oplus c'_i = m_i \oplus m'_i$, which is essentially a running key cipher (see Example 1.2 and Section 8.4) and easy to break.

Modern one-time pads have the keys written on magnetic or optical media that are destroyed after use. In advance of the communication, two copies of the random key stream must be created and distributed to the sender and receiver. Assuming the key is not reused, the one-time pad achieves perfect secrecy because, given any M and C, there is always a key stream K that will encipher M as C, so that every C occurs with equal probability, assuming the keys are equally likely. Hence, $p(M = m | C = c) = p(M = m)$ and so $H(M|C) = H(M)$.

If a cipher has perfect secrecy, then there must be at least as many keys as plaintexts. Otherwise, there would be some pairs m, c with no key to decipher c into m. Then $p(M = m | C = c) = 0$ for these particular m and c, and so $H(M|C) < H(M)$, which would violate the definition of perfect secrecy.

8.3   Unicity Distance

How much information can be contained in a string of n letters of English? If we allow any string of n letters, then there are 26^n possible strings. If they are equally likely, then the entropy of such a string is $\log_2 26^n = n \log_2 26$. Thus, the amount of information per letter is $R = \log_2 26 \approx 4.7$. This is called the absolute rate of English. In general, the **absolute rate of a language** with a letters in its alphabet is the maximum number of bits per letter in a string, namely, $R = \log_2 a$. The absolute rate would be higher if we counted spaces and punctuation as "letters" of the alphabet.

How much information is contained per letter in a *meaningful* string of letters of English? If we could list all meaningful n-letter strings X of English and determine the probability of each, then we could compute $H(X)$ and the number of bits of information per letter would be $H(X)/n$. Finally, we could define the rate of English to be $\lim_{n\to\infty} H(X)/n$. For any language, we define the **rate of the language for messages** X **of length** n as $r_n = H(X)/n$ and the **rate of the language** to be $r = \lim_{n\to\infty} r_n$. This is the average number of bits of entropy per letter in meaningful messages. Although we cannot compute $H(X)/n$ for n of any interesting length, Shannon [108] proposed a way to estimate r for English and found that $r \approx 1$ bit per letter.

The **redundancy of a language** is defined to be $D = R - r$. The redundancy of English is about 3.7 bits per letter. With these definitions we see that there are $2^{Rn} = 26^n$ n-letter messages, of which 2^{rn} are meaningful and $2^{Rn} - 2^{rn}$ are meaningless.

A cipher is **unconditionally secure** if $H(K|C)$ does not go to 0 as the length of C increases without bound. For example, a one-time pad is unconditionally secure. Let us consider ciphers that do not have this property.

DEFINITION 8.4   *If the conditional entropy $H(K|C)$ goes to 0 as the length of C increases, then the cipher is **theoretically breakable**, and the* **unicity distance** *is the shortest length n of C for which $H(K|C) \leq 1$.*

If $H(K|C) \leq 1$, then there is no more than 1 bit of uncertainty about the key, that is, the key has one of two possible values. Then, any given ciphertext can be deciphered in at most two different ways, and a cryptanalyst aware of the nature of the communication should be able to decide which of the two possible plaintexts was sent.

For most ciphers we can only estimate the unicity distance. We now derive a useful approximation to it.

We assume that all 2^{rn} meaningful n-letter messages have equal probability 2^{-rn}, and that all meaningless messages have probability 0. Here we are assuming the equally-likely case, which maximizes entropy and is the worst case.

We assume that there are $2^{H(K)}$ keys, and they are equally likely. That is, $p(K = k) = 2^{-H(K)}$ for each key k.

A **random cipher** is one in which the decipherment $D_K(C)$ is an independent random variable uniformly distributed over all 2^{Rn} messages, both meaningful and meaningless. This means that for a given k and C, $D_k(C)$ is as likely to produce one plaintext message as any other. Actually the decipherments are not completely independent because a given key must uniquely encipher a given message, so that $D_k(C) \neq D_k(C')$ for $C \neq C'$, that is, the deciphering function D_k is one-to-one for each key.

Assume we have a random cipher and suppose $C = E_k(M)$. A **spurious**

key decipherment or **false solution** of C is either $C = E_{k'}(M)$ or $C = E_{k''}(M')$ where M' is meaningful. (We are not concerned with meaningless false solutions, as they are easily detected.) In the first case ($C = E_{k'}(M)$), the key k' may or may not decipher other C enciphered with k. For every correct decipherment there are $2^{H(K)} - 1$ other keys, each with probability

$$q = 2^{rn}/2^{Rn} = 2^{-Dn}$$

of giving a false solution. Let F be the number of false solutions. Then $F = (2^{H(K)} - 1)q \approx 2^{H(K)-Dn}$. When n is large enough so that $F \le 1$, we have enough ciphertext to break the cipher. At the borderline case where $F = 1$, we have $H(K) = Dn$. Thus $n = H(K)/D$ is approximately the unicity distance.

Example 8.5

DES is a block cipher with 56-bit keys and 64-bit blocks of plain and cipher text. Now 64 bits is 8 characters. For English, $D = 3.7$, so $n = H(K)/D = 56/3.7 = 15.1$ characters, or about two blocks.

8.4 Some Obsolete Ciphers

Kahn [55] tells the fascinating history of cryptography up to 1967, including tales about many of the ciphers mentioned in this section.

We compute the standard approximation to the unicity distance for several simple ciphers, and mention techniques for breaking them.

Recall that transposition ciphers rearrange characters or bits. They have a fixed period, d, say. If we assume that all $d!$ permutations are equally likely, which is the worst case, then the unicity distance is

$$n = \frac{H(K)}{D} = \frac{\log_2(d!)}{D} \approx \frac{d\log_2(d/e)}{3.2} \approx 0.3d\log_2(d/e),$$

where we have used Stirling's formula to approximate $d!$.

For example, with a 3×9 matrix we have $d = 27$ and $n = 27.9$.

THEOREM 8.1 Stirling's formula
For $n \ge 1$, $n! \approx \sqrt{2\pi n}(n/e)^n$ and $\log(n!) \approx n\log(n/e)$.

See Feller [43], page 50, for a proof.

Use the frequency distribution of pairs or triples of letters to break transposition ciphers. The process is called **anagramming**.

Use frequency counts to distinguish transposition ciphers from substitution ciphers. With transposition ciphers, the letters of the alphabet have their normal frequency; with substitution ciphers, they do not.

Recall that substitution ciphers replace (blocks of) characters by other characters. One classification lists four types of substitution ciphers. They are

1. Simple: Replace each m_i by c_i.

2. Homophonic: Replace m_i by a random one of many possible c_i.

3. Polyalphabetic: Use multiple maps from the plaintext alphabet to the ciphertext alphabet.

4. Polygram: Make arbitrary substitutions for groups of characters.

1. **Simple substitution ciphers** replace each m_i by c_i. Write the enciphering function as $f(m) = c$. For example, the Caesar cipher rotates the alphabet: $f(m) = (m + k) \bmod n$, where n is the alphabet size. For English, the unicity distance is $H(K)/D = (\log_2 26)/3.2 \approx 1.5$ letters.

If all $n!$ permutations of the alphabet are equally likely (the worst case for the cryptanalyst) in a simple substitution cipher, then the unicity distance is $\log_2(n!)/D$. For English, $n = 26$ and the unicity distance would be $\log(26!)/3.2 \approx 27.6$.

These ciphers may be broken with frequency analysis and trial and error. Some are published in newspapers as puzzles to amuse readers.

In an **affine cipher**, $f(m) = (am + b) \bmod n$. Break it by guessing some two-letter pairs and solving two congruences in the two unknowns a and b. An exercise gives an example of finding the unicity distance. Remember that a and n must be relatively prime in order for messages to be deciphered. Therefore, there are $\phi(n)$ choices for a and n choices for b.

2. **Homophonic substitution ciphers** replace m_i by a random one of many possible c_i.

To confound the frequency analysis that succeeds so well for simple substitution ciphers, one might use a ciphertext alphabet larger than the plaintext alphabet and assign each plaintext letter a to a subset (**homophone**) $f(a)$ of the ciphertext alphabet. To permit deciphering, we require that $f(a)$ and $f(b)$ be disjoint when $a \neq b$. Encipher each m_i in the plaintext as a randomly chosen $c_i \in f(m_i)$.

Usually, the ciphertext alphabet is much larger than the plaintext alphabet and the size of $f(a)$ is proportional to the frequency of occurrence of a in English. Then the letters of the ciphertext alphabet have a uniform distribution in the ciphertext. Use the frequency distribution of pairs of letters to break.

One can define f via a standard text using the number of an instance of the letter as its cipher.

One can encipher two plaintext messages of equal length together using a 26×26 matrix of ciphers. One cannot tell which message it is without the key.

3. **Polyalphabetic substitution ciphers** use multiple maps f_i from the plaintext alphabet to the ciphertext alphabet.

Suppose we encipher $M = m_0 m_1 \ldots$ as $C = f_0(m_0) f_1(m_1) \ldots$. Let n be the length of the alphabet. The sequence $\{f_i\}$ may be periodic, perhaps defined by a **keyword** $K = k_0 \ldots k_{d-1}$.

For example, the Vigenère cipher uses $f_i(a) = (a + k_{i \bmod d}) \bmod n$ and the Beaufort cipher uses $f_i(a) = (k_{i \bmod d} - a) \bmod n$. If the period of the key (the number of letters in the keyword) is d, then the unicity distance is $H(K)/D = \log_2(n^d)/D = (d/D) \log_2 n$. For English, this is $d \log_2(26)/3.2 \approx 1.47d$.

There are two basic methods to find the period of periodic polyalphabetic substitution ciphers, which is the first step in breaking them. The Kasiski method, due to F. W. Kasiski, looks for repetitions in cipher text. They might occur at multiples of the period d; so, the period might be a divisor of the gcd of several of the differences.

W. Friedman [44] invented the **Index of Coincidence Method**, which measures frequency variations of letters to guess the approximate size of the period d. Let $\{a_0, a_1, \ldots, a_{n-1}\}$ be the (plain or ciphertext) alphabet. Let F_i be the number of times a_i occurs in a ciphertext of length N. Define the Index of Coincidence as

$$IC = \left(\sum_{i=0}^{n-1} \frac{F_i(F_i - 1)}{2} \right) \bigg/ \left(\frac{N(N-1)}{2} \right).$$

Then IC represents the probability that two letters chosen at random in the ciphertext are the same.

One can estimate IC theoretically in terms of the period d. See Section 2.7 of Barr [10] for the derivation. For English and a polyalphabetic cipher with period d, the expected value of IC is

$$IC = \frac{0.065(N - d) + 0.038N(d - 1)}{d(N - 1)}.$$

Solving for d gives the estimate

$$d \approx \frac{0.027N}{0.065 - IC + N(IC - 0.038)}.$$

For large N, we have $d \approx 0.027/(IC - 0.038)$.

Example 8.6

Suppose that a Kasiski analysis suggests that the period d of a polyalphabetic substitution cipher is a divisor of 15 and that the Index of Coincidence of a large ciphertext sample is 0.043.

The divisors of 15 are 1, 3, 5 and 15. An IC of 0.043 implies a period of $d = 5.4$, so d is near 5 or 6. Therefore, the period is probably 5.

Once the period d is determined, the cipher may be broken using frequency analysis and trial and error. Think of the cipher as d interwoven simple substitution ciphers.

Polyalphabetic ciphers can also have nonperiodic mapping functions from the plaintext to ciphertext alphabets. **Running key** substitution ciphers use a known text (in a standard book, say) as a key. Encrypt as for a Caesar or Vigenère cipher, except that the key is not constant or periodic. Since the key is as long as the message, this cipher may seem to be unbreakable, like the one-time pad, but it is not if the key is redundant, as in English text. Roughly speaking, this is so because a large proportion of letters in both key and plaintext will be high frequency letters (ETAONISRHDL).

Rotor machines produce running key substitution ciphers with large period d. If there are 26 letters in the alphabet and t rotors, we have $d = 26^t$. The Enigma was a rotor machine with four rotors used by the Germans in World War II and broken by Alan Turing using group theory.

The UNIX[1] crypt(1) command is a (software) rotor machine with one rotor having 256 positions. See Reeds and Weinberger [95] for its cryptanalysis.

A one-time pad is another example of a nonperiodic polyalphabetic substitution cipher.

4. **Polygram substitution ciphers** make arbitrary substitutions for groups of characters. One example is the **Hill cipher**, due to Hill [54], which codes blocks of n letters into column vectors of dimension n. It enciphers a block of n letters by multiplying it by an $n \times n$ matrix to get a vector of n ciphertext letters. The matrix must be invertible modulo the alphabet size to permit deciphering.

For example, suppose $n = 2$ and we encode the alphabet as A= 0, B= 1, etc. Then the plaintext AT would be encoded as $\begin{bmatrix} 0 \\ 19 \end{bmatrix}$ and the plaintext NO would be encoded as $\begin{bmatrix} 13 \\ 14 \end{bmatrix}$. Suppose the key matrix is $K = \begin{bmatrix} 3 & 18 \\ 21 & 11 \end{bmatrix}$. Then AT would be enciphered as

$$K \begin{bmatrix} 0 \\ 19 \end{bmatrix} = \begin{bmatrix} 3 & 18 \\ 21 & 11 \end{bmatrix} \begin{bmatrix} 0 \\ 19 \end{bmatrix} = \begin{bmatrix} 342 \\ 209 \end{bmatrix} \equiv \begin{bmatrix} 4 \\ 1 \end{bmatrix} \quad (\text{mod } 26),$$

which may be converted back into the letters EB. Similarly, NO would be enciphered as $\begin{bmatrix} 5 \\ 11 \end{bmatrix}$ or FL in letters.

Someone who knew the key matrix K could decrypt ciphertext by multiplying the vectors by K^{-1} (mod 26). The matrix may be inverted by the usual techniques of linear algebra, keeping in mind that any division by d must be done by multiplying by the multiplicative inverse of d modulo 26. The methods are similar to those used in Example 5.7. We will illustrate a different method by inverting K modulo 26 by Cramer's rule. The determinant of K is $3 \cdot 11 - 18 \cdot 21 \equiv 19 \pmod{26}$. Now $\phi(26) = 12$, so the inverse of 19 is

---

[1]UNIX is a trademark of Bell Labs.

$19^{11} \equiv 11 \pmod{26}$ by Corollary 6.2. Then by Cramer's rule,

$$K^{-1} \equiv 11 \begin{bmatrix} 11 & -18 \\ -21 & 3 \end{bmatrix} \equiv \begin{bmatrix} 17 & 10 \\ 3 & 7 \end{bmatrix} \pmod{26}.$$

The ciphertext FL would be deciphered as

$$K^{-1} \begin{bmatrix} 5 \\ 11 \end{bmatrix} = \begin{bmatrix} 17 & 10 \\ 3 & 7 \end{bmatrix} \begin{bmatrix} 5 \\ 11 \end{bmatrix} \equiv \begin{bmatrix} 13 \\ 14 \end{bmatrix} \pmod{26},$$

or NO in letters.

One can break the Hill cipher easily with a known-plaintext attack. If one knows n plaintext-ciphertext blocks, then one can determine K through linear algebra. Suppose we didn't know K, but we did know that $K \begin{bmatrix} 0 \\ 19 \end{bmatrix} \equiv \begin{bmatrix} 4 \\ 1 \end{bmatrix} \pmod{26}$ and $K \begin{bmatrix} 13 \\ 14 \end{bmatrix} \equiv \begin{bmatrix} 5 \\ 11 \end{bmatrix} \pmod{26}$. These matrix equations are equivalent to the single equation

$$K \begin{bmatrix} 0 & 13 \\ 19 & 14 \end{bmatrix} \equiv \begin{bmatrix} 4 & 5 \\ 1 & 11 \end{bmatrix} \pmod{26},$$

which is easy to solve for K using linear algebra.

A ciphertext-only attack is harder. Cryptanalysis based on letter frequency does not work because a Hill cipher encrypts blocks of n letters together. If n is small, one could use the frequency of n-letter blocks to guess n of the blocks and then proceed as in the known-plaintext attack above.

8.5 The Entropy of Number Theoretic Ciphers

Plaintext and ciphertext are encoded as numbers when number theoretic ciphers are used. These numbers are grouped into large blocks which hold the codes of many letters and which form numbers modulo some large integer m. These numbers modulo m are enciphered by computing some function modulo m. The key is typically a number about the size of m in these ciphers and its entropy $H(K)$ is roughly $\log_2 m$. The key is chosen large enough so that one cannot try all the keys, making a brute force ciphertext-only attack infeasible.

Known-plaintext attacks on number theoretic ciphers are generally thwarted by making the cryptanalyst solve a hard problem of number theory. When exponentiation modulo m is used as the enciphering function, the cryptanalyst must solve a discrete logarithm problem to effect a known-plaintext attack.

It is amusing to note that the key entropy for all public-key ciphers is zero because one can always compute the secret key from public data. Public-key ciphers do not rely on large key entropy for their secrecy, but rather on the difficulty of computing the secret key from public data. The cryptanalyst must solve a hard number theory problem, like the discrete logarithm problem, to deduce the secret key from public data.

8.6   *Exercises*

1. Let X be an integer variable represented with 24 bits. Suppose that the probability is $1/2$ that X is in the range $[0, 2^{11} - 1]$, with all such values being equally likely, and $1/2$ that X is in the range $[2^{11}, 2^{24} - 1]$, with all such values being equally likely. Compute the entropy $H(X)$.

2. Suppose that X is one of two messages. Use calculus to prove that the entropy $H(X)$ is maximal when the two messages are equally likely. When is the entropy minimal?

3. Prove the four properties of entropy listed at the end of Section 8.1.

4. Let M be a 6-digit number in the range $[0, 10^6 - 1]$ enciphered with a Caesar-type shifted substitution cipher with key K in the range $0 \le K \le 9$. For example, if $K = 2$, then $M = 214759$ is enciphered as $C = 436971$. Compute $H(M)$, $H(M|C)$ and $H(K|C)$, assuming that all values of M and K are equally likely.

5. Suppose that meaningful English language plaintext messages 1000 letters long are enciphered using keys that are strings of letters. (Here "letter" means one of the 26 letters A, B, ..., Z.) Explain why perfect secrecy can be achieved with keys shorter than 1000 letters long, and compute the minimum length of keys if perfect secrecy is desired.

6. Let M be a secret message revealing the name of a spy. There are five suspects: two females, Alice and Bethany, and three males, Chuck, Dennis and Edgar. Exactly one of the five suspects is the spy. The message M is correct. Alice, Bethany and Chuck each have probability 0.25 of being the spy while Dennis and Edgar each have probability 0.125 of being the spy.

   a. Compute the entropy $H(M)$.

   b. Let S be a message telling whether the spy is male or female. Compute $H(M|S)$.

7. A secret message was enciphered using the affine substitution cipher $E(x) = (3x + 24) \bmod 26$. The ciphertext is RTOLK TOIK. Find the plaintext.

8. Consider an affine substitution cipher using the transformation $f(m) = (k_1 m + k_0) \bmod 26$. It is suspected that the plaintext letter E ($= 4$) corresponds to the ciphertext letter F ($= 5$) and that the plaintext letter H ($= 7$) corresponds to the ciphertext letter W ($= 22$). Assuming these correspondences are correct, break the cipher by finding k_1 and k_0.

9. The people on the island of Cobol speak Cobolese. The Cobolese alphabet has 45 letters and the written language has a rate of $r = 2.0$ bits

per letter. For its diplomatic communications, the government of Cobol uses affine ciphers of the form $f(a) = (ak_1 + k_0) \bmod 45$. Naturally, the keys k_0 and k_1 are chosen so that the deciphering function f^{-1} is a well defined function and so that all such keys are equally likely. Determine the standard approximation to the unicity distance of these ciphers.

10. One hundred characters of ciphertext from a suspected Beaufort cipher were intercepted by one of your agents. Here is the frequency distribution of the letters of the alphabet in this sample of ciphertext:

A	B	C	D	E	F	G	H	I	J	K	L	M
2	10	2	5	3	8	1	2	2	5	1	3	1

N	O	P	Q	R	S	T	U	V	W	X	Y	Z
2	10	1	8	1	8	5	2	1	3	5	1	8

a. Compute the Index of Coincidence IC for this sample.

b. What do you think is the period of the key?

11. Suppose that a Kasiski analysis of ciphertext from a Vigenère cipher identifies these six pairs of repeated sequences of ciphertext letters:

Location of start of						
first occurrence	10	21	37	49	58	72
second occurrence	34	65	109	105	162	132

What can you conclude about the period of the Vigenère cipher? Explain your answer.

12. Consider a synchronous stream cipher (from Shamir [103]) whose i-th key block is $k_i = (i+1)^d \bmod n$, where the large integer n is public and d is secret. The i-th message block m_i is enciphered as $c_i = m_i \oplus k_i$. Show that this cipher is vulnerable to a known-plaintext attack. Specifically, show how to compute k_3 and k_5 from the two pairs (m_1, c_1) and (m_2, c_2). Given many plaintext-ciphertext pairs, can a cryptanalyst determine d?

13. Consider a synchronous stream cipher (from Shamir [103]) whose i-th key block is $k_i = S^{1/d_i} \bmod n$, where $n = pq$, and the large primes p and q are secret, S is secret and relatively prime to n, the d_i are pairwise relatively prime and also relatively prime to $\phi(n)$, and $S^{1/d_i} \bmod n$ is the d_i-th root of S modulo n.

Show how to compute the keys from p, q, S, and the d_i's. Explain why this technique cannot be used to find the square root of S modulo n.

14. A message is enciphered using a product cipher which consists of one Hill cipher followed by (composed with) another Hill cipher. Each of these Hill ciphers uses a 2×2 matrix which is invertible modulo 26. Does

the product cipher have a well defined inverse (deciphering) function? If so, is the product cipher more secure, less secure or just as secure as a single Hill cipher? Justify your answer.

Chapter 9

Groups, Rings and Fields

This chapter considers some topics from modern algebra that have important uses in cryptography. We begin with group theory. Many cryptographic functions are computations in groups. Then we study rings, which generalize the structure of the integers modulo m. We consider fields, which generalize the integers modulo a prime p. We investigate polynomials and then make a brief incursion into algebraic number theory, which we need to describe the number field sieve integer factoring algorithm. Other books that cover the same material as this chapter are [78] and [53].

9.1 Groups

Operations like addition, multiplication and exponentiation, which combine two numbers and produce a third number, are called **binary operations**. A group is a set with a binary operation satisfying certain properties. In this section only, the symbol $\oplus$ represents a generic binary operation rather than exclusive-or, which is its meaning in the rest of this book.

DEFINITION 9.1   *A **group** G is a set of elements together with a binary operation $\oplus$ such that*

1. The set is closed under the operation, that is, for every a, b in G, $a \oplus b$ is a unique element of G.

*2. The **associative law** holds, that is, for all a, b, c in G,*

$$a \oplus (b \oplus c) = (a \oplus b) \oplus c.$$

*3. The set has a unique **identity element** e such that $a \oplus e = e \oplus a = a$ for every element a of G.*

*4. Every element a of G has a unique **inverse** a^{-1} in G, with the property $a \oplus a^{-1} = a^{-1} \oplus a = e$.*

*A group is called **commutative** or **abelian** if $a \oplus b = b \oplus a$ for every pair of elements a, b of G.*

*A group is **finite** if it has only a finite number of elements. The number of elements of a finite group is called the **order** of the group. If a group has an infinite number of elements, it is an **infinite** group.*

Nonabelian groups played an important role in breaking the German Enigma cipher during World War II. (See [95].) Most groups arising in number theory are abelian. All groups studied in this book are abelian.

The set of all integers $\{\ldots, -2, -1, 0, 1, 2, \ldots\}$ forms an infinite abelian group with addition $(+)$ for the binary operation, 0 for the identity, and $-a$ for the inverse of a. However, this set does not form a group with multiplication as the operation because 1 would have to be the identity and elements other than ± 1 lack inverses in the set.

If m is a positive integer, a complete set of residues modulo m forms an abelian group with addition modulo m as the binary operation. The identity is the residue class containing 0. The inverse of the residue class containing a is the residue class containing $-a$. The associative law is inherited from the integers, that is, $a + (b + c) = (a + b) + c$ implies $a + (b + c) \equiv (a + b) + c \pmod{m}$. This group is called the **additive group modulo** m. Different CSR's modulo m produce additive groups modulo m with different appearance but the same structure under the addition operation. Two such groups are essentially the same; only the elements have been renamed. These groups are called isomorphic.

DEFINITION 9.2   *We call two groups, G with operation $\oplus$ and G' with operation $\otimes$, **isomorphic** and write $G \cong G'$ if there is a one-to-one correspondence between the elements of G and those of G' such that if $a \in G$ corresponds to $a' \in G'$, then $a \oplus b$ corresponds to $a' \otimes b'$.*

We will regard isomorphic groups as being the same group. The discussion above proves this theorem.

THEOREM 9.1 Integers modulo m are a group under addition
A CSR modulo m for a group with addition modulo m as the operation. Any two CSR's modulo m form isomorphic groups.

At this point we will drop the notation $\oplus$ for the binary operation of a group. When we discuss groups abstractly, we will write the operation as multiplication and write 1 for the identity element. We write ab for $a \oplus b$, abc for $a \oplus (b \oplus c) = (a \oplus b) \oplus c$, a^2 for $a \oplus a$, etc. We write a^i for the product of i a's. However, when a group inherits its operation from another group, then we will write the operation as in the other group. For example, we will continue to write $+$ for the addition operation in the group of integers modulo m under addition. When $+$ is the group operation, we will write $2a$ for $a + a$, etc., and use 0 for the identity element.

A CSR modulo m does not form a group under multiplication because 0 has no inverse. In addition, if m is composite, then proper factors of m lack inverses, too.

THEOREM 9.2 RSR modulo m is a group under multiplication
Let $m > 1$ be an integer. Then any RSR modulo m forms a group with multiplication modulo m as operation. This group has order $\phi(m)$. Different RSR's modulo m produce isomorphic groups.

This group, denoted R_m, is called the **multiplicative group modulo** m.

PROOF Write $n = \phi(m)$ and let $r_1, \ldots, r_n$ be an RSR modulo m. Theorem 3.10 shows that the set is closed under multiplication modulo m. The associative property is inherited from the integers, that is, $a(bc) = (ab)c$ implies $a(bc) \equiv (ab)c \pmod{m}$. The identity is the element $r_i \equiv 1 \pmod{m}$. Inverses exist because the congruence $r_j x \equiv r_i \pmod{m}$ has a unique solution by Theorem 5.6. Two different RSR's are congruent, element by element, modulo m, and this correspondence gives an isomorphism between the two groups. ∎

9.2 Simple Properties of Groups

THEOREM 9.3 Cancellation in group equations
In any group, if $ab = ac$, then $b = c$. If a is an element of a finite group with identity 1, then there is a unique smallest positive integer i with $a^i = 1$.

PROOF Multiply $ab = ac$ by a^{-1} to get $a^{-1}(ab) = a^{-1}(ac)$. The associative law and the properties of inverse and identity yield $1b = 1c$ and $b = c$. Consider the powers of a: $1, a, a^2, a^3, \ldots$. Since the group is finite, there must be a repeated power of the form $a^u = a^v$, where $u < v$. Write this as $a^u 1 = a^u a^{v-u}$. By the cancellation property just proved, $1 = a^{v-u}$. Hence, $a^i = 1$ for some positive integer, namely, $v - u$, and so there must be a smallest positive integer with this property. ∎

DEFINITION 9.3 *Let a be an element of a group. If there is a positive integer i with $a^i = 1$, then a is said to have **finite order** (even if G is not a finite group). If a has finite order, then the **order** of a is the smallest positive integer i with $a^i = 1$. The element a has **infinite order** if there is no positive integer i with $a^i = 1$. A **cyclic group** is one that contains an element a whose* *powers a^i and a^{-i} make up the entire group. An element a with this property is called a **generator** of the group and is said to **generate** the group.*

By Theorem 9.3, every element of a finite group has finite order. The identity element 1 has finite order in every group.

The set of all integers with $+$ for the operation is a cyclic group of infinite order. It is generated by 1. The "powers" of 1 are $0, \pm 1, \pm 2, \ldots$. Every element $a \neq 0$ of this group has infinite order.

The integers modulo $m > 0$ with $+$ for the operation form a cyclic group of order m. The residue class of 1 is a generator.

If m is a positive integer, then all cyclic groups of order m are isomorphic. If a and b generate two cyclic groups of order m, then the one-to-one correspondence makes a^i correspond with b^i for each integer i. Let C_m denote a cyclic group of order m.

The multiplicative group modulo m, R_m, of Theorem 9.2 may or may not be cyclic. The order of a in this group is the same as the order of a we defined in Definition 6.7. A generator for R_m is the same as a primitive root modulo m. Theorem 6.13 said that $m > 1$ has a primitive root if and only if $m = 2$, 4, p^e or $2p^e$, where p is an odd prime and $e \geq 1$. The group R_m is cyclic for the same set of m. When R_m is cyclic we have $R_m \cong C_{\phi(m)}$.

THEOREM 9.4 Lagrange's theorem

The order of an element of a finite group divides the order of the group. If n is the order of the group, then $a^n = 1$ for every element a of the group.

PROOF   Let a have order i. A proof like that of the second part of Theorem 9.3 shows that the members of $A = \{1, a, a^2, \ldots, a^{i-1}\}$ are i distinct elements of the group. If A is not the whole group, then the group has another element a_2. We show that the set $B = \{a_2, a_2 a, \ldots, a_2 a^{i-1}\}$ contains i new elements different from the ones in A. First of all, if $a_2 a^j = a_2 a^k$, then $a^j = a^k$ by Theorem 9.3, contrary to A having distinct elements. Also, if $a_2 a^j = a^k$, then $a_2 = a^{k-j}$ and a_2 would be in A. If $A \cup B$ is not the whole group, then the group has another element a_3, and one can show that $C = \{a_3, a_3 a, \ldots, a_3 a^{i-1}\}$ contains i new elements not in $A \cup B$. Since the group is finite, this process of obtaining new elements a_j must terminate with a last batch, say the batch of i elements including a_k. Then the order of the group must be $n = ik$, and the order i of a divides n. Finally, $a^n = (a^i)^k = 1^k = 1$. ∎

The theorem just proved implies Euler's theorem. The group R_m has order $n = \phi(m)$. The elements of R_m are integers a relatively prime to m. Lagrange's theorem says that $a^n = 1$, that is, $a^{\phi(m)} \equiv 1 \pmod{m}$. Lagrange's theorem has this corollary, which generalizes Corollary 6.2.

COROLLARY 9.1

If a is an element of a group of order n, then a^{n-1} is the inverse of a.

PROOF We have $aa^{n-1} = a^{n-1}a = a^n = 1$ by Lagrange's theorem. ∎

The fast exponentiation algorithm works in any group, and computes a^n with only $O(\log n)$ group operations. Corollary 9.1 provides an efficient method of finding inverses in any group whose order is known.

If G and H are two groups, we can define a group operation on the set of ordered pairs (g, h) of elements of the two groups by $(g_1, h_1) \cdot (g_2, h_2) = (g_1 g_2, h_1 h_2)$, where $g_i \in G$ and $h_i \in H$. The set of ordered pairs with this operation forms a group $G \otimes H$, called the **direct product** of G and H. The identity in the direct product is $(1_G, 1_H)$ and the inverse of (g, h) is (g^{-1}, h^{-1}), with the obvious notation. In a similar way, we may form the direct product of three or even more groups by defining a group operation on the set of ordered triples or quadruples, etc.

A theorem of group theory says that every finite abelian group is isomorphic to a direct product of cyclic groups. We can find this direct product for the group R_m. Let $m = p_1^{e_1} \cdots p_k^{e_k}$ be the standard factorization of m. The Chinese remainder theorem implies that

$$R_m \cong R_{p_1^{e_1}} \otimes \cdots \otimes R_{p_k^{e_k}}.$$

If p is an odd prime, then $R_{p^e} \cong C_{\phi(p^e)}$ is cyclic. When $p = 2$, we have $R_2 \cong C_1$ and $R_4 \cong C_2$. One can show that $R_{2^e} \cong C_2 \otimes C_{2^{e-2}}$ for $e \geq 3$. (In fact, (-1) generates C_2 and 5 generates $C_{2^{e-2}}$. That is, one can show that every odd number is $\equiv \pm(5^k) \pmod{2^e}$ for some k.) Thus, R_m is expressed as the direct product of k or $k + 1$ cyclic groups, depending on the power of 2 dividing m.

DEFINITION 9.4 A **subgroup** *of a group is a subset of the group that forms a group with the same binary operation.*

The associative law holds automatically for a subset of a group. To verify that a subset S is a subgroup, one must check that the identity element is in S, that a^{-1} is in S whenever a is in S and that ab is in S whenever a and b are in S. A subgroup of an abelian group is automatically abelian. It is easy to see that a subgroup of a cyclic group is cyclic.

The real Lagrange's theorem states that the order of a subgroup of a group G divides the order of G. Theorem 9.4 is the special case for the cyclic subgroup generated by a. The real Lagrange's theorem may be proved by a slightly more complicated argument than the proof of Theorem 9.4.

9.3   The Baby-Step-Giant-Step Algorithm

When we say we are "given a group," we mean that we have a way to represent group elements by strings of symbols, we know which string represents the identity and we have algorithms to decide whether two strings represent the same group element, to compute the string which represents the product of elements represented by two given strings and to find the inverse of any given element. We shall assume these algorithms are efficient. This is true for groups used in cryptography. The complexity of other group algorithms is measured in units of these group operations.

This section considers the following problem. Given a finite cyclic group G with generator g and an element b of G, find the smallest integer k so that $g^k = b$. This problem generalizes the discrete logarithm problem for groups R_m when m has a primitive root. That is why it is called the **discrete logarithm problem** for groups.

The simplest algorithm for solving this problem is to compute successively, $g, g^2, g^3, \ldots$, and compare each power of g with b, stopping at the first equality. If G has order n, then this algorithm takes $O(n)$ group operations and $O(1)$ space.

If one wished to compute discrete logarithms of group elements very quickly, one could precompute a table of the discrete logarithm of every element in G and simply look up the discrete logarithm of a when it was needed. The precomputation time is $O(n \log n)$ group operations to form the n powers of g and sort the pairs (g^i, i). The main computation takes no group operations but $O(n)$ space. The time for table lookup is probably $O(\log n)$, depending on the representation of group elements.

Shanks' [105] **baby-step-giant-step algorithm** computes the discrete logarithm of an element of a finite cyclic group with a complexity about midway (on a logarithmic scale) between those of the two simple algorithms just stated. It does not require that we know the order of the group exactly; an upper bound on the size is good enough. A slightly different version of this algorithm appears as Algorithm 5.3.1 in Crandall and Pomerance [33].

[Baby-step-giant-step algorithm for discrete logarithms in a group]
Input: A finite cyclic group G, a generator g, an upper bound n on the order of G, and an element b of G.
Output: An integer k for which $g^k = b$.

```
Precomputation:
L = ⌈√n⌉
a = 1
for (i = 0 to L − 1) {
        store (a, i) in a Table A
        a = a * g
        }
sort Table A in order of its first components
```

```
Main computation:
h = (g⁻¹)ᴸ
a = b
for (j = 0 to L − 1) {
          if (a is the first component of a
                      pair (a, i) in Table A) {
                   write "Log of b is i + jL" and exit
                   }
      a = a * h
      }
write "Error:  n was too small."
```

The algorithm is called the baby-step-giant-step algorithm because the variable a takes baby steps of length 1 through powers of g in the first `for` loop and giant steps of length L through powers of g in the second `for` loop.

THEOREM 9.5 Complexity of the baby-step-giant-step algorithm

The baby-step-giant-step algorithm correctly finds the discrete logarithm of an element b of a finite cyclic group with generator g and order $\leq n$. The complexity of this algorithm is $O(\sqrt{n} \log n)$ group operations and $O(\sqrt{n})$ space.

PROOF   Suppose $g^k = b$. Then $0 \leq k < n$ and so k is a two-digit number in base $L = \lceil \sqrt{n} \rceil$, that is $k = i + jL$ for some $0 \leq i, j < L$. This means that $b = g^k = g^{i+Lj} = g^i(g^L)^j$, so $g^i = bh^j$, where $h = (g^L)^{-1} = (g^{-1})^L$. Table A contains pairs (g^s, s) for $0 \leq s < L$. The second `for` loop forms $a = bh^t$ and searches for this group element as first component of a pair in Table A. It will certainly find such a pair when $t = j$ because the pair (g^i, i) is in Table A and $g^i = bh^j$.

The group element $h = (g^{-1})^L$ may be computed in $O(\log n)$ group operations by fast exponentiation. We are assuming that we can compute g^{-1} in one group operation even if we don't know the exact order of the group. The `for` loops clearly take $O(\sqrt{n} \log n)$ group operations because $L < 1 + \sqrt{n}$ and $\log n$ comparisons are needed to seek each a in the sorted Table A in the second loop. Sorting Table A of length L requires $O(\sqrt{n} \log n)$ comparisons of strings representing group elements. The only large data structure is Table A, and it occupies $O(\sqrt{n})$ space.   ∎

If we wish to compute discrete logarithms modulo a prime p, the baby-step-giant-step algorithm roughly doubles the length of p for which we can do the calculation, as compared to the first algorithm in this section.

9.4   *Rings and Fields*

DEFINITION 9.5   A **ring** *is a set of at least two elements with two binary operations, addition* $(+)$ *and multiplication* $(\times)$*, which is an abelian group with identity zero* (0) *under addition and whose multiplication is associative* $(a \times (b \times c) = (a \times b) \times c)$ *and distributive over addition* $(a \times (b + c) = (a \times b) + (a \times c)$ *and* $(b + c) \times a = (b \times a) + (c \times a))$*. A ring is* **commutative** *if* $a \times b = b \times a$ *for every* a *and* b*. If the elements of a ring, other than 0, form a commutative group under* $\times$*, then the ring is called a* **field**.

All rings in this book will be commutative and the multiplication will have an identity which we will write 1 and call the **unity** of the ring. We will write multiplication in rings in the usual way and omit the $\times$. The set **Z** of all integers with the usual operations is a commutative ring with unity. It is not a field because most integers have no inverses under multiplication. The set of all rational numbers **Q** is a field, as are the set **R** of all real numbers and the set **C** of all complex numbers.

THEOREM 9.6   The integers modulo m are a ring
The set $Z_m = \{0, 1, \ldots, m - 1\}$*, with arithmetic defined modulo* m*, forms a commutative ring for every integer* $m > 1$*. This ring is a field if and only if* m *is prime.*

PROOF   By Theorem 9.1, the set Z_m is a group under addition modulo m. Multiplication modulo m inherits its associative, commutative and distributive properties from the integers. (For example, $a(b + c) \equiv ab + ac \pmod{m}$ because $a(b + c) = ab + ac$.) This shows that Z_m is a commutative ring.

Theorem 9.2 shows that any RSR modulo m forms an abelian group under multiplication modulo m. If m is prime, then the elements of Z_m other than 0 are an RSR, and therefore a commutative group. Thus, Z_m is a field if m is prime. If m is not prime, then $m = ij$ for some $1 < i \leq j < m$ and the congruence $ix \equiv 1 \pmod{m}$ has no solution, by Theorem 5.6. This shows that the element i of Z_m has no multiplicative inverse in Z_m; so, the set of nonzero elements of Z_m is not a group under multiplication, and Z_m is not a field.   ∎

The set of all 2×2 matrices with integer entries is a commutative ring with unity I, the identity matrix. Likewise, the set of all 2×2 matrices with entries modulo $m > 1$ is a commutative ring with unity I.

Let R and S be two rings. A **homomorphism** from R to S is a function f from R into S which preserves addition and multiplication. This means that $f(0) = 0$, $f(1) = 1$, $f(a + b) = f(a) + f(b)$ and $f(ab) = f(a)f(b)$ for all a and b in R. If f is onto, then the ring S is called the **homomorphic**

image of R. For each integer $m > 1$ there is a homomorphism from $\mathbf{Z}$ onto Z_m defined by $f(a) = (a \bmod m)$, that is, $f(a)$ is the congruence class of a modulo m. An **isomorphism** from R to S is a homomorphism which is one-to-one and onto, that is, a one-to-one correspondence between rings that preserves addition and multiplication. We say R and S are **isomorphic** if there is an isomorphism from one to the other.

9.5 *Polynomials*

Let F be a ring. A **polynomial with coefficients in** F is an expression $f(x) = a_n x^n + a_{n-1} x^{n-1} + \cdots + a_1 x + a_0$, where the coefficients a_i are in F. The set of all such polynomials is denoted $F[x]$. The **degree** of a polynomial $f(x)$ is the exponent on the highest power of x having a nonzero coefficient. If the **leading coefficient**, a_n, is $\neq 0$ in the expression above for $f(x)$, then the polynomial $f(x)$ has degree n. The coefficient a_0 of x^0 is the **constant term**. Constant polynomials have degree 0, except for the zero polynomial 0, which has no degree. A polynomial is **monic** if its leading coefficient is 1. Two polynomials are **equal** if they have the same degree and all corresponding coefficients are equal.

Polynomials in $F[x]$ may be added and multiplied in a natural way. If $f(x) = a_n x^n + a_{n-1} x^{n-1} + \cdots + a_1 x + a_0$ and $g(x) = b_n x^n + b_{n-1} x^{n-1} + \cdots + b_1 x + b_0$, then their sum is

$$f(x) + g(x) = (a_n + b_n)x^n + \cdots (a_1 + b_1)x + (a_0 + b_0)$$

and their product is

$$f(x)g(x) = (a_n b_n)x^{2n} + \cdots + (a_2 b_0 + a_1 b_1 + a_0 b_2)x^2 + (a_1 b_0 + a_0 b_1)x + (a_0 b_0).$$

With these operations, $F[x]$ is a commutative ring with unity. The zero element is the polynomial 0 and the unity is the constant polynomial 1.

From now on in this section, we assume F is a field.

THEOREM 9.7 Division algorithm for polynomials

Let $f(x)$ and $g(x)$ be two polynomials in $F[x]$, where F is a field. If $g(x)$ is not the zero polynomial, then there exist polynomials $q(x)$ and $r(x)$ in $F[x]$ with $f(x) = q(x)g(x) + r(x)$ and either $r(x)$ is the zero polynomial or else the degree of $r(x)$ is less than the degree of $g(x)$.

The proof is a statement of the long division algorithm you learned for polynomials in high school.

A **zero** of a polynomial $f(x)$, or a **root** of $f(x) = 0$, is a quantity α, belonging either to F or to a larger field, for which $f(\alpha) = 0$.

THEOREM 9.8 The factor theorem

If $\alpha \in F$ is a zero of a polynomial $f(x)$ in $F[x]$, where F is a field, then there is a polynomial $q(x)$ in $F[x]$ for which $f(x) = (x - \alpha)q(x)$.

PROOF   By Theorem 9.7 with $g(x) = (x - \alpha)$, we can write $f(x) = (x - \alpha)q(x) + r(x)$, for some polynomials $q(x)$ and $r(x)$ in $F[x]$. If $r(x)$ has a degree, then it is less than 1, that is, $r(x)$ must be constant. Substituting $x = \alpha$ shows that this constant must be 0, and the theorem is proved.   ∎

COROLLARY 9.2

The number of zeros of a polynomial in $F[x]$, where F is a field, is no more than its degree.

PROOF   Use induction on the degree. Compare with the proof of Theorem 5.8.   ∎

It is essential that F be a field in this corollary. We saw in Theorem 7.17 that the polynomial $x^2 - 1$ has four zeros in Z_8. By Theorem 7.18, if a is a quadratic residue modulo pq, then $x^2 - a$ has four zeros in Z_{pq}.

The division algorithm for polynomials allows us to define divisibility and greatest common divisors for polynomials, just as we did for integers in Chapter 3. Let F be a field. If $f(x)$ and $g(x)$ are in the polynomial ring $F[x]$ and $f \neq 0$, then $f(x)$ is called a **divisor** of $g(x)$ if there is a polynomial $q(x)$ in $F[x]$ with $g(x) = q(x)f(x)$. We write $f(x)|g(x)$ if this is so. If $a \in F$ and $a \neq 0$, then the constant polynomial a divides every polynomial in $F[x]$. Also if $a \in F$ and $a \neq 0$, then $f(x)|g(x)$ if and only if $(af(x))|g(x)$.

A **greatest common divisor** of two polynomials $f(x)$ and $g(x)$, not both 0, in $F[x]$ is a monic polynomial $d(x)$ of highest degree which divides both $f(x)$ and $g(x)$. We write $d(x) = \gcd(f(x), g(x))$ in this case.

THEOREM 9.9 GCD is a linear function of polynomials

If the two polynomials $f(x)$ and $g(x)$ in $F[x]$ are not both 0, and $d(x)$ is a greatest common divisor of $f(x)$ and $g(x)$, then there are polynomials $a(x)$ and $b(x)$ in $F[x]$ such that $a(x)f(x) + b(x)g(x) = d(x)$.

The theorem may be proved in the same way as Theorem 3.9. A consequence of Theorem 9.9 is that the greatest common divisor of two polynomials is unique. The analogue of the extended Euclidean algorithm for integers may be used to compute the greatest common divisor of two polynomials. The only difference here is that, if the last nonzero remainder is not a monic polynomial, then we must multiply it by the inverse of its leading coefficient to make it monic.

A polynomial $p(x)$ of degree at least one in $F[x]$ is called **irreducible** (over F) if it cannot be written as the product of two nonconstant polynomials in $F[x]$ of lower degree. Irreducible polynomials are analogues of prime numbers. The analogue of Lemma 4.2 is this statement: If $p(x)$ is irreducible and $p(x)|f(x)g(x)$, then either $p(x)|f(x)$ or $p(x)|g(x)$. This can be used to prove the analogue of Theorem 4.1.

THEOREM 9.10 Factorization of polynomials
Every nonconstant polynomial $f(x)$ in $F[x]$ can be written in the form

$$f(x) = ap_1(x) \cdots p_k(x),$$

where the $p_i(x)$ are irreducible polynomials in $F[x]$.

The irreducible (over **Q**) factors of the polynomials $x^n - 1$ are called the **cyclotomic polynomials**. There is exactly one irreducible monic polynomial $\Phi_n(x)$, called the n-th cyclotomic polynomial, which divides $x^n - 1$ but no $x^k - 1$ for $1 \leq k < n$. The coefficients of each $\Phi_n(x)$ are integers. The degree of $\Phi_n(x)$ is $\phi(n)$. One can show that $x^n - 1 = \prod_{d|n} \Phi_d(x)$ (and so $\sum_{d|n} \phi(d) = n$). The zeros of $\Phi_n(x)$ are the n-th roots of 1 in the complex numbers **C** which are not k-th roots of 1 for any $1 \leq k < n$. These numbers are called the **primitive n-th roots of unity**. They are the complex numbers $e^{2\pi i j/n}$, where $1 \leq j \leq n$ and $\gcd(j, n) = 1$. The first few cyclotomic polynomials are

$$\Phi_1(x) = x - 1$$
$$\Phi_2(x) = x + 1$$
$$\Phi_3(x) = x^2 + x + 1$$
$$\Phi_4(x) = x^2 + 1$$
$$\Phi_5(x) = x^4 + x^3 + x^2 + x + 1$$
$$\Phi_6(x) = x^2 - x + 1$$

The notion of divisibility for polynomials allows us to define congruence of polynomials. Let $f(x)$ be a nonzero polynomial in $F[x]$. Let $a(x)$ and $b(x)$ be two polynomials in $F[x]$. We say $a(x)$ is **congruent to $b(x)$ modulo $f(x)$** and write $a(x) \equiv b(x) \pmod{f(x)}$ if $f(x)$ divides $a(x) - b(x)$. Congruence defines an equivalence relation on $F[x]$. Let $[a(x)]$ be the congruence class containing the polynomial $a(x)$, that is, $[a(x)] = \{b(x) \in F[x]; b(x) \equiv a(x) \pmod{f(x)}\}$. Then the formulas

$$[a(x)] + [b(x)] = [a(x) + b(x)]$$
$$[a(x)][b(x)] = [a(x)b(x)]$$

are well defined rules for addition and multiplication of congruence classes. The set of congruence classes with these operations forms a commutative ring with unity. The ring is denoted $F[x]/(f(x))$. The zero element of the ring is the class $[f(x)]$ and the unity is the class $[1]$.

If $f(x)$ has degree $d > 0$, then it is easy to prove, using Theorem 9.7, that the distinct congruence classes of $F[x]/(f(x))$ are the classes $[a(x)]$, where $a(x)$ is an arbitrary polynomial in $F[x]$ of degree $< d$.

We will construct new fields using the following theorem.

THEOREM 9.11 Fields correspond to irreducible polynomials
Let $p(x)$ be a polynomial in $F[x]$, where F is a field. Then $p(x)$ is irreducible if and only if the ring $F[x]/(p(x))$ is a field.

PROOF   Suppose $p(x)$ is irreducible. Let $[a(x)]$ be a nonzero element of $F[x]/(p(x))$. Then $p(x)$ does not divide $a(x)$ and so $\gcd(a(x), p(x)) = 1$. By Theorem 9.9, there are polynomials $b(x)$ and $c(x)$ so that $b(x)a(x)+c(x)p(x) = 1$. Hence, $a(x)b(x) \equiv 1 \pmod{p(x)}$, which means $[a(x)][b(x)] = [1]$. Thus, $[a(x)]$ has a multiplicative inverse, and so $F[x]/(p(x))$ is a field.

Conversely, if $p(x)$ were not irreducible, then the congruence classes of its factors would not have inverses, so $F[x]/(p(x))$ would not be a field. ∎

Theorem 9.11 may be used to construct finite fields of size p^e elements for every prime p and every positive integer e. We know from Theorem 9.6 that $\mathbf{F}_p = Z_p$ is a field for every prime p. It contains p elements. Let $f(x)$ be an irreducible polynomial of degree e in $\mathbf{F}_p[x]$. Then $\mathbf{F}_{p^e} = \mathbf{F}_p[x]/(f(x))$ is a field with exactly p^e elements, and all finite fields arise this way. One can prove that there is such an irreducible polynomial for every prime p and every positive integer e. Let n_e denote the number of monic irreducible polynomials of degree e over $\mathbf{F}_p$, where p is prime. Then one can show that $n_e = (p^e - \sum_{1 \le d < e, d \mid e} dn_d)/e$.

Example 9.1

Use $f(x) = x^8 + x^4 + x^3 + x + 1$ in $\mathbf{F}_2[x]$ to construct a field with 2^8 elements.

We first show that $f(x)$ is irreducible over $\mathbf{F}_2$, the integers modulo 2. Since the degree of $f(x)$ is 8, if $f(x)$ were not irreducible, then $f(x)$ would be divisible by an irreducible polynomial of degree 1, 2, 3 or 4. If $f(x)$ had a linear factor $x + a$, then $f(a) = 0$. But $f(0) = f(1) = 1 \ne 0$. It is easy to see that the only irreducible polynomial in $\mathbf{F}_2[x]$ of degree two is $x^2 + x + 1$. (Note that $x^2 + 1 = (x + 1)^2$.) Similarly, the only irreducible cubic polynomials in $\mathbf{F}_2[x]$ are $x^3 + x^2 + 1$ and $x^3 + x + 1$. Long division shows that none of these three trinomials divide $f(x)$. We leave it to the reader to find the three irreducible polynomials in $\mathbf{F}_2[x]$ of degree 4 and show they do not divide $f(x)$.

By Theorem 9.11, $\mathbf{F}_{2^8} = \mathbf{F}_2[x]/(f(x))$ is a field. Its 2^8 elements are 0 and the polynomials of degree < 8 with coefficients in $\mathbf{F}_2$. The 8 coefficients of such a polynomial are bits, and the polynomials correspond in a natural way to 8-bit bytes. Addition in $\mathbf{F}_{2^8}$ corresponds to the exclusive-or of bytes. To multiply two polynomials in $\mathbf{F}_{2^8}$, first multiply them as ordinary polynomials (over $\mathbf{Z}$) and reduce the coefficients modulo 2. Then divide the product by $f(x)$. The remainder is the product of the two polynomials in $\mathbf{F}_{2^8}$.

One can show that two different irreducible polynomials $f(x)$ and $g(x)$ in $\mathbf{F}_p[x]$ with the same degree e produce isomorphic fields $\mathbf{F}_p[x]/(f(x))$ and $\mathbf{F}_p[x]/(g(x))$, which we regard as the same field $\mathbf{F}_{p^e}$.

Note that, when $e > 1$, $\mathbf{F}_{p^e}$ is not isomorphic to Z_{p^e}. Both rings contain p^e elements, but Z_{p^e} is not a field because, for example, the congruence class of p has no multiplicative inverse.

Let p be prime and $q = p^e$ for some positive integer e. The set of nonzero elements of $\mathbf{F}_q$ forms a group of order $q - 1$. By Lagrange's theorem, if $a \in \mathbf{F}_q$ and $a \neq 0$, then $a^{q-1} = 1$. In fact, there is an element of order $q - 1$ in this group.

THEOREM 9.12 Multiplicative group of a finite field is cyclic
The multiplicative group of a finite field is cyclic.

PROOF   Let F be the finite field and let G be its multiplicative group. By Corollary 9.2, for every $n \geq 1$ the equation $x^n = 1$ has at most n roots in F. Let a be an element of G with largest order N. Let b be any element of G, and call n its order. We will show that G is cyclic by proving that $b = a^j$ for some integer j.

If n does not divide N, then there is a prime p and a power $q = p^s$ of p so that q divides n but not N. It is easy to see that the order of $ab^{n/q}$ is $\mathrm{lcm}(N, q) > N$, which contradicts the definition of N as the largest order of any element of G. Therefore n divides N.

The equation $x^n = 1$ has the n distinct roots $a^{iN/n}$ in G, with $0 \leq i < n$. Since b satisfies $b^n = 1$, it must be one of these roots, that is $b = a^j$ with $j = iN/n$ for some $0 \leq i < n$. ∎

9.6   Algebraic Number Theory

In this section, we give a brief introduction to algebraic number theory needed to understand the number field sieve factoring algorithm.

Let $\mathbf{Z}[x]$ denote the ring of polynomials with integer coefficients.

DEFINITION 9.6   An **algebraic integer of degree** d *is the zero in the complex numbers* $\mathbf{C}$ *of a monic polynomial of degree* d *in* $\mathbf{Z}[x]$ *which is not the zero of such a polynomial with lower degree.*

For example, $\sqrt{5}$ and $i = \sqrt{-1}$ are algebraic integers of degree 2. They are the zeros of the polynomials $x^2 - 5$ and $x^2 + 1$, respectively, but not the zero of any linear polynomial with integer coefficients. If a and b are integers, then $a + bi$ is an algebraic integer of degree 2 because it is the zero of $x^2 - 2ax + (a^2 + b^2)$. Let $\mathbf{Z}[i]$ denote the set of all $a + bi$, where a and b are integers. This set is called the **Gaussian integers**. It contains the integers $\mathbf{Z}$ as the subset with $b = 0$ and shares many properties with $\mathbf{Z}$.

A major portion of algebraic number theory studies how algebraic integers factor. In the Gaussian integers, 3 and 7 cannot be factored, but $2 = (1 + i)(1 - i)$ and $13 = (3 + 2i)(3 - 2i)$ can be factored, even though they are primes in $\mathbf{Z}$. The Gaussian integers 3, 7, $1 + i$ and $3 - 2i$ are primes in $\mathbf{Z}[i]$.

The primes in $\mathbf{Z}$ are called **rational primes** to distinguish them from the Gaussian integer primes. Likewise, the elements of $\mathbf{Z}$ are sometimes called **rational integers** to distinguish them from algebraic integers not in $\mathbf{Z}$.

DEFINITION 9.7   A **unit** of a commutative ring R with unity 1 is an element having a multiplicative inverse in R. An **irreducible element** of R is a nonzero, nonunit element α whose only factorizations in R are the trivial ones $\alpha = u\beta$ with one factor u being a unit. If $\alpha = u\beta$, where u is a unit, then α and β are called **associates**. An algebraic integer α has **unique factorization** in R if any two factorizations of α into the product of irreducibles and units are the same except for replacing irreducibles by their associates and using different units.

The units in $\mathbf{Z}[i]$ are $+1, -1, +i$ and $-i$. The units in $\mathbf{Z}$ are $+1$ and -1. The irreducible elements in $\mathbf{Z}$ are the primes p and their associates $-p$. In $\mathbf{Z}[i]$, $2 + i$ is irreducible and has the associates $2 - i$, $1 + 2i$ and $1 - 2i$. The Gaussian integers have unique factorization. The number 5 can be factored as $(2 + i)(2 - i)$, as $(1 + 2i)(1 - 2i)$ and also as $(-i)(2 + i)(1 - 2i)$. All three of the factorizations are considered the same.

Now let $\mathbf{Z}[\sqrt{-6}]$ denote the set of all numbers of the form $a + b\sqrt{-6}$, where a and b are integers. This set forms a commutative ring with unity under addition and multiplication. Define the **norm** of $a + b\sqrt{-6}$ to be $N(a + b\sqrt{-6}) = a^2 + 6b^2$. We say $a + b\sqrt{-6}$ is factored if we can write

$$a + b\sqrt{-6} = (c + d\sqrt{-6})(e + f\sqrt{-6})$$

with $N(c + d\sqrt{-6}) > 1$ and $N(e + f\sqrt{-6}) > 1$. This restriction avoids trivial factorings. The norm function is completely multiplicative, that is, if α and β are in $\mathbf{Z}[\sqrt{-6}]$, then $N(\alpha\beta) = N(\alpha)N(\beta)$. It follows that $a + b\sqrt{-6}$ is factored if $a + b\sqrt{-6} = \alpha\beta$ with $1 < N(\alpha) < N(a + b\sqrt{-6}) = a^2 + 6b^2$ and $1 < N(\beta) < N(a + b\sqrt{-6})$. These inequalities show that a number in $\mathbf{Z}[\sqrt{-6}]$ can break up into only a finite number of factors. Note also that $N(a + b\sqrt{-6}) \geq 6$ if $b \neq 0$. This shows that 2 and 5 are irreducibles; they do not factor in $\mathbf{Z}[\sqrt{-6}]$. Now 10 can be factored in two different ways:

$$10 = 2 \cdot 5 = (2 + \sqrt{-6})(2 - \sqrt{-6}). \tag{9.1}$$

The ring $\mathbf{Z}[\sqrt{-6}]$ does not have unique factorization.

DEFINITION 9.8   *A nonzero algebraic integer α* **divides** *an algebraic integer β (written $\alpha|\beta$) if there is an algebraic integer γ so that $\beta = \alpha\gamma$. A nonzero algebraic integer α is* **prime** *if it is not a unit and whenever $\alpha|\beta\gamma$, either $\alpha|\beta$ or $\alpha|\gamma$.*

In the integers $\mathbf{Z}$, every irreducible is prime, by Lemma 4.2. In any ring in which every irreducible is prime, one can prove that factorization into irreducibles (or primes) is unique; our proof of Theorem 4.1 shows this. Not every irreducible in $\mathbf{Z}[\sqrt{-6}]$ is prime. The four factors of 10 in Equation (9.1) are irreducible but not prime.

DEFINITION 9.9   *An* **algebraic number** *of degree d is the zero in the complex numbers $\mathbf{C}$ of a polynomial of degree d in $\mathbf{Z}[x]$ which is not the zero of such a polynomial with lower degree.*

This definition is the same as Definition 9.6 of algebraic integer except that the word "monic" is dropped.

DEFINITION 9.10   *If $E \subset F$ are two fields, we call E a* **subfield** *of F, and F an* **extension field** *of E.*

DEFINITION 9.11   *An* **algebraic number field** *is an extension field of $\mathbf{Q}$ that contains only algebraic numbers. If α is an algebraic number of degree d, then the* **algebraic number field of degree d over $\mathbf{Q}$ generated by α** *is the smallest extension field $\mathbf{Q}(\alpha)$ of $\mathbf{Q}$ containing α.*

One can show that $\mathbf{Q}(\alpha)$ is the intersection of all algebraic number fields containing α. The monic polynomial in $\mathbf{Q}[x]$ of degree d satisfied by α is the **minimal polynomial** of α over $\mathbf{Q}$. When considered as a vector space over $\mathbf{Q}$, $\mathbf{Q}(\alpha)$ has dimension d over $\mathbf{Q}$. It is known that every extension field E of $\mathbf{Q}$ that is a finite-dimensional vector space over $\mathbf{Q}$ has the form $\mathbf{Q}(\alpha)$ for some algebraic number α. The elements of $\mathbf{Q}(\alpha)$ are all sums $\sum_{j=0}^{d-1} a_j\alpha^j$, where the a_j are in $\mathbf{Q}$. Define $\mathbf{Z}(\alpha)$ to be the set of all sums $\sum_{j=0}^{d-1} a_j\alpha^j$, where the a_j are in $\mathbf{Z}$.

If α is an algebraic number of degree d, then its **conjugates** are the d roots in $\mathbf{C}$ of its minimal polynomial. The **norm** of α, $N(\alpha)$, is the product of its conjugates, and equals $(-1)^d$ times the constant term of the minimal polynomial of α. The norm satisfies $N(\alpha\beta) = N(\alpha)N(\beta)$. The norm of an algebraic integer is an integer. The minimal polynomial of an algebraic integer has integer coefficients.

Let α be an algebraic number. The set of all algebraic integers in $\mathbf{Q}(\alpha)$ forms a ring $\mathcal{I}$ called the **ring of integers** in $\mathbf{Q}(\alpha)$. This ring $\mathcal{I}$ always contains the ring $\mathbf{Z}(\alpha)$, and may be equal to it.

Example 9.2

The ring $\mathbf{Z}(\sqrt{13})$ consists of all numbers $a + b\sqrt{13}$, where a and b are integers. However, the ring of integers in $\mathbf{Q}(\sqrt{13})$ consists of all numbers $a+b(1+\sqrt{13})/2$, where a and b are integers. The number $(1 + \sqrt{13})/2$ is an algebraic integer because it is a zero of the polynomial $x^2 - x - 3$. Clearly, it is in $\mathbf{Q}(\sqrt{13})$.

If the ring $\mathcal{I}$ of integers in $\mathbf{Q}(\alpha)$ has unique factorization, then $\mathcal{I}$ is called a **unique factorization domain**.

9.7   Exercises

1. Which of the following are groups?

    a. The even numbers with addition as the operation.

    b. The integers with subtraction as the operation.

    c. The odd numbers with multiplication as the operation.

    d. The rational numbers a/b with $b = 1$ or 2, with addition as the operation.

    e. The rational numbers a/b with $b = 1$, 2 or 3, with addition as the operation.

2. Show that the groups R_5 and R_8 have the same size, but that they are not isomorphic.

3. Show that the groups C_6 and R_9 are isomorphic.

4. Let G be a finite cyclic group generated by g. Let b be an element of G. Suppose two numbers $A < B$ are known, with $B - A$ small compared to the order of G, such that there is an integer k in $A \leq k < B$ for which $g^k = b$. Modify Shanks' baby-step-giant-step algorithm to create an algorithm that will discover this k in $O(\sqrt{B - A}\log(B - A))$ group operations and $O(\sqrt{B - A})$ space.

5. Alice and Bob debate whether Shanks' baby-step-giant-step algorithm works because of the birthday paradox of Theorem 2.4. Does it?

6. Do the real numbers of the form $x + y\sqrt[3]{2}$, where x and y are rational numbers, form a ring with the usual addition and multiplication? If so, is this ring a field?

7. Let $F = \mathbf{F}_3$, the field with three elements. Let $f(x) = 2x^2 + 1$ and $g(x) = x^3 + x^2 + 2$ be two polynomials in $F[x]$. Compute $d(x) = \gcd(f(x), g(x))$ and find polynomials $a(x)$ and $b(x)$ so that $a(x)f(x) + b(x)g(x) = d(x)$.

8. In Example 9.1, find the three irreducible polynomials in $\mathbf{F}_2[x]$ of degree 4 and show that they do not divide $f(x) = x^8 + x^4 + x^3 + x + 1$.

9. Find explicit formulas for the cyclotomic polynomials $\Phi_p(x)$ and $\Phi_{2p}(x)$, where p is any odd prime.

10. According to Theorem 9.12, the multiplicative group of the field $\mathbf{F}_{2^8}$ constructed in Example 9.1 is cyclic. Find a generator for it.

Chapter 10

Exponential Methods of Factoring Integers

This chapter introduces methods of factoring integers that are slower than the fastest known ones. They require time $O(n^c)$, where $c > 0$ is a constant, to factor n. They are called "exponential algorithms" because their time complexity is exponential in $\log n$, the length of the input, since $n^c = e^{c \ln n}$. We study them because they are fairly simple, some are used as procedures in faster factoring methods, and because, since they sometimes work surprisingly quickly, we have to avoid numbers they can factor when choosing cryptographic keys that must not be factored. See the books by Crandall and Pomerance [33], Cohen [28], and Riesel [96] for more about these factoring algorithms.

The trial division algorithm from Chapter 4 is an excellent way to factor fairly small numbers. Example 4.5 shows that it has little chance of factoring large integers completely, although it almost always finds some small prime factors of random large integers. Half of all integers have a factor of 2. About 92% of large odd integers have a prime divisor below 1,000,000. Example 4.5 shows that if we don't want someone to factor our secret key, then it must not have a small prime factor.

Throughout this chapter, n will be the odd composite number to factor.

10.1   Fermat's Difference of Squares Method

This is the second oldest factoring method, after trial division. Fermat tried to express n as a difference of two squares, $x^2 - y^2$, with the pair x, y different from $(n + 1)/2, (n - 1)/2$. Any other representation of n as $x^2 - y^2$ gives a nontrivial factorization $n = (x - y)(x + y)$. Clearly, $x \geq \sqrt{n}$.

We illustrate the algorithm with $n = 481$ in the following table. The variable x begins with $\lfloor\sqrt{n}\rfloor$ rather than $\lceil\sqrt{n}\,\rceil$ to check whether n is a square.

x	$t = 2x + 1$	x^2	$r = x^2 - n$
21	43	441	-40
22	45	484	3
23	47	529	48
24	49	576	95
25	51	625	$144 = 12^2$

The last line of the table shows that $481 = 25^2 - 12^2 = (25 - 12)(25 + 12)$.

Why is there a column for $2x + 1$ in the table? After the numbers in the first row have been computed, we can find the new x^2 by adding the old $2x+1$ to the old x^2, since $(x+1)^2 = x^2 + (2x+1)$. Likewise, the next r can be found by adding the old $2x + 1$ to the old r. The column for x is needed only at the end. It need not be computed at all because it can be found easily from $2x + 1$ when the algorithm finishes. This suggests the following algorithm.

[Fermat's difference of squares factoring algorithm]
Input: An odd composite positive integer n to factor.
Output: The factors a and b of n.

```
x = ⌊√n⌋
t = 2x + 1
r = x² − n
while (r is not a square) {
        r = r + t
        t = t + 2
        }
x = (t − 1)/2
y = ⌊√r⌋
    return the factors x − y and x + y
```

In two lines of the algorithm we must find the integer part of the square root of an integer. A good way to do this is with a modification of Newton's method. The initial value of x in the algorithm below can be any integer $\geq \sqrt{n}$, the closer to $\sqrt{n}$ the better. The value in the first line of the algorithm is easy to compute on a binary computer.

[Integer part of the square root of a positive integer]
Input: A positive integer n.
Output: $x = \lfloor\sqrt{n}\rfloor$.

```
x = 2^⌈(log₂ n)/2⌉
y = ⌊(x + ⌊n/x⌋)/2⌋
while (y < x) {
        x = y
        y = ⌊(x + ⌊n/x⌋)/2⌋
```

```
      }
   return x
```

It is an easy exercise to show that this algorithm is correct and finishes in $O(\log \log n)$ iterations.

Let us return to Fermat's difference of squares factoring algorithm. The condition in the `while` loop may be tested as, "Is $r \neq (\lfloor \sqrt{r} \rfloor)^2$?", where the square root is computed by the algorithm just given. However, the rest of the loop contains only two additions. The integer square root algorithm uses several divisions and would dominate the time for the loop. Fermat, working by hand with decimal numbers, solved this problem by recognizing possible squares by their low-order digits. Every square has last decimal digit 0, 1, 4, 5, 6 or 9. In the example in the table above, $r = 3$ and 48 cannot be squares because their last digits are not in the list. Only 22 two-digit numbers may occur as the last two decimal digits of a square. A binary computer can test whether r might be a square with the logical operation $(r\&63)$ to find $(r \bmod 64)$, followed by looking up the remainder in a table of the twelve possible squares modulo 64. If r passes this test, then look up $(r \bmod p^e)$ in a table of possible squares modulo p^e for a few small odd prime powers p^e. Only in case r passes all these tests need one check "$r \neq (\lfloor \sqrt{r} \rfloor)^2$?". Tricks like these amortize evaluation of the `while` condition to a cost comparable to the cost of the two addition operations inside the loop.

THEOREM 10.1 Complexity of Fermat's factoring algorithm

Let the odd composite positive integer $n = ab$, where a is the largest divisor of n which is $\leq \sqrt{n}$. Let $k = a/\sqrt{n}$, so that $0 < k \leq 1$. Then the `while` loop in Fermat's difference of squares factoring algorithm is executed

$$1 + (1 - k)^2 \sqrt{n}/(2k)$$

times.

PROOF   If $a = b$, then n is a square, $k = 1$ and the `while` loop ends after the first iteration. In any case, when the algorithm ends, $x - y = a$ and $x + y = b$, that is, $x = (a+b)/2$ and $y = (b-a)/2$. At this time, $x = (t-1)/2$, so $t = 1 + a + b = 1 + a + n/a$. The variable t increases by 2 at each iteration, begins at the first odd number $> 2\sqrt{n}$ and stops at $1 + a + n/a$. Hence, the `while` loop is executed

$$1 + \frac{1}{2}\left(\left(a + \frac{n}{a}\right) - 2\sqrt{n}\right) = 1 + \frac{(\sqrt{n} - a)^2}{2a} = 1 + \frac{(1-k)^2}{2k}\sqrt{n}$$

times.   ∎

Theorem 10.1 does not say that the time complexity of Fermat's algorithm is $O(\sqrt{n})$ to factor n. In the worst case, $n = 3p$, for some prime p, we have $k = 3/\sqrt{n}$ and the `while` loop is performed essentially $n/6$ times.

If $\sqrt{n} - a$ is $O(n^{1/4})$, then the theorem shows that Fermat's algorithm takes about $\sqrt{n} - a$ steps. The cryptographic significance of this result is that if $n = pq$ is public but the primes p and q must stay secret, then one must not choose p and q too close to each other. If $p < q$ and $q - p$ is small enough so that an attacker could perform $q - p$ simple operations, then n can be factored by Fermat's method. As one can build special hardware devices (called **sieves**) capable of executing the `while` loop at high speed, it is best to require $q - p > 10^{25}$. See Lehmer [62], [63] and Williams [128] for information about the construction and use of sieves.

10.2   *Pollard's Rho Method*

This method is also called the Monte Carlo factoring method because it constructs a sequence of random numbers and the city of Monte Carlo is well known for randomness.

Let n be the composite number to factor and let p be an unknown prime factor of n. Pollard [85] proposed choosing a random function f from the set $\{0, 1, \ldots, n-1\}$ into itself, picking a random starting number s in the set and iterating f:

$$s, f(s), f(f(s)), f(f(f(s))), \ldots.$$

If we consider these numbers modulo the unknown prime p, we get a sequence of integers in the smaller set $\{0, 1, \ldots, p-1\}$. We know from the birthday paradox, Theorem 2.4, that some number in the smaller set will be repeated after about $\sqrt{p}$ iterations of f. If u, v were the iterates of f with $u \equiv v \pmod{p}$, then probably $\gcd(u-v, n) = p$ because p divides $u-v$ and n and it is unlikely that any other prime divisor of n divides $u - v$. But how can we detect this repeated value when it happens? We don't know p and must iterate f modulo n. Suppose that $f(u) \equiv f(v) \pmod{p}$ whenever $u \equiv v \pmod{p}$.

Write s_i for the i-fold iterate of f starting at s. That is, $s_0 = s$ and $s_i = f(s_{i-1})$ for $i > 0$. If $s_i \equiv s_j \pmod{p}$, then p divides $s_i - s_j$ and also $\gcd(s_i - s_j, n)$. However, we can't compute a gcd for every pair $i, j < \sqrt{p}$ because there would be about $\frac{1}{2}(\sqrt{p})^2 = p/2$ pairs and we might as well use trial division to find p.

There is a beautiful solution to this problem due to Floyd. (See Exercise 6b in Section 3.1 of [56].) The **Floyd cycle-finding algorithm** computes two iterates of f together in the same loop, with one instance running twice as fast as the other. This trick generates s_m and s_{2m} together and forms $\gcd(s_{2m} - s_m, n)$, hoping to find p. Here is why the trick works. Suppose $s_i \equiv s_j \pmod{p}$ for some $i < j$. By the birthday paradox, Theorem 2.4, the first j for which this congruence holds for some $i < j$ is $O(\sqrt{p})$. Let $k = j - i$.

Then for any $m \geq i$ and $t \geq 0$, $s_m \equiv s_{m+tk} \pmod{p}$. When $m = k\lceil i/k \rceil$ and $t = \lceil i/k \rceil$ we have $s_m \equiv s_{2m} \pmod{p}$ and $m \leq j$, so m is $O(\sqrt{p})$.

What is a good choice for the random function $f(x)$? It must be easy to compute modulo n. Low degree polynomials with integer coefficients come to mind first. They satisfy $f(u) \equiv f(v) \pmod{p}$ whenever $u \equiv v \pmod{p}$, by Corollary 5.1.

One should not use a linear polynomial because they are not random enough. Consider first $f(x) = (x + b) \bmod n$. Then $s_i \equiv s + ib \pmod{p}$. We may suppose that p does not divide b, since otherwise $\gcd(b, n) = p$ factors n. In this case, we have $s_i \equiv s_j \pmod{p}$ if and only if $i \equiv j \pmod{p}$ by Theorem 5.5. This means that the sequence $\{s_i \bmod p\}$ has period p, and trial division of n would find p sooner.

Now suppose $f(x) = (ax + b) \bmod n$. We may suppose that p does not divide $a - 1$, for otherwise $\gcd(a - 1, n) = p$ factors n. In this situation, $1 - a$ has a multiplicative inverse modulo p. An easy induction shows that $s_i \equiv sa^i + b/(1 - a) \pmod{p}$ for $i \geq 0$. Then $s_i \equiv s_j \pmod{p}$ if and only if $a^i \equiv a^j \pmod{p}$. If a happened to be a primitive root modulo p, then this congruence would hold if and only if $i \equiv j \pmod{p - 1}$, by Theorem 6.15. Hence the sequence $\{s_i \bmod p\}$ has period $p - 1$, and trial division of n would find p sooner. The same difficulty would arise whenever a happened to have a large order modulo p, which happens often by Theorem 6.14.

The next simplest functions to compute modulo n are quadratic polynomials. One should avoid $f(x) = x^2 \bmod n$ because this choice gives $s_i \equiv s^{2^i} \pmod{p}$. Suppose s is a primitive root modulo p, which has a reasonable chance of happening. Then $s_i \equiv s_j \pmod{p}$ if and only if $2^i \equiv 2^j \pmod{p - 1}$ by Theorem 6.15. Write $p - 1 = d2^h$ with d odd. Suppose $i > h$ and $j > h$. The last congruence is equivalent to $2^{i-h} \equiv 2^{j-h} \pmod{d}$. Then $s_i \bmod p$ would have a period equal to the order of 2 modulo d, which might easily be much larger than $\sqrt{p}$.

Another quadratic polynomial to avoid is $f(x) = (x^2 - 2) \bmod n$. By Theorem 7.3, there is a 50% chance that $s^2 - 4$ is a quadratic residue modulo p. If this happens, then we can solve the congruence $r^2 - sr + 1 \equiv 0 \pmod{p}$, which has discriminant $s^2 - 4$, and which is equivalent to $s \equiv r + (1/r) \pmod{p}$. A simple induction shows that $s_i \equiv r^{2^i} + r^{-(2^i)} \pmod{p}$. Then $s_i \equiv s_j \pmod{p}$ if $2^i \equiv 2^j \pmod{p - 1}$, and we have the same long period problem as with $f(x) = x^2 \bmod m$.

Although no one has proved that any polynomials $f(x) = (x^2 + b) \bmod n$ are random mappings when $b \neq 0$ or -2, experiments suggest that these are good choices. We avoid terms ax in these polynomials because they make f harder to evaluate. Here is the algorithm.

[Pollard rho factorization algorithm]
Input: A composite number n to factor.
Output: A proper factor of n, or else "give up."

```
Choose a random b in 1 ≤ b ≤ n − 3
Choose a random s in 0 ≤ s ≤ n − 1
A = B = s
Define a function f(x) = (x² + b) mod n
g = 1
while (g = 1) {
        A = f(A)
        B = f(f(B))
        g = gcd(A − B, n)
        }
if (g < n) { write "g is a proper factor of n" }
else { either give up or try again with new s and/or b }
```

If we reach the last line of the algorithm, it means that $g = n$, that is, we have found all prime factors of n together. There is a fair chance that we will separate them, and find just one of them, if we restart the algorithm with new random b and s.

The factor g of n written in the next-to-last line is not guaranteed to be prime. It is possible that we may find two or more prime factors of n together. One should always test g for primality.

As noted above, assuming f is a random mapping, the complexity of the Pollard rho method is $O(\sqrt{p})$ steps, where p is the smallest prime factor of n. Since $p \leq \sqrt{n}$, this complexity is $O(n^{1/4})$.

Example 10.1

Try Pollard rho factorization of $n = 9271$ with $s = b = 1$.

The first 12 iterates are: $s_0 = 1$, $s_1 = 2$, 5, 26, 677, 4051, $s_6 = 932$, 6422, 4677, 4041, 3451, 5438, $s_{12} = 6626$. We have $\gcd(s_{12} - s_6, n) = \gcd(6626 - 932, 9271) = 73$. The (hidden) values of $s_i \bmod 73$ are shown in the figure below. The shape is the reason for the algorithm's name.

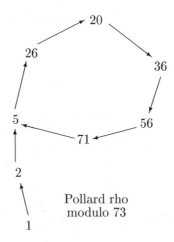

Pollard rho
modulo 73

The most expensive step in the `while` loop is the gcd. Its cost may be amortized by adding a new variable C, initialized at 1, replacing the gcd by the instruction $C = C(A - B) \bmod n$, and computing $g = \gcd(C, n)$ occasionally. One strategy performs the gcd only when the iteration number is a Fibonacci number. This causes gcd's to be done less and less frequently.

The Pollard rho factoring algorithm roughly doubles the size of prime factors we can discover, as compared to trial division. The Monte Carlo method will find a 20-digit prime factor of n with about the same work needed to find a 10-digit factor by trial division.

10.3 *Pollard's $p-1$ Method*

Pollard invented two factoring methods in the 1970's. One was the rho method and the other [84] was the $p-1$ method. The $p-1$ method is based on Fermat's little theorem (Theorem 6.1), which says that $a^{p-1} \equiv 1 \pmod{p}$ when p is a prime which does not divide a. Thus $a^L \equiv 1 \pmod{p}$ for any multiple L of $p - 1$. If also $p|n$, then p divides $\gcd(a^L - 1, n)$. Of course, we cannot compute $a^L \bmod p$ because p is an unknown prime factor of n. However, we can compute $a^L \bmod n$. Pollard's idea is to let L have many divisors of the form $p - 1$ and thus try many potential prime factors p of n at once.

The number $p - 1$, where p is a large prime, seems to factor in the same way as a random integer of about the same size. In particular, the statements about smooth numbers in Section 4.4 seem to apply to numbers of the special form $p-1$. If $p-1$ is B-smooth, that is, the largest prime factor of $p-1$ is $\leq B$, then $p-1$ will divide L if L is the product of all primes $\leq B$, with appropriate multiplicity. If a prime $q \leq B$ divides $p - 1$, then q cannot divide $p - 1$ more than $\log_q p - 1 = (\log p / \log q) - 1$ times. This number is an upper bound on the "appropriate multiplicity" of q in L. However, large primes rarely divide large random integers more than one time. A reasonable compromise for L is to choose a bound B, which tells how much work one is willing to do in an effort to factor n, and define L to be the least common multiple of the positive integers up to B. One can show that this $L = \prod q^e$, where q runs over all primes $\leq B$ and, for each q, q^e is the largest power of q which is $\leq B$. Typically, B is in the millions and L is enormous. There is no need to compute L. As each q^e is formed, one computes $a = a^{q^e}$. Here is the algorithm.

[Simple Pollard $p - 1$ factorization method, first stage]
Input: A composite positive integer n to factor and a bound B.
Output: A proper factor p of n, or else give up.

```
Find the primes p₁ = 2, p₂, ..., pₖ ≤ B
a = 2
for (i = 1 to k) {
        e = ⌊(log B)/ log pᵢ⌋
        f = pᵢᵉ
```

$$a = a^f \bmod n$$
$$\}$$
$$g = \gcd(a - 1, n)$$
```
if 1 < g < n { print "g divides n" }
else { give up }
```

The primes may be found by the sieve of Eratosthenes. Exponentiation is done by the fast exponentiation algorithm. The gcd should be computed once every few thousand primes rather than just once at the end, with the `for` loop continuing if $g = 1$. If $g = 1$ at the end, one can either give up or try the second stage described below. If $g = n$, then all prime divisors p of n have been discovered together. When this happens, if one has saved the value of a at the previous gcd, one can return to it and compute a gcd after each exponentiation in an effort to separate the prime divisors p of n. But even this trick won't work in case $p - 1$ has the same largest prime divisor q for every prime factor p of n. This happens, for example, when one tries to factor $1247 = 29 \cdot 43$, since $29 - 1 = 2^2 \cdot 7$ and $43 - 1 = 2 \cdot 3 \cdot 7$.

If we use the Pollard $p - 1$ algorithm with bound B to try to factor n, and n has a prime factor p, then the probability that we will find p is roughly the probability that a number near p is B-smooth, which is $\rho((\log p)/\log B)$ by Theorem 4.9. But if $p - 1$ has a prime factor $> B$, then we will fail. We could fail to find a prime factor p as small as $p = 2q + 1$, where q is the first prime $> B$ (or $> B_2$, if the second stage is used). On the other hand, the $p - 1$ algorithm occasionally has a spectacular success, like the 30-digit prime divisor $p = 1744633866657191516033932614401$ of $2^{740} + 1$ found by R. Baillie. He used $B = 500,000$ and succeeded because

$$p - 1 = 2^8 \cdot 5^2 \cdot 17 \cdot 37 \cdot 1627 \cdot 5387 \cdot 68111 \cdot 152081 \cdot 477361.$$

The second stage of the algorithm chooses a second bound $B_2 > B$, perhaps $B_2 = 100B$, and seeks a factor p of n for which the largest prime factor of $p - 1$ is $\leq B_2$ and the second largest prime factor is $\leq B$. In other words, $p - 1$ is 1-semismooth with respect to B_2 and B in the terminology of Section 4.4. Theorem 4.11 predicts that the probability of finding p is about $\rho_1((\log p)/\log B, (\log p)/\log B_2)$.

Here is one version of the second stage. At the end of the first stage (the algorithm above), a has the value $2^L \pmod{n}$. Let $q_1 < q_2 < \ldots < q_t$ be the primes between B and B_2. The idea is to compute successively $2^{Lq_i} \pmod{n}$ and then $\gcd(2^{Lq_i} - 1, n)$ for $1 \leq i \leq k$. The first power $2^{Lq_1} \pmod{n}$ is computed directly. The differences $q_{i+1} - q_i$ are even numbers and much smaller than the q_i themselves. Precompute $2^{Ld} \pmod{n}$ for $d = 2, 4, \ldots$ up to a few hundred. To find $2^{Lq_{i+1}} \pmod{n}$ from $2^{Lq_i} \pmod{n}$, multiply the latter by $2^{Ld} \pmod{n}$, where $d = q_{i+1} - q_i$. The amortized cost of computing $2^{Lq_i} \pmod{n}$ for $1 \leq i \leq k$ is a single multiplication modulo n. We can save time on the gcd's by multiplying several values of $(2^{Lq_i} \bmod n) - 1$

modulo n and taking the gcd of the product with n. R. Brent found the factor $p = 49858990580788843054012690078841$ of $2^{977} - 1$ with this method. Since

$$p - 1 = 2^3 \cdot 5 \cdot 13 \cdot 19 \cdot 977 \cdot 1231 \cdot 4643 \cdot 74941 \cdot 1045397 \cdot 11535449,$$

he must have used $B \geq 1045397$ and $B_2 \geq 11535449$. Many tricks and variations speed this algorithm, especially its second stage.

The cryptographic significance of Pollard's $p - 1$ algorithm is that, if we don't want an adversary to be able to factor a large composite number n, then each prime factor p of n must have the property that $p - 1$ contains a prime factor q so large that it is not feasible to perform q operations.

There is a complementary algorithm, due to Williams [123] and called the $p + 1$ factoring method, which discovers a prime divisor p of n provided $p + 1$ is smooth. Therefore, if a cryptographic key n must not be factored, then $p + 1$ must have a large prime factor for each prime factor p of n.

10.4 Square Form Factorization

A **binary quadratic form** is a function $ax^2 + bxy + cy^2$ of x and y. We require the coefficients a, b, c to be integers. It is customary to suppress the variables and write (a, b, c) for $ax^2 + bxy + cy^2$. Two quadratic forms are **equivalent** if a linear change of variables with determinant 1 changes one form into the other. A quadratic form (a, b, c) **represents** an integer m if there exist integers x and y so that $ax^2 + bxy + cy^2 = m$. The **discriminant** of (a, b, c) is $D = b^2 - 4ac$. Equivalent forms have the same discriminant and represent the same set of integers. The theory of quadratic forms was developed by Gauss [45].

In order to determine whether two quadratic forms with the same discriminant are equivalent, it is convenient to select one form from each equivalence class and call it **reduced**. Then two forms would be equivalent if and only if their reduced forms are the same. When $D < 0$, one can define a unique reduced form in each equivalence class via simple inequalities on the coefficients. There is a polynomial-time algorithm for computing the reduced form equivalent to any given quadratic form.

However, when $D > 0$, it is not possible to define a unique reduced form in each equivalence class *and* have an efficient algorithm for finding the reduced form equivalent to any given one. Instead, simple inequalities define reduced forms with given discriminant $D > 0$. (A form (a, b, c) is reduced if $|\sqrt{D} - 2|a|| < b < \sqrt{D}$.) Most equivalence classes contain many reduced forms. Usually, a class contains about $\sqrt{D}$ different reduced forms, but occasionally this set is quite small. There is an efficient algorithm for finding a reduced form equivalent to any given one. The reduced forms in each equivalence class are arranged in a cycle. There is an efficient algorithm for computing the "next" reduced form in the cycle after a given one. To decide whether two quadratic forms with discriminant $D > 0$ are equivalent, compute the reduced forms equivalent to each. Beginning with one of the reduced forms, cycle through the reduced forms, using the "next" algorithm, until you either find the other reduced form or return to the first one. In the first case, the original forms are equivalent; in the second case, they are not. All the reduced forms with discriminant $D > 0$ have coefficients $0 \leq |a|,\, b,\, |c| < \sqrt{D}$.

A **square form** is a quadratic form (a, b, c) in which a is the square of an integer: $a = r^2$. If one visits the reduced forms of an equivalence class, the expected number of iterations of the "next" algorithm between square forms is typically $O(D^{1/4})$. When a square form is found, it often leads to a factorization of D as follows. One constructs a form (r, s, t), where $a = r^2$ in the square form, and visits the successive reduced forms in the equivalence class of (r, s, t) until one reaches two consecutive reduced forms (u, v, w), (j, v, k), with the same middle coefficient v. Then u divides D.

D. Shanks devised the following factoring algorithm called the SQUare FOrm Factorization algorithm, or SQUFOF. Given an odd positive integer n, let $h = \lfloor \sqrt{n} \rfloor$ and $c = h^2 - n$. If $c = 0$, then $n = h^2$ is a square and we are done. Otherwise, $c < 0$ and the form $(1, 2h, c)$ is reduced and has discriminant $D = (2h)^2 - 4c = 4h^2 - 4(h^2 - n) = 4n$. Visit the reduced forms equivalent to this one until a square form is found. Detect squares a efficiently as in Fermat's difference of squares method. Use the square form as above to find a factor u of $D = 4n$. If u is odd, it divides n. If u is even, $u/2$ divides n.

The number of forms visited in the equivalence class of (r, s, t) after the square form is located is always very close to one-half of the number of forms tested in the first equivalence class to find the square form. Since the latter number is $O(D^{1/4})$, or $O(n^{1/4})$, the total number of quadratic forms visited is $O(n^{1/4})$, and this is the complexity of SQUFOF.

The algorithm can fail in two ways. The equivalence class of the initial form may be so small that it contains no reduced square form. In the rare event that this happens, one can apply the algorithm to $3n$ or $5n$, etc., which have completely different, and probably larger, equivalence classes.

The other failure possibility is that the factor of n produced by the algorithm may be trivial, with u or $u/2 = 1$. Shanks found a way to distinguish between square forms that lead to trivial factors of n and those that lead to proper factors. If the square form (r^2, b, c) leads to a trivial factor, then there was a form (r, i, j) visited earlier in the equivalence class, in fact, about half way from the beginning to (r^2, b, c). Since $r^2 < 2\sqrt{n}$, we can detect unproductive square forms by maintaining a list of $a < \sqrt{2\sqrt{n}}$ which occur in forms (a, b, c) and ignoring square forms (r^2, b, c) with r on the list. This is how the algorithm is usually implemented (see the versions in Cohen [28], Algorithm 8.7.2 and Riesel [96], pages 190–192), although some square forms with r on the list do lead to proper factors of n. In unpublished work, Shanks found necessary and sufficient conditions for a square form to lead to a proper factor of n.

SQUFOF has the remarkable property that, after the first couple of steps in factoring n, all arithmetic is performed on integers $< 2\sqrt{n}$. Although the complexity of SQUFOF is $O(n^{1/4})$, the implied constant is tiny. The main loop uses only a handful of arithmetic operations to pass from one quadratic form to the next. (See the program in Riesel [96].) If $\lfloor 2\sqrt{n} \rfloor$ fits in a single-precision integer variable, then each pass through the main loop takes only about a dozen machine cycles. On a 32-bit one-gigahertz machine, SQUFOF can factor almost any 18-digit number in less than a millisecond. In contrast to the factoring algorithms mentioned earlier in this chapter, the complexity depends only on the size of n and not on the size of its prime factors. While SQUFOF is useless in a direct assault on a cryptographic key, it has an important use in factoring auxiliary numbers arising in more powerful factoring algorithms, such as the quadratic and number field sieves.

10.5 Exercises

1. Show that the algorithm for the integer part of the square root of a positive integer is correct and takes $O(\log \log n)$ iterations.

2. Find the 22 two-digit numbers that may be the last two decimal digits of a square.

3. Find the twelve square residues modulo 64.

4. Factor 18779 by Fermat's difference of squares method.

5. Factor 18779 by Pollard's rho method.

6. Factor 18779 by Pollard's $p - 1$ method.

7. Experiment with simple, easily evaluated random functions other than $f(x) = (x^2 + b) \bmod n$ in Pollard's rho method.

8. Show that the least common multiple L of the positive integers up to B is $L = \prod p^e$, where p runs over all primes $\leq B$ and, for each p, p^e is the largest power of p that is $\leq B$.

9. Show that equivalent binary quadratic forms have the same discriminant and represent the same set of integers.

10. Prove that if a, b and c are integers, $D = b^2 - 4ac > 0$ and $\left| \sqrt{D} - 2|a| \right| < b < \sqrt{D}$, then a and c have opposite signs, $b < \sqrt{D}$ and $|a| + |c| < \sqrt{D}$, so that both $|a|$ and $|c|$ are $< \sqrt{D}$.

Chapter 11

Finding Large Primes

Many cryptographic algorithms require prime numbers of a certain size. If the prime need not be secret, then one can get one from a book or web site. There are thousands of primes in the Cunningham Project electronic book [18] or the web site with the full tables from that work, `http://www.cerias.purdue.edu/homes/ssw/cun/third/index.html`. Alternatively, one can form a random large prime by one of the methods for finding secret primes.

One needs a source of random numbers to generate secret random primes. Some methods for finding them are described in Chapter 15.

Number theorists who identify large primes distinguish between "primality testing" and "primality proving." There are simple and swift algorithms for testing large odd numbers for being "probably prime." When used properly, these "probable primality tests" are nearly infallible, but could say that a composite number is prime. They never assert that a prime number is composite. Numbers that pass these tests are called "industrial-grade primes." When a rigorous proof of the primality of a large probable prime is desired, one must resort to slow, complicated algorithms unless the prime has a special form which facilitates its primality proof.

Every prime has a short, simple proof of its primality, but it is usually difficult to discover such a proof when the prime is large. See Theorem 11.16.

There are three ways to find large secret primes for cryptographic use.

1. Test random large numbers and choose the first probable prime. In other words, use industrial-grade primes.

2. Test random large numbers for being probably prime. When you find one, prove rigorously that it is prime.

3. Use random numbers to construct a large prime having special form which permits an easy rigorous proof of its primality.

We consider the first method in the next two sections, and the second and third methods in the two following sections.

11.1   *Stronger Probable Prime Tests*

Recall that Theorem 6.1, Fermat's little theorem, says that if p is prime and p does not divide the integer a, then $a^{p-1} \equiv 1 \pmod{p}$. The fast exponentiation algorithm makes the arithmetic of $a^{p-1} \bmod p$ easy. Also in Chapter 6, we defined a probable prime to base a to be an odd integer n with $a^{n-1} \equiv 1 \pmod{n}$. Thus, every prime is a probable prime to every base it does not divide. We defined a pseudoprime to be a composite probable prime. We noted after Definition 6.1 that if we had a list of all pseudoprimes to some base a up to some limit L, then we could devise a simple, fast primality test: An integer $n < L$ is prime if and only if it is a probable prime to base a and it is not on the list. One difficulty with this test is that lists of pseudoprimes, to base 2, say, do not reach high enough to encompass the range of primes of cryptographic interest. A second problem is that there are too many pseudoprimes to any particular base; the list of all of them would be too long. One might try to solve this problem with pseudoprime tests to multiple bases. However, this proposed solution does not work because there are lots of Carmichael numbers, which are pseudoprimes to every possible base.

In Chapter 7, we devised a more discriminating probable prime test. An Euler probable prime to base a was defined as an integer n for which $\gcd(a, n) = 1$ and $a^{(n-1)/2} \equiv (a/n) \pmod{n}$. The Jacobi symbol $(a/n) = \pm 1$ because $\gcd(a, n) = 1$. An Euler pseudoprime to base a is a composite Euler probable prime to base a. We proved in Theorem 7.12 that every Euler probable prime is a probable prime (to the same base).

Our goal in this section and the next one is to find even more discriminating, but still rapid, probable prime tests. The first one was inspired by the fast exponentiation algorithm to compute $a^{n-1} \bmod n$.

DEFINITION 11.1   *An odd positive integer n, with $n - 1 = 2^s d$, where d is odd, is a **strong probable prime to base** a if either $a^d \equiv 1 \pmod{n}$ or $a^{d \cdot 2^r} \equiv -1 \pmod{n}$ for some $0 \le r < s$. A **strong pseudoprime to base** a is a composite strong probable prime to base a.*

The left to right variation of fast exponentiation computes $a^{n-1} \bmod n$ by first finding $a^d \bmod n$, and then squaring the result s times modulo n. Thus, fast exponentiation automatically produces the remainders, which are compared to $+1$ or -1 in the definition.

Every prime p is a strong probable prime to every base a it does not divide because $a^{p-1} \equiv 1 \pmod{p}$, by Theorem 6.1, and the only solutions to $x^2 \equiv 1 \pmod{p}$ are $x \equiv \pm 1 \pmod{p}$, by Theorem 7.1.

One can show that there are infinitely many strong pseudoprimes to every base $a \ge 1$. However, they are rarer than Euler pseudoprimes.

The bases $+1$ and -1 are not interesting because every odd composite integer n is a pseudoprime, an Euler pseudoprime and a strong pseudoprime to both of these bases.

It is easy to see that every strong probable prime is a probable prime to the same base, because the definition says that we will get ± 1 at some step before the last step in computing $a^{n-1} \bmod n$ by fast exponentiation, and this number will be squared at least once.

THEOREM 11.1 Strong probable primes are Euler probable primes
Every strong probable prime is an Euler probable prime to the same base. Every strong pseudoprime is an Euler pseudoprime to the same base.

Since every prime not dividing a is both a strong probable prime to base a and an Euler probable prime to base a, the two statements are equivalent. For a proof of the second statement, see Theorem 3 of [89] or Theorem 9.12 of [99].

In some cases, one can prove that Euler pseudoprimes must be strong pseudoprimes.

THEOREM 11.2 Euler pseudoprimes $\equiv 3 \pmod 4$ are strong
If $n \equiv 3 \pmod 4$ is an Euler pseudoprime to base a, then n is a strong pseudoprime to base a.

PROOF Since $n \equiv 3 \pmod 4$, we have $n - 1 = 2d$, where d is odd. Because n is an Euler pseudoprime, we have $\gcd(a, n) = 1$ and $a^d \equiv (a/n) = \pm 1 \pmod n$. Therefore, n satisfies one of the two cases of the definition of strong pseudoprime, depending on the sign ± 1. ∎

THEOREM 11.3 Euler and $(a/n) = -1$ imply strong
If n is an Euler pseudoprime to base a and $(a/n) = -1$, then n is a strong pseudoprime to base a.

PROOF If $n - 1 = 2^s d$, with d odd, then $a^{2^{s-1}d} = a^{(n-1)/2} \equiv (a/n) = -1 \pmod n$ because n is an Euler pseudoprime to base a. Then the second case of the definition of strong pseudoprime applies to n. ∎

Recall that R_n denotes the multiplicative group of congruence classes relatively prime to n.

THEOREM 11.4 Pseudoprime bases form a subgroup of R_n
Let n be an integer greater than 1. The set of all bases $1 \leq a < n$ to which n is a pseudoprime forms a subgroup of R_n under multiplication modulo n.

The set of all bases $1 \leq a < n$ to which n is an Euler pseudoprime forms a subgroup of R_n.

PROOF   By Theorem 3.10, if $\gcd(a, n) = \gcd(b, n)$, then $\gcd(ab, n) = 1$. If n is a pseudoprime to bases a and b, then n is a pseudoprime to base ab because $(ab)^{n-1} = a^{n-1}b^{n-1} \equiv 1 \cdot 1 = 1 \pmod{n}$. If n is an Euler pseudoprime to bases a and b, then n is an Euler pseudoprime to base ab because $(ab)^{(n-1)/2} = a^{(n-1)/2}b^{(n-1)/2} \equiv (a/n)(b/n) = (ab/n) \pmod{n}$ by Part 2 of Theorem 7.9. ∎

For a Carmichael number n, the group of all pseudoprime bases is all of R_n. One can prove that for every composite $n > 1$ there is at least one a in $1 < a < n$ with $\gcd(a, n) = 1$ so that n is *not* an Euler pseudoprime to base a. Hence, the group of all Euler pseudoprime bases for n is always a proper subgroup of R_n. Since the order of a subgroup divides the order of the whole group, by Lagrange's theorem, the number of Euler pseudoprime bases for n must be $\leq$ half the size of R_n, and we have this theorem.

THEOREM 11.5 Number of Euler pseudoprime bases
If n is an odd composite positive integer, then the number of bases a in $1 \leq a < n$ with $\gcd(a, n) = 1$ to which n is an Euler pseudoprime is $\leq \phi(n)/2$.

This theorem yields the following probabilistic primality test.

[Solovay-Strassen probabilistic primality test]
Input: Two integers $n > 1$, which is odd, and $k \geq 1$.
Output: Either "n is prime" or "n is composite."

```
for (i = 1 to k) {
        Choose a random integer a in 1 < a < n − 1
        if (gcd(a, n) > 1) { return "n is composite" }
        if (a^(n−1)/2 ≢ (a/n) (mod n))
                { return "n is composite" }
        }
return "n is prime"
```

THEOREM 11.6 Solovay-Strassen probabilistic primality test
If n is an odd prime, then the algorithm returns "n is prime." If n is odd and composite, then the algorithm returns "n is composite" with probability at least $1 - 2^{-k}$. The time complexity of the algorithm is $O((\log n)^3)$ bit operations.

PROOF   If n is prime, then $\gcd(a, n) = 1$ because $1 < a < n$, and so

$a^{(n-1)/2} \equiv (a/n) \pmod{n}$ by Euler's criterion. Therefore, the `for` loop will finish and the algorithm will return "n is prime."

Now suppose n is composite. If $\gcd(a, n) > 1$ for some chosen a, then the algorithm returns "n is composite." Otherwise, $\gcd(a, n) = 1$ for every such a, and each chosen a is in R_n. By Theorem 11.5, for each a in R_n the probability is $\leq 1/2$ that n is an Euler pseudoprime to base a. Hence, the probability that n is an Euler pseudoprime for every one of the k random bases a is $\leq (1/2)^k$. So the algorithm returns "n is composite" with probability at least $1 - 2^{-k}$.

The complexity follows from Corollary 3.1 and Theorems 6.2 and 7.11. ∎

We can construct a better probabilistic test of primality by using strong probable primes in place of Euler probable primes in the test above.

THEOREM 11.7 Number of strong pseudoprime bases
For each odd composite integer n, the number of bases to which n is a strong pseudoprime is $\leq (n-1)/4$.

For each odd composite integer $n > 9$, the number of bases to which n is a strong pseudoprime is $\leq \phi(n)/4$.

For a proof, see Theorem 5.10 of Rosen [99] or Theorem 3.4.4 of Crandall and Pomerance [33]. The theorem was first proved independently by Monier [73] and Rabin [93]. Earlier, Miller [72] had proposed a similar but slightly more complicated test. The set of all bases to which n is a strong pseudoprime usually does not form a subgroup of R_n. The idea of the proof of the theorem is to show that this set is a subset of a proper subgroup of the group of Euler pseudoprime bases for n.

This theorem gives the following improved probabilistic primality test.

[Miller-Rabin probabilistic primality test]
Input: Two integers $n > 1$, which is odd, and $k \geq 1$.
Output: Either "n is prime" or "n is composite."

```
for (i = 1 to k) {
        Choose a random integer a in 1 < a < n − 1
        if (a is not a strong probable prime to base a)
                { return "n is composite" }
        }
return "n is prime"
```

One can prove the following theorem in the same way as Theorem 11.6, but using Theorem 11.7 in place of Theorem 11.5.

THEOREM 11.8 Miller-Rabin probabilistic primality test
If n is an odd prime, then the algorithm returns "n is prime." If n is odd

and composite, then the algorithm returns "n is composite" with probability at least $1 - 4^{-k}$. The time complexity of the algorithm is $O((\log n)^3)$ bit operations.

DEFINITION 11.2   If n is an odd composite integer and $1 \leq a < n$, then a is called a **witness for** n if n is not a strong pseudoprime to base a.

In other words, a is a witness to the compositeness of n; a can testify, via a strong probable prime test, that n is composite. Theorem 11.7 says that at least three-fourths of the integers a in $1 \leq a < n$ are witnesses for n. How hard is it to find one witness? Let $W(n)$ be the least witness for n. If we could prove that $W(n) < C$ for all composite n and some constant C, then we would have a very simple and fast primality test. Unfortunately, Alford et al. [3] prove that this is not so.

THEOREM 11.9 The least witness may be large
For infinitely many odd composite n we have

$$W(n) > (\ln n)^{1/(3 \ln \ln \ln n)} = \exp\left(\frac{\ln \ln n}{3 \ln \ln \ln n}\right).$$

On the other hand, if you believe the extended Riemann Hypothesis, then this theorem of Bach [7] is useful.

THEOREM 11.10 The least witness isn't too large
Assuming the extended Riemann Hypothesis, $W(n) < 2(\ln n)^2$ for every odd composite integer n.

If $n \approx 10^{100}$, then $2(\ln n)^2 \approx 106038$. Therefore, you can prove that a 100-digit odd number is prime by doing about 100,000 strong pseudoprime tests on it, assuming the extended Riemann Hypothesis is valid. This is not a reasonable way to find a 100-digit prime. Keep reading.

11.2   Lucas Probable Prime Tests

In this section, we develop the theory of binary linear recurrences and a probable prime test using them. The test is based on a generalization of the following theorem, which we will prove later. See Williams [124] for much more about primality tests developed by Lucas.

THEOREM 11.11 Divisibility of Fibonacci numbers by primes
If n is prime, u_i is the i-th Fibonacci number and $(n/5)$ is the Legendre symbol, then n divides $u_{n-(n/5)}$.

Example 11.1

Since $(3/5) = -1$, 3 divides $u_4 = 3$. Since $(11/5) = +1$, 11 divides $u_{10} = 55$.
Since $(5/5) = 0$, 5 divides $u_5 = 5$.

DEFINITION 11.3 *The* **Lucas sequences with parameters** P **and**
Q *are the two sequences* $\{u_n\}$ *and* $\{v_n\}$ *defined by* $u_0 = 0$, $u_1 = 1$, $v_0 = 2$,
$v_1 = P$, *and* $u_n = Pu_{n-1} - Qu_{n-2}$, $v_n = Pv_{n-1} - Qv_{n-2}$, *for* $n \geq 2$. *We
sometimes write* $u_n = u_n(P, Q)$ *and* $v_n = v_n(P, Q)$ *to show the dependence on
the parameters* P *and* Q. *Let* $x^2 - Px + Q$ *be the* **recurrence polynomial**
associated to the Lucas sequences, let $D = P^2 - 4Q$ *be the discriminant of
this polynomial and let* α *and* β *be the two zeros of the polynomial.*

To get the Fibonacci numbers u_n, let $P = 1$ and $Q = -1$. In that case,
$v_n = v_n(1, -1)$ are called the **Lucas numbers**. The recurrence polynomial
is $x^2 - x - 1$, with discriminant $D = 5$ and roots $\alpha, \beta = (1 \pm \sqrt{5})/2$.

In this section, the parameters P and Q will always be integers. In this
situation, all the u_n and v_n are integers. Usually, we will also assume that
$D = P^2 - 4Q$ is not a square. This implies that $D \neq 0$, so $\alpha \neq \beta$. From the
equation $(x - \alpha)(x - \beta) = x^2 - Px + Q$, we see that $\alpha + \beta = P$ and $\alpha\beta = Q$.
If we let $\alpha = (P + \sqrt{D})/2$ and $\beta = (P - \sqrt{D})/2$, then $\alpha - \beta = \sqrt{D}$. It is easy
to show by induction on n that

$$u_n = \frac{\alpha^n - \beta^n}{\alpha - \beta} = \frac{\alpha^n - \beta^n}{\sqrt{D}} \quad \text{and} \quad v_n = \alpha^n + \beta^n$$

for $n \geq 0$. These formulas are called the generalized **Binet equations**, and
provide an alternate definition of the Lucas sequences.

There is a natural way to compute Lucas sequences using 2×2 matrices.
Define $L = \begin{bmatrix} P & -Q \\ 1 & 0 \end{bmatrix}$ and, for $n \geq 0$, $A_n = \begin{bmatrix} u_{n+1} & v_{n+1} \\ u_n & v_n \end{bmatrix}$. Then $A_0 =$
$\begin{bmatrix} 1 & P \\ 0 & 2 \end{bmatrix}$. A simple induction shows that $A_n = L^n A_0$ for $n \geq 0$, where L^0
means the identity matrix. This is not just a pretty formula. It provides a
quick way to compute u_n and v_n when n is huge. The fast exponentiation
algorithm of Chapter 6 applies to anything we can multiply associatively,
including matrices. Thus, Theorem 6.2 says that we can compute L^n in our
formula with only $O(\log n)$ matrix multiplications. If we wish to compute
$u_n \bmod m$ or $v_n \bmod m$, we should reduce each matrix entry modulo m as it
is computed. This will keep the numbers small ($< m^2$) even if n has hundreds
of digits.

We need the formulas in the next theorem to prove the generalization of
Theorem 11.11.

THEOREM 11.12 Lucas sequences in terms of binomial coefficients
For integers $n \geq 0$ we have

$$2^{n-1}u_n = \sum_{\substack{i=0 \\ i \text{ odd}}}^{n} \binom{n}{i} P^{n-i} D^{(i-1)/2}$$

$$\text{and } 2^{n-1}v_n = \sum_{\substack{i=0 \\ i \text{ even}}}^{n} \binom{n}{i} P^{n-i} D^{i/2}.$$

PROOF   Begin with the formula for u_n in terms of the two roots of the recurrence polynomial.

$$u_n = \frac{\alpha^n - \beta^n}{\alpha - \beta} = \frac{(P + \sqrt{D})^n - (P - \sqrt{D})^n}{2^n \sqrt{D}}.$$

Apply the binomial theorem to the two binomial powers and get

$$2^n u_n = \frac{1}{\sqrt{D}} \sum_{i=0}^{n} \binom{n}{i} P^{n-i} \left(\left(\sqrt{D}\right)^i - \left(-\sqrt{D}\right)^i \right).$$

When i is even, the terms $\left(\sqrt{D}\right)^i - \left(-\sqrt{D}\right)^i$ cancel, but they add when i is odd. Hence

$$2^n u_n = \frac{2}{\sqrt{D}} \sum_{\substack{i=0 \\ i \text{ odd}}}^{n} \binom{n}{i} P^{n-i} \left(\sqrt{D}\right)^i.$$

We obtain the first formula when we cancel one $\sqrt{D}$ and divide by 2. The second formula is proved the same way, starting from $v_n = \alpha^n + \beta^n$. ∎

The next theorem generalizes Theorem 11.11 and proves it.

THEOREM 11.13 Divisibility properties of Lucas sequences
If p is an odd prime not dividing PQ, then

$$u_{p-(D/p)} \equiv \quad 0 \quad \pmod{p},$$
$$u_p \equiv (D/p) \quad \pmod{p} \text{ and}$$
$$v_p \equiv v_1 = P \quad \pmod{p}.$$

If also $\gcd(p, D) = 1$, then

$$v_{p-(D/p)} \equiv 2Q^{(1-(D/p))/2} \pmod{p}.$$

PROOF   First let $n = p$ in the formula for u_n in Theorem 11.12. Note that since p is prime, it divides every binomial coefficient $\binom{p}{i}$ with $1 \leq i \leq n - 1$.

The only remaining term with odd i in the sum is the one with $i = p$. Also, $2^{p-1} \equiv 1 \pmod{p}$ by Fermat's little theorem. We find

$$u_p \equiv D^{(p-1)/2} \equiv (D/p) \pmod{p},$$

by Euler's criterion. This proves the second formula.

To prove the first one, let $n = p+1$ in the formula for u_n in Theorem 11.12. Since p is prime, it divides the binomial coefficients $\binom{p+1}{i}$ with $2 \le i \le p-1$. The only odd i not in this interval are $i = 1$ and $i = p$, so the sum reduces to two terms. We have $2^p \equiv 2 \pmod{p}$ by Fermat's little theorem. We find

$$2u_{p+1} \equiv (p+1)P^p D^0 + (p+1)P^1 D^{(p-1)/2} \equiv P(1 + (D/p)) \pmod{p},$$

where we have again used Euler's criterion and also $P^p \equiv P \pmod{p}$ by Fermat's little theorem. If $(D/p) = -1$, we see immediately that p divides $u_{p+1} = u_{p-(D/p)}$. If $(D/p) = +1$, then we have $2u_{p+1} \equiv 2P \pmod{p}$, so $u_{p+1} \equiv P \pmod{p}$. By the second formula, which we proved above, $u_p \equiv (D/p) = +1 \pmod{p}$. Substituting into the recurrence formula, $u_{p+1} = Pu_p - Qu_{p-1}$, we find $P \equiv P(+1) - Qu_{p-1} \pmod{p}$. This yields $Qu_{p-1} \equiv 0 \pmod{p}$. Since $\gcd(p, Q) = 1$ we can divide by Q and find that p divides $u_{p-1} = u_{p-(D/p)}$. The other two congruences are proved the same way, using the formula for v_n in Theorem 11.12. ∎

Two matrices are **congruent** modulo n if their corresponding entries are congruent modulo n. Let I denote the 2×2 identity matrix.

THEOREM 11.14 Fermat's little theorem for Lucas sequences
Let $L = \begin{bmatrix} P & -Q \\ 1 & 0 \end{bmatrix}$ be the matrix used to compute the Lucas sequences with parameters P and Q. Let $D = P^2 - 4Q$. Let p be a prime not dividing $2PQD$. If $(D/p) = +1$, then $L^{p-1} \equiv I \pmod{p}$. In any case, $L^{p^2-1} \equiv I \pmod{p}$.

PROOF Suppose $(D/p) = +1$. Then Theorem 11.13 says that

$$A_{p-1} = \begin{bmatrix} u_p & v_p \\ u_{p-1} & v_{p-1} \end{bmatrix} \equiv \begin{bmatrix} 1 & P \\ 0 & 2 \end{bmatrix} = A_0 \pmod{p}.$$

But also $A_{p-1} = L^{p-1}A_0$. Since A_0 has determinant 2, it is invertible modulo the odd prime p. Therefore, $L^{p-1} \equiv I \pmod{p}$. We have $L^{p^2-1} = (L^{p-1})^{p+1} \equiv I^{p+1} = I \pmod{p}$.

Now suppose $(D/p) = -1$. In this case, Theorem 11.13 says that

$$A_p = L^P A_0 = \begin{bmatrix} u_{p+1} & v_{p+1} \\ u_p & v_p \end{bmatrix} \equiv \begin{bmatrix} 0 & 2Q \\ -1 & P \end{bmatrix} \pmod{p}.$$

Since A_0 is invertible modulo p, with inverse $A_0^{-1} \equiv \frac{1}{2} \begin{bmatrix} 2 & -P \\ 0 & 1 \end{bmatrix}$ (mod p), we find that $L^p \equiv \begin{bmatrix} 0 & Q \\ -1 & P \end{bmatrix}$ (mod p) and $L^{p+1} = LL^p \equiv \begin{bmatrix} Q & 0 \\ 0 & Q \end{bmatrix} = QI$ (mod p). Then $L^{p^2-1} = (L^{p+1})^{p-1} \equiv Q^{p-1}I \equiv I$ (mod p) by Fermat's little theorem.

∎

Why did we call this theorem, "Fermat's little theorem for Lucas sequences?" Fermat's little theorem says that if you raise a, relatively prime to a prime p, to the power $p - 1$ modulo p, you will get the identity element 1 of R_p. The theorem says that if you raise L, which describes a Lucas sequence, to the power $p^2 - 1$ modulo the prime p, relatively prime to $2PQD$, you will get the identity element I in the cyclic group of powers of L modulo p. Note that $p^2 - 1$ is the order of the multiplicative group of the field $\mathbf{F}_{p^2}$ with p^2 elements, and this group is also cyclic by Theorem 9.12. If the matrix L were an element of $\mathbf{F}_{p^2}$, the last statement of the theorem would follow from Lagrange's theorem. The connection between Lucas sequences and this field is shown on page 132 of [33], where the first formula of Theorem 11.13 is proved as Theorem 3.5.3.

The four congruences of Theorem 11.13 are valid at least for all primes p not dividing $2PQD$. In fact, when p is allowed to be composite, but $\gcd(p, 2PQD) = 1$, any two of the congruences imply the other two. Baillie and Wagstaff [9] found that they seldom hold when p is an odd composite number. They focussed on the first congruence when they made this definition.

DEFINITION 11.4   A **Lucas probable prime with parameters P and Q** is an integer $n > 1$ with $\gcd(n, 2PQD) = 1$ and $u_{n-(D/n)} \equiv 0$ (mod n), where $D = P^2 - 4Q$. A **Lucas pseudoprime with parameters P and Q** is a composite Lucas probable prime with the same parameters.

Baillie and Wagstaff [9] showed that Lucas pseudoprimes are rare and defined Lucas analogues of Euler and strong pseudoprimes.

The bases $a = \pm 1$ are avoided in probable prime tests because every odd number is a probable prime to these bases. Likewise, the parameters $(P, Q) = (1, 1)$ and $(-1, 1)$ must be avoided in Lucas probable prime tests because every odd n satisfies $u_{n-(D/n)} \equiv 0$ (mod n) with either of these choices.

We mentioned that D should not be a square. In fact, D should not even be a quadratic residue modulo n in a Lucas probable prime test on n. For if $D \equiv b^2$ (mod n), then $(D/n) = +1$, $P \equiv b + 2$ (mod n), $Q \equiv b + 1$ (mod n), $\alpha \equiv Q$ (mod p), $\beta \equiv 1$ (mod p), and $u_{n-1} \equiv (Q^{n-1} - 1)/b$ (mod n); so, the Lucas test is an ordinary probable prime test in disguise. The complexity of a Lucas probable prime test is several times that of a probable prime test; so, one might as well perform a probable prime test with base $a = b + 1$ rather than a Lucas probable prime test with $D \equiv b^2 \pmod{n}$.

Selfridge [89] proposed the following method of choosing the parameters for a Lucas probable prime test that avoids the problem of D being a quadratic residue modulo n. Let D be the first member of the sequence $5, -7, 9, -11, 13, -15, \ldots$ for which the Jacobi symbol $(D/n) = -1$. Let $P = 1$ and $Q = (1 - D)/4$. It is known (page 1416 of [9]) that the expected number of D's which must be tried, before a suitable one is found, is about 1.8. When $n \equiv 2$ or 3 $\pmod{5}$, the first discriminant, $D = 5$, is chosen and the Lucas sequence is the Fibonacci numbers.

Pinch [82] has computed the pseudoprimes to base 2 up to 10^{13}. With Selfridge's parameter choices for the Lucas sequence, not a single known strong pseudoprime to base 2 is also a Lucas pseudoprime. In fact, Pomerance, Selfridge and Wagstaff [89] made this conjecture.

CONJECTURE

No odd composite positive integer is both a strong pseudoprime to base 2 and a Lucas pseudoprime with Selfridge's choice of parameters P and Q.

In 1980, they [89] offered \$30 for a proof or disproof of the conjecture, and have since raised this reward to \$620. The conjecture is certainly true for all integers $< 10^{13}$.

A simplified version of the conjecture asserts that there is no composite number n whose last decimal digit is 3 or 7, which is strong pseudoprime to base 2 and which divides the Fibonacci number u_{n+1}.

Those cryptographers satisfied with "industrial-grade primes" should select strong probable primes to base 2 which are also Lucas probable primes, as in the Conjecture. The tests are simple, elegant and provide the added benefit that if you are the first to detect a failure of the conjecture, then you will collect \$620.

11.3 Rigorous Proof of Primality

Recall the Lucas-Lehmer primality test.

THEOREM 11.15 Lucas-Lehmer $n - 1$ primality test = Theorem 6.10
Let $n > 1$ and a be integers such that $a^{n-1} \equiv 1 \pmod{n}$. If $a^{(n-1)/p} \not\equiv 1 \pmod{n}$ holds for every prime p dividing $n - 1$, then n is prime.

Suppose we had proved that n is prime via this theorem and we wished to convince someone else that n is prime. How little information can we provide and still make the verification easy? We would certainly provide the primitive root a for n and reveal the prime factorization of $n - 1$. But the certificate of primality for n would not be complete until we gave certificates for the primality of each of the prime factors of $n - 1$ as well. This certificate would have a tree structure. Each node would contain a prime n and a primitive root a for n. Each node (n, a) would have a child node, with the same format, for each distinct prime divisor of $n - 1$. As we must stop somewhere, let us assume that everyone knows that 2 is prime. How large could the tree be? Let $N(n)$ be the number of nodes in the tree that certifies the primality of n. Let $M(n)$ be the total number of multiplications modulo a number $\leq n$ a verifier would have to perform to check the certificate using Theorem 6.10. Pratt proved the following theorem.

THEOREM 11.16 Every prime has a succinct certificate
With the notation above, for every odd prime n we have $N(n) < \log_2 n$ and $M(n) < 2(\log_2 n)^2$.

For details of the proof, see Pratt [90] or Bach and Shallit [8].

Theorem 6.10 may be used iteratively to construct large, random primes.

[Really simple large prime generation algorithm]

Begin with a prime p_1. Let $i = 1$. Repeat Steps 1 through 5 until p_i is large enough.

1. Let k be a random small integer and let $n = 2kp_i + 1$.

2. If $2^{n-1} \not\equiv 1 \pmod{n}$, then n is composite by Fermat's little theorem, so return to Step 1.

3. Otherwise, n is probably prime, so try to prove n is prime using the Lucas-Lehmer theorem just stated. Note that $n - 1 = 2kp_i$ is easy to factor completely because it has the known prime factor p_i, which should be removed first, and because k is small. Try the primes < 30 for possible values of a.

4. If you succeed in finding an a which satisfies the conditions of the theorem, then n is proved prime. Let $p_{i+1} = n$ and $i = i + 1$ and go to Step 1.

5. Otherwise, try a new random k. (Go to Step 1.)

During the construction of the last p_i one may have to restrict the size of k to produce a prime of the required size. Typical sizes for k before the last p_i might be 10 or 15 decimal digits—small enough to factor easily by trial division.

There are several enhancements to Theorem 6.10 that accelerate this algorithm. The first is that one can use different a's for different prime factors p of $n - 1$, provided one checks $a^{n-1} \equiv 1 \pmod{n}$ once for each a used. The proof of this version of the theorem is the same as for Theorem 6.10. Note that if more than a is used, then no a is guaranteed to be a primitive root modulo n. If you want to construct a large prime p_i *and* a primitive root a for it, then you must use the original version of the Lucas-Lehmer theorem to show p_i is prime, although you may use the more flexible version in the prime proofs of p_j with $j < i$.

Suppose the prime p_i must be secret, and is a factor of a public key. The algorithm has the advantage that p_i will be immune to discovery by the Pollard $p - 1$ method, because $p_i - 1$ has the large prime factor p_{i-1}.

However, the algorithm is slow because it builds up primes little by little. The next theorem, which may be viewed as another enhancement of Theorem 6.10, allows one to jump ahead to larger primes much faster because it requires only a partial factorization of $n - 1$.

THEOREM 11.17 Pocklington-Lehmer theorem
Let n be odd and $n - 1 = FR$, where the complete factorization of F is known. Suppose that for every prime p dividing F there is an integer a such that $a^{n-1} \equiv 1 \pmod{n}$ and $\gcd(a^{(n-1)/p} - 1, n) = 1$. Then every prime factor of n is $\equiv 1 \pmod{F}$.
If also $F \geq \sqrt{n}$, then n is prime.

PROOF Let $F = \prod p_i^{e_i}$ be the standard factorization of F, and let a_i be the integer a for p_i in the hypothesis. Let p be a prime divisor of n and let f_i be the order of a_i modulo p. Then f_i divides $p - 1$. Since $a_i^{n-1} \equiv 1 \pmod{n}$, we have f_i divides $n - 1$. But $\gcd(a_i^{(n-1)/p_i} - 1, p) = 1$, so f_i does not divide $(n - 1)/p_i$. Hence, $p_i^{e_i}$ divides f_i. Therefore, $p_i^{e_i}$ divides $p - 1$ for each i, and so F must divide $p - 1$.
Suppose $F \geq \sqrt{n}$. Then every prime factor p of n must be greater than $F \geq \sqrt{n}$, and so n is prime by Theorem 4.8. ∎

Of course, the a's in the theorem need not be primitive roots modulo n in case n turns out to be prime.

This theorem allows us to construct a new prime with about twice as many digits as the previous one.

[Doubling the size of a random prime]
Input: A prime p.
Output: A prime n near p^2.

```
repeat {
        let k be a random integer between p/2 and p.
        n = 2kp + 1
        if 2^{n-1} ≢ 1 (mod n) restart this loop.
        try to prove n is prime via Theorem 11.17.
        if you succeed, end the loop.
        } until n is prime
```

By the prime number theorem, the expected number of iterations of the loop needed to find a prime n is about $\ln p$.

In applying Theorem 11.17 in the algorithm above, let $F = p$ and $R = 2k$. It may seem strange to put the known factor 2 into R, but it would take longer to check the hypotheses of Theorem 11.17 if we put the 2 in F. For the integer a of the theorem, try the ten primes < 30.

To construct a large prime near X, begin with a known prime near the 2^i-th root of X, for some convenient i, and apply the algorithm i times with the known prime as the first input, and each subsequent input equal to the previous output. Adjust k in the final iteration of the loop to make the last n just the right size. The large prime p will have a rigorous proof of its primality and $p - 1$ will have a large prime factor to make p immune to discovery by the Pollard $p - 1$ method. If you wish to make p impossible to find by the $p + 1$ method, then try to factor $p + 1$ and reject p unless you can factor it completely and show that $p + 1$ has just one large probable prime factor.

There are theorems analogous to Theorems 6.10 and 11.17 in which n can be proved prime provided $n + 1$ can be factored completely or partly. Lucas sequences replace powers of a in these results.

In the following theorems, if n is the odd number to be proved prime, then we let $\mathcal{U}_n$ be the set of all Lucas sequences $\{u_n\}$ for which the Jacobi symbol $(D/n) = -1$. The theorems are proved in [17]. Theorem 11.19 was first proved by Morrison [76].

THEOREM 11.18 Primality test with $n + 1$ completely factored
Let $n > 1$ be odd. If for each prime p dividing $n + 1$, there exists a Lucas sequence $\{u_n\}$ in $\mathcal{U}_n$ for which n divides u_{n+1} but not $u_{(n+1)/p}$, then n is prime.

THEOREM 11.19 Primality test with $n + 1$ partly factored
Let n be odd and $n + 1 = FR$, where the complete factorization of F is known. Suppose that for every prime p dividing F there is a Lucas sequence $\{u_n\}$ in $\mathcal{U}_n$ for which n divides u_{n+1} and $\gcd(u_{(n+1)/p}, n) = 1$. Then every prime factor q of n is $\equiv (D/q) \pmod{F}$.
If also $F > \sqrt{n} + 1$, then n is prime.

Theorems like 11.17 and 11.19 can be enhanced by letting F be smaller when a lower bound is known on the prime divisors of R. Each theorem has another version which proves that n is prime provided $F > n^{1/3}$. These two theorems may be combined into a theorem that asserts that if we have a completely factored divisor $F > n^{1/3}$ of $n^2 - 1$, then we can rigorously decide in polynomial time whether n is prime. See [17] for details. These theorems suffice to give a quick proof of primality of almost any prime $< 10^{50}$. See Appendix B of the Cunningham Project [18] for thousands of examples.

Note that $n - 1$ and $n + 1$ are the first two cyclotomic polynomials evaluated at $x = n$. Williams and his associates [126], [127], [125], [121] have generalized the theorems above to some higher cyclotomic polynomials, proving that one can rigorously decide in polynomial time whether n is prime, given a sufficiently large completely factored divisor of

$$(n - 1)(n + 1)(n^2 + 1)(n^2 - n + 1)(n^2 + n + 1).$$

11.4 Prime Proofs for Arbitrary Large Integers

There are two types of practical algorithms for proving primality of large primes without special form. These algorithms can show that a 100-digit prime is prime in a few seconds or less. They take a few hours for 1000-digit primes, which are larger than primes currently used in cryptography. However, all of these algorithms are complicated.

One collection of these algorithms generalizes the theorems of Williams, mentioned at the end of the previous section, to even higher cyclotomic polynomials. Adleman, Pomerance and Rumely [1] invented an algorithm of this sort that correctly decides whether n is prime in $< (\ln n)^{c \ln \ln \ln n}$ steps for some constant $c > 0$. They offer a simple, practical, probabilistic version as well as a more complicated deterministic version. The probabilistic version always gives the correct answer; its random choices affect only the running time. Both versions were soon improved. Lenstra [66] simplified the deterministic version, while Cohen and Lenstra [29] made the probabilistic version faster. These algorithms, which use cyclotomy, all have superpolynomial complexity, although the exponent on $\ln n$ grows so slowly $(\ln \ln \ln n)$ that the algorithms nearly run in polynomial time. Because of their complexity, the cyclotomy algorithms lost favor in the 1990's when practical elliptic curve prime proving algorithms were invented. In 1998, Mihăilescu [71] found further improvements in cyclotomy prime proving algorithms. Although his new algorithm still has superpolynomial running time, it is faster than elliptic curve methods for primes having no more than a few thousand decimal digits. Prime proofs using cyclotomy do not have useful succinct certificates. One must redo most of the calculation to verify a prime proof of this type.

Elliptic curve prime proving algorithms can prove n is prime in expected polynomial time $O((\log n)^6)$ and are described in Chapter 12. Elliptic curve prime proofs do have succinct certificates. Such a proof can be verified in much less time than it took to discover it.

Note: As I write this, a new prime proving algorithm has just been announced by Manindra Agrawal, Neeraj Kayal, and Nitin Saxena. If it turns out to be correct, its complexity will be provably polynomial time. Their algorithm uses simple mathematics compared to cyclotomy or elliptic curves.

11.5 Exercises

1. The year is 2020. As the Chief Scientist of a large computer security company, you are implementing a new cryptosystem that uses 1000-digit primes as keys. The algorithm chooses a random 1000-digit integer R and then tests $R + 1$, $R + 2$, ..., for being prime until it finds the first prime number $R + k$ greater than R. What is the average (or expected) number of integers $R + i$ the algorithm tests for being prime?

2. Show that if n is an Euler pseudoprime to base 2 and $n \equiv 5 \pmod 8$, then n is a strong pseudoprime to base 2.

3. Show that if n is an Euler pseudoprime to base 3 and $n \equiv 5 \pmod{12}$, then n is a strong pseudoprime to base 3.

4. Find a composite number n and two bases a and b so that n is a strong pseudoprime to base a and to base b, but not to base ab.

5. Prove that if n is a strong pseudoprime to base a, then n is a strong pseudoprime to base a^i for every integer i.

6. Prove the formula for v_n in Theorem 11.12.

7. Prove the formulas for v_n in Theorem 11.13.

8. Prove that every odd n satisfies $u_{n-(D/n)} \equiv 0 \pmod n$ with either of the parameter sets $(P, Q) = (1, 1)$ and $(-1, 1)$.

9. Prove that if you are given a large odd composite integer n and an integer a so that n is a pseudoprime to base a, but not a strong pseudoprime to base a, then you can factor n in polynomial time.

10. Show that if the conjecture of Section 11.2 is true and n is a composite number with last decimal digit 3 or 7, then either n is not a strong pseudoprime to base 2 or n does not divide the Fibonacci number u_{n+1}.

11. Many cryptographic algorithms require primes 2^m bits long for some particular m. For large k, not a power of 2, there are almost certainly many primes with k bits in their binary representation, exactly two of which are 0 bits. (See Wagstaff [117].) Prove that for all $m \geq 1$ there is no prime with 2^m bits in its binary representation, exactly two of which are 0 bits.

Chapter 12

Elliptic Curves

Elliptic curves are abelian groups created by defining a binary operation on the points of the graph of certain polynomial equations in two variables. These groups have several properties that make them useful in cryptography. One can test equality and perform the group operation on pairs of points efficiently. When the coefficients of the polynomial are integers, we can study the points whose coordinates are also integers, if any. Reducing the coefficients and points modulo a prime p produces an interesting finite abelian group whose order is near p. Choosing random coefficients results in groups with random orders near p. There is an integer factoring algorithm that finds the unknown factor p provided the order of an elliptic curve group is smooth, just as the Pollard $p - 1$ algorithm finds p when $p - 1$ is smooth. There is a probabilistic algorithm for proving p is prime that generalizes the Lucas-Lehmer primality test by replacing $p - 1$ with the order of an elliptic curve group modulo p. Finally, an elliptic curve group may be used directly in cryptographic algorithms in many of the same ways the multiplicative group of integers modulo p can be used. In these applications, the discrete logarithm problem is harder for elliptic curve groups than for the integers modulo p, permitting smaller parameters and faster algorithms. We will say more about this version of the discrete logarithm problem in Chapter 14.

See [58], [111] and [78] for alternate treatments of elliptic curves, with applications to factoring, prime proving and cryptography.

12.1 Definitions and Examples

Let $f(x, y)$ be a polynomial in two variables. The **degree** of a term $cx^m y^n$ with constant $c \neq 0$ is $m + n$. The **degree** of the polynomial $f(x, y)$ is the highest degree of any of its terms. Let $f(x, y)$ have degree d. A straight line $ax + by + c = 0$ intersects the graph of $f(x, y) = 0$ in at most d points because if we solve for x in the equation for the line and substitute for x in $f(x, y) = 0$, we get a polynomial equation of degree d in y, and it has at most d zeros, each of which gives a unique value for x when substituted into the equation for the line.

We wish to define a group operation on the graph of $f(x, y) = 0$. First we must give a rule for specifying a third point on the graph from two given points. One way of defining such a rule is to draw a straight line through the two given points and look for a third point of intersection of the line and the graph. From the discussion of intersections of straight lines and graphs above, we see that this is most likely to work when the degree of $f(x, y)$ is 3. In that case, the straight line through two points of the graph will intersect it in at most one more point.

We can ensure that there is exactly one more point of intersection if we count the intersections with multiplicity. If the straight line and the graph are tangent at point P, then P counts as two points of intersection. If the straight line and the graph are tangent at a point P of inflection of the graph, then P counts as three points of intersection. For $ax + by + c = 0$ to be the equation of a straight line, at least one of a, b must be $\neq 0$. Suppose the straight line $ax + by + c = 0$ with $b \neq 0$ intersects the graph of $f(x, y) = 0$ in two points P and Q. Then $f(x, (-ax - c)/b) = 0$ is a cubic equation in x and we know it has two real roots, namely, the x-coordinates of P and Q. In this situation, the cubic equation has exactly one more root, the x-coordinate of a third point of intersection R. We can find the y-coordinate of R from the equation of the straight line.

The binary operation $\oplus$ defined by $P \oplus Q = R$ turns out not to define a group on the graph, but a simple modification does work. To keep the formulas for the binary operation simple, we will restrict $f(x, y)$ to have the form $y^2 - (x^3 + ax + b)$, which is called the **Weierstrass form** of the elliptic curve.

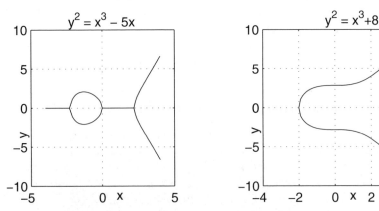

Figure 12.1   Graphs of $y^2 = x^3 - 5x$ and $y^2 = x^3 + 8$.

DEFINITION 12.1   *An **elliptic curve** is the graph E or $E_{a,b}$ of an equation $y^2 = x^3 + ax + b$, where x, y, a and b are real numbers, rational numbers or integers modulo $m > 1$. The set E also contains a **point at infinity**, denoted ∞.*

The point ∞ is not a point on the graph of $y^2 = x^3 + ax + b$. It will be the identity of the elliptic curve group. The points of E, other than ∞, look like one of the graphs in Figure 12.1.

The discriminant $b^2 - 4ac$ vanishes when the quadratic equation $ax^2 + bx + c = 0$ has a repeated root. For the cubic equation $x^3 + ax + b = 0$, the discriminant is $4a^3 + 27b^2$. It vanishes when the cubic has a repeated root. We will assume that that this discriminant is $\neq 0$, so that the cubic does not have a repeated root. Thus, we are excluding elliptic curves like those in Figure 12.2, which have a "double point" and a "cusp."

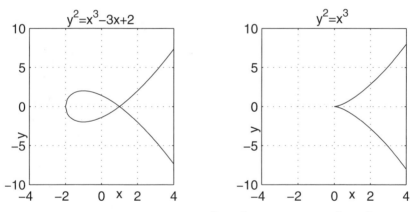

Figure 12.2   Graphs of $y^2 = x^3 - 3x + 2$ and $y^2 = x^3$.

If $P = (x, y)$ lies on the graph of $y^2 = x^3 + ax + b$, we define $-P = (x, -y)$, that is, $-P$ is P reflected in the x-axis.

Given two points P and Q, on the graph but not on the same vertical line, define $P + Q = -R$, where R is the third point on the straight line through P and Q.

If P and Q are distinct points on the graph and on the same vertical line, then they must have the form $(x, \pm y)$, that is, $Q = -P$, and we define $P + Q = \infty$, the identity element of the group.

Also, $P+\infty = \infty + P = P$ for any element P of the elliptic curve (including ∞).

To add a point $P \neq \infty$ to itself, draw the tangent line to the graph at P. If the tangent line is vertical, then $P = (x, 0)$ and we define $P + P = \infty$. If the tangent line is not vertical, then it intersects the graph in exactly one more point R, and we define $P + P = -R$. (If P is a point of inflection, then $R = P$.)

The addition rule may be expressed as $P + Q + R = \infty$ if and only if P, Q, R are on the same straight line.

THEOREM 12.1 An elliptic curve is a group

An elliptic curve E with the addition operation $+$ forms an abelian group with identity ∞. The inverse of P is $-P$.

PROOF   The operation $+$ is well defined and assigns an element $P+Q$ of E to every pair or elements P, Q of E. It is easy to check that ∞ is the identity, that the inverse of P is $-P$ and that $P + Q = Q + P$. The associative law $(P + Q) + R = P + (Q + R)$ may be verified by a long and tedious calculation using the addition formulas that follow.   ∎

There are several short proofs of the associative law, but each requires knowledge of some hard mathematics, like algebraic geometry. See Silverman [111] for a proof.

Let E be defined by $y^2 = x^3 + ax + b$. Let $P = (x_1, y_1)$ and $Q = (x_2, y_2)$. If $x_1 = x_2$ and $y_1 = -y_2$, then $P = -Q$ and $P + Q = \infty$. Otherwise, let s be the slope $s = (y_2 - y_1)/(x_2 - x_1)$ of the line through P and Q when $P \neq Q$, and let s be the slope $s = (3x_1^2 + a)/(2y_1)$ of the tangent line to $y^2 = x^3 + ax + b$ at P when $P = Q$. Then $P + Q = (x_3, y_3)$, where $x_3 = s^2 - 2x_1$ and $y_3 = s(x_1 - x_3) - y_1$.

Example 12.1

On the elliptic curve $y^2 = x^3 - 5x$, add the points $P = (-1, 2)$ and $Q = (0, 0)$.

Using the formula above, we find that the slope is $s = (0 - 2)/(0 - (-1)) = -2$. Then $x_3 = (-2)^2 - (-1) - 0 = 5$ and $y_3 = (-2)(-1 - 5) - 2 = 10$, so $P + Q = (5, 10)$. One should check the arithmetic by verifying that the sum is a point on the curve. Here the check is $10^2 = 5^3 - 5 \cdot 5$.

Example 12.2

On the elliptic curve $y^2 = x^3 + 8$, compute $P + P$, where $P = (1, 3)$.

We use the second formula for the slope because $P = Q$. We have $s = (3 \cdot 1^2 + 0)/(2 \cdot 3) = 1/2$, $x_3 = (1/2)^2 - 1 - 1 = -7/4$ and $y_3 = (1/2)(1 - (-7/4)) - 3 = -13/8$, so $P + P = (-7/4, -13/8)$.

Note that if a and b and the coordinates of points P and Q on the elliptic curve $E_{a,b}$ are rational numbers, then the coordinates of $P+Q$ will be rational numbers (unless $P + Q = \infty$). Therefore, if a and b and the coordinates of points P and Q on the elliptic curve $E_{a,b}$ are integers modulo m, then the coordinates of $P+Q$ will be integers modulo m, unless $P+Q = \infty$, provided that any division needed to add points is by a number relatively prime to m. The modulus m cannot be even because we have to divide by 2 in the formula for the slope s when $P = Q$. The condition on the discriminant becomes $4a^3 + 27b^2 \not\equiv 0 \pmod{m}$. Of course, the graph is not a curve in the plane; it is just a set of pairs of numbers modulo m.

Let us look at the points of the elliptic curve $y^2 \equiv x^3 + 3x + 4 \pmod 7$.

x	$(x^3 + 3x + 4) \bmod 7$	y
0	4	2, 5
1	1	1, 6
2	4	2, 5
3	5	none
4	3	none
5	4	2, 5
6	0	0

There are ten points on this elliptic curve, counting ∞.

Example 12.3

Add the points $(1, 1) + (2, 5)$ on the curve whose points were just listed.

We have $s = (5 - 1)/(2 - 1) = 4$, $x_3 = 4^2 - 1 - 2 = 13 \equiv 6 \pmod 7$ and $y_3 \equiv 4(1 - 6) - 1 \equiv 0 \pmod 7$, so the sum is $(6, 0)$.

Example 12.4

Double the point $(2, 2)$ on the same curve.

We must add $(2, 2) + (2, 2)$. We have $s = (3 \cdot 2^2 + 3)/(2 \cdot 2) \equiv 2 \pmod 7$, $x_3 = 2^2 - 2 - 2 \equiv 0 \pmod 7$ and $y_3 \equiv 2(2 - 0) - 1 \equiv 2 \pmod 7$, so the sum is $(0, 2)$.

There is a simple formula for the number of points on an elliptic curve modulo a prime.

THEOREM 12.2 The number of points on an elliptic curve

The number N of points on the elliptic curve $y^2 \equiv x^3 + ax + b \pmod p$ is $N = p + 1 + \sum_{x=0}^{p-1}((x^3 + ax + b)/p)$, where (r/p) is the Legendre symbol.

PROOF   Each x between 0 and $p - 1$ gives one value $x^3 + ax + b$. The number of y between 0 and $p - 1$ with $y^2 \equiv x^3 + ax + b \pmod p$ is 0, 1, or 2 according as $x^3 + ax + b$ is a quadratic nonresidue, is $\equiv 0$, or is a quadratic residue, all modulo p, by Part 1 of Theorem 7.5. Counting ∞, we have

$$N = 1 + \sum_{x=0}^{p-1} \left(1 + \left(\frac{x^3 + ax + b}{p}\right)\right) = p + 1 + \sum_{x=0}^{p-1} \left(\frac{x^3 + ax + b}{p}\right).$$

∎

Theorem 7.3 says that there are as many quadratic residues as quadratic nonresidues in the interval $1 \leq r \leq p - 1$. Thus the Legendre symbol in Theorem 12.2 will be $+1$ about as often as it will be -1. Hence, we expect the number of points on a random elliptic curve modulo p to be close to $p + 1$. H. Hasse proved that this is so.

THEOREM 12.3 Hasse's theorem
Let the elliptic curve E modulo a prime p have N points. Then

$$p + 1 - 2\sqrt{p} < N < p + 1 + 2\sqrt{p}.$$

We omit the proof, which is hard. See Section V.1 of Silverman [111] for a proof. Deuring [38] proved that every integer N in $p + 1 - 2\sqrt{p} < N < p + 1 + 2\sqrt{p}$ actually occurs as the order of an elliptic curve $y^2 \equiv x^3 + ax + b \pmod{p}$ for some pair a, b. Lenstra [67] showed that the orders of these elliptic curves are well distributed in the interval when random pairs a, b are chosen.

In certain cases, we can determine the order of an elliptic curve without computation. For example, if $p \equiv 3 \pmod 4$ and $b = 0$, then the curve $E_{a,0}$ modulo p has exactly $p + 1$ points.

Now that we have a rich family of abelian groups modulo primes, we can ask about their structure. A theorem of Cassels [24] implies that elliptic curve groups modulo a prime p are either cyclic or the direct product of two cyclic groups. In the latter case, the order of the smaller cyclic group divides both $p - 1$ and the order of the larger cyclic group.

When P is a point on an elliptic curve and k is a positive integer we write kP for the sum $P + P + \cdots + P$ of k P's. We also define $0P = \infty$ and $kP = (-k)(-P)$ when k is a negative integer. The fast exponentiation algorithm, with multiplication replaced by addition of points of an elliptic curve, provides a speedy way to compute kP. It takes $O(\log|k|)$ group operations to find kP when $k \neq 0$.

12.2   *Factoring with Elliptic Curves*

In 1985, H. W. Lenstra, Jr. invented an ingenious new factoring algorithm which uses elliptic curves. Recall that Pollard's $p - 1$ factoring algorithm performs a calculation $(a^L \bmod n)$ in the integers modulo n which mirrors a hidden calculation ($a^L \bmod p$) in the multiplicative group R_p. The factor p of n is discovered when the order $p-1$ of the group R_p divides L. The $p-1$ algorithm fails to find p if $p-1$ happens to have a large prime divisor. Lenstra's idea is to replace the group R_p with an elliptic curve group $E_{a,b}$ modulo p. By Hasse's theorem, the two groups have roughly the same size. But there is only one group R_p and there are many elliptic curve groups modulo p. If the order of R_p has a large prime factor, we are stuck. But if the order of $E_{a,b}$ modulo p has a large prime factor, we just change a and b and try another elliptic curve group.

In Pollard's $p-1$ algorithm, we raise a random number a to the L power modulo n and the factor p appears as $\gcd(a^L - 1, n)$. In Lenstra's **elliptic curve method**, or ECM, we multiply a random point P on an elliptic curve times the integer L, that is, we add P to itself L times, and the factor p appears when we use the extended Euclidean algorithm to try to compute a multiplicative inverse modulo n as part of the slope calculation. Here is the algorithm.

[Simple elliptic curve factorization method, first stage]
Input: A composite positive integer n to factor and a bound B.
Output: A proper factor p of n, or else give up.

```
Find the primes p₁ = 2, p₂, . . . , pₖ ≤ B
Choose a random elliptic curve E_{a,b} modulo n
          and a random point P ≠ ∞ on it
g = gcd(4a³ + 27b², n)
if (g = n) choose a new curve and point P
if (g > 1) report the factor g of n and stop
for (i = 1 to k) {
          e = ⌈(log B)/ log pᵢ⌉
          P = (pᵢᵉ)P or else find a factor g of n
          }
Give up or try another random curve
```

Whenever we compute a multiple hP we reduce the coordinates modulo n. Imagine that the coordinates are also reduced modulo p, a prime divisor of n. Here is why the algorithm works. If the order of the elliptic curve modulo p divides L, then $LP = \infty$ by Lagrange's theorem. Since $P \neq \infty$, at some point during the calculation we must have $P_1 + P_2 = \infty$ for two points $P_1, P_2 \neq \infty$ working with coordinates modulo p. According to the formulas for addition of points, the only way this could happen is if $P_1 = -P_2$, that is, $P_1 = (x, y)$ and $P_2 = (x, -y)$, working with x and y modulo p. It could happen that $P_1 = -P_2$ with coordinates modulo n, but only if that equation held with coordinates modulo q for every prime factor q of n, which is unlikely if n has more than one large prime factor. It is much more likely that $P_1 \neq -P_2$ with coordinates modulo n. Write $P_1 = (x_1, y_1)$ and $P_2 = (x_2, y_2)$ with coordinates modulo n.

We have $x_1 \equiv x_2 \pmod{p}$ because $P_1 = -P_2$ with modulo p coordinates. If $x_1 \neq x_2$, then we will try to compute the slope $s = (y_2 - y_1)/(x_2 - x_1)$ and discover the factor p when we attempt to find the inverse of $x_2 - x_1$ modulo n because $\gcd(x_2 - x_1, n) > 1$. And if $x_1 = x_2$, but $y_1 \neq -y_2$, then $y_1 = y_2$, and so $P_1 = P_2$. In this case we are doubling ("squaring") the point P_1 during the fast "exponentiation." We must have $y_1 \neq 0$ because $y_1 \neq -y_2 = -y_1$. However, $P_1 = -P_2$ with modulo p coordinates. Since also $P_1 = P_2$, we must have $y_1 \equiv 0 \pmod{p}$. But $y_1 \neq -y_2$ and we use the formula $s = (3x_1^2 + a)/(2y_1)$ to compute the slope. When we use the extended Euclidean algorithm to try to find an inverse of $2y_1$, we will discover p because p divides both y_1 and n. See Proposition VI.3.1 of Koblitz [58] for more details of why the algorithm works.

Technically, the "elliptic curve modulo n" is really not an elliptic curve because the addition of points is not defined for every pair of points. This is why we use the condition $\gcd(4a^3 + 27b^2, n) = 1$ in the algorithm. This condition ensures that $4a^3 + 27b^2 \not\equiv 0 \pmod{q}$ for each prime divisor q of n, so that $E_{a,b}$ is a valid elliptic curve (without repeated zeros of $x^3 + ax + b$) modulo each q.

If we make some reasonable assumptions, we can determine the complexity of the elliptic curve method. To factor large numbers n, the basic algorithm for one curve, given above, is repeated until a factor p is found. If the probability is $1/m$ that p will be discovered by one instance of the algorithm, then the expected number of curves that must be tried is m. The factor p is discovered by the algorithm if the order N of the elliptic curve modulo p is B-smooth. (This is actually not quite true because some prime $p_i \leq B$ might divide N to a higher power than the exponent e tried in the algorithm. But such bad luck is very rare and the statement is essentially true.) The Hasse interval $p + 1 - 2\sqrt{p} < N < p + 1 + 2\sqrt{p}$ is too short to prove that any N in it is B-smooth. The principal assumption we make is that the probability that N is B-smooth is u^{-u}, where $u = (\ln p)/\ln B$, which would follow from a theorem of Canfield et al. [23], if N could be chosen from a longer interval $(p/2 < N < 3p/2$, say) than the Hasse interval.

We will also assume that the optimal value of B is used in the algorithm. This is a problem because the optimal value depends on p, which is unknown. However, if the algorithm is used with slowly increasing values of B, the effect is the same as if the optimal B were used. This is how people actually use the algorithm. See Silverman and Wagstaff [112] for advice about how to increase B gradually.

THEOREM 12.4 Complexity of elliptic curve method

Let n be a positive integer with an unknown prime factor p. Let B be the optimal bound for finding p by the elliptic curve algorithm. Assume that a random elliptic curve $E_{a,b}$ modulo p with $4a^3 + 27b^2 \not\equiv 0 \pmod{p}$ has a B-smooth order with probability u^{-u}, where $u = (\ln p)/\ln B$. Define

$L(x) = \exp(\sqrt{(\ln x)\ln\ln x})$.

Then $B = L(p)^{\sqrt{2}/2}$. *The expected total number of group additions performed when the elliptic curve algorithm is used to discover p is $L(p)^{\sqrt{2}}$. The expected total work needed to discover one prime factor of n is $L(n)$ group additions.*

PROOF   By the prime number theorem, there are about $B/\ln B$ primes $\leq B$. For nearly all of these primes q, the largest power $q^e \leq B$ has $e = 1$. The fast "exponentiation" used to compute $(q^e)P$ takes about $\log B$ group additions. Hence the total count of group additions per curve with bound B is about $(B/\ln B)\ln B = B$.

Since the probability of finding p with any single curve is u^{-u}, the expected number of curves required is $1/u^{-u} = u^u$. Therefore, if we use bound B in the algorithm, the total number of group operations needed to find p is $f(B) = Bu^u$. We must find the B which minimizes $f(B)$.

Let $a = (\ln B)/\ln L(p)$ so that $B = (L(p))^a$. We will express $f(B)$ in terms of a. We have $\ln B = a\ln L(p) = a\sqrt{(\ln p)\ln\ln p}$, so

$$u = \frac{\ln p}{\ln B} = \frac{\ln p}{a\sqrt{(\ln p)\ln\ln p}} = \frac{1}{a}\sqrt{\frac{\ln p}{\ln\ln p}}$$

and $\ln u = \frac{1}{2}\ln\ln p - \frac{1}{2}\ln\ln\ln p - \ln a \approx \frac{1}{2}\ln\ln p$ since the other two terms are small compared to $\frac{1}{2}\ln\ln p$. Hence,

$$u\ln u \approx \frac{1}{a}\sqrt{\frac{\ln p}{\ln\ln p}}\cdot\frac{1}{2}\ln\ln p = \frac{1}{2a}\ln L(p).$$

Therefore, $u^u = e^{u\ln u} \approx L(p)^{1/(2a)}$ and the function we seek to minimize is

$$f(B) = Bu^u \approx L(p)^a L(p)^{1/(2a)} = L(p)^{a+1/(2a)}.$$

Since $L(p)$ is a positive constant (for $p > e^e$), the minimum of $f(B)$ will occur when $a + 1/(2a)$ is minimal. It is an easy calculus exercise to show that the minimum of $a + 1/(2a)$ occurs when $a = \sqrt{2}/2$, and the minimum value is $\sqrt{2}$. Therefore, the optimal B is $L(p)^{\sqrt{2}/2}$. With this B, the expected total number of group additions is $f(B) = L(p)^{\sqrt{2}}$.

Let p be the smallest prime factor of n. Then $p \leq \sqrt{n}$, $\ln p \leq (1/2)\ln n$ and $\ln\ln p < \ln\ln n$, so the expected total number of group additions is

$$L(p)^{\sqrt{2}} = \exp(\sqrt{2(\ln p)\ln\ln p}) < \exp(\sqrt{(\ln n)\ln\ln n}) = L(n).$$

∎

Like the Pollard $p - 1$ algorithm that inspired it, Lenstra's elliptic curve algorithm admits a second stage. The second stage of the algorithm chooses a second bound $B_2 > B$, perhaps $B_2 = 100B$. At the end of the first stage (the algorithm above), the variable P is the point Q equal to L times the original point P. Let $q_1 < q_2 < \ldots < q_t$ be the primes between B and B_2. The idea is to compute successively $(Lq_i)P$ for $i = 1, 2, \ldots t$, where P is the original point. The first point $(Lq_1)P$ is computed directly as $(q_1)Q$. The differences $q_{i+1} - q_i$ are all even numbers and much smaller than the q_i themselves. Precompute $(Ld)P = dQ$ for $d = 2, 4, \ldots$ up to a few hundred. To find $(Lq_{i+1})P$ from $(Lq_i)P$, add the latter to $(Ld)P = dQ$, where $d = q_{i+1} - q_i$. The amortized cost of computing each $(Lq_i)P$ for $1 \leq i \leq k$ is a single addition of two points on $E_{a,b}$.

The second stage finds a factor p of n when the order of the elliptic curve is 1-semismooth with respect to B_2 and B in the terminology of Section 4.4. The probability of finding p is about $\rho_1((\log p)/\log B, (\log p)/\log B_2)$ per curve, according to Theorem 4.11.

The cryptographic significance of the elliptic curve method is that, if we don't want an adversary to be able to factor a large composite number n, then each prime factor p of n must have the property that there are no B-smooth integers between $p + 1 - 2\sqrt{p}$ and $p + 1 + 2\sqrt{p}$, where the adversary is able to perform $O(B)$ operations. As this goal is impossible to achieve, the best we can do is make the smallest prime divisor of n as large as possible. If n must be composite, but we are free to choose the number of its prime divisors, we should opt for only two of them.

Of the many enhancements of the elliptic curve method, let me mention just one. Recall that in the Pollard $p - 1$ method, we may delay computing the greatest common divisor with n. In fact, the version of the algorithm we stated had just one gcd at the end. On the other hand, the elliptic curve method has a gcd hidden in every addition of points. This step slows the algorithm significantly. We may avoid it by using a different coordinate system. The system of representing points (x, y) on the curve $y^2 = x^3 + ax + b$ is called the "affine coordinate system" and (x, y) an "affine" point. With "homogeneous" or "projective coordinates," one lets $x = X/Z$ and $y = Y/Z$ and clears the denominators, obtaining the equation $Y^2 Z = X^3 + aXZ^2 + bZ^3$. Points in this system are triples $[X, Y, Z]$. When $Z \neq 0$, the point $[X, Y, Z]$ corresponds to the affine point $(X/Z, Y/Z)$. Any point $[X, Y, 0]$ with $Z = 0$ represents ∞. One can derive formulas for adding points $[X_1, Y_1, Z_1]$ and $[X_2, Y_2, Z_2]$ in the new system as follows. If either $Z_1 = 0$ or $Z_2 = 0$, the sum is the other point $(P + \infty = P)$. Otherwise, there are formulas for the projective coordinates of the sum $[X_3, Y_3, Z_3]$ which involve only addition, subtraction and multiplication modulo p, and which may be derived this way. Formally add the points $(X_1/Z_1, Y_1/Z_1)$ and $(X_2/Z_2, Y_2/Z_2)$ using the affine rules given above. Leave the fractions as fractions. Replace each condition like $X_1/Z_1 = X_2/Z_2$ by a condition $X_1 Z_2 = X_2 Z_1$, which avoids division. In the two cases $P+Q$ and $P+P$ not involving ∞, write the formulas for the affine coordinates as fractions with a common denominator. Let Z_3 be the common denominator and let X_3 and Y_3 be the two numerators. If one stores intermediate results like $X_1 Z_2 - X_2 Z_1$ to be used later, one can reduce the addition of two points to only 13 or 14 multiplications, a handful of additions and subtractions and *no* divisions modulo p. Other coordinate systems reduce the computational labor of adding points even further. See Section 7.2 of [33] for more details. In these systems, one computes $\gcd(Z, n)$ once at the end, or more often, to see whether n has been factored (yet).

12.3 Primality Proving with Elliptic Curves

Elliptic curve prime proving (ECPP) algorithms are the only ones that can prove a prime is prime in polynomial time. They can prove n is prime in expected time $O((\log n)^6)$. The first such algorithms were published by Goldwasser and Kilian [47]. Atkin and Morain [6] made substantial improvements and made the algorithm practical.

We will describe the original algorithm by Goldwasser and Kilian. The next theorem is an elliptic curve analogue of Theorem 11.17. (The variables m, s and P in the next theorem correspond to $n-1$, F and a in the Pocklington-Lehmer theorem.) The words "elliptic curve" are in quotes in the statement because we don't know that it really is an elliptic curve until after n is proved prime.

THEOREM 12.5 Goldwasser-Kilian ECPP

Let n be a positive integer relatively prime to 6. Let s and m be positive integers with $s|m$. Let E be an "elliptic curve" modulo n. Suppose there is a point P of E such that we can perform the curve operations to compute mP and find $mP = \infty$, and for every prime p dividing s we can perform the curve operations to compute $(m/p)P$ and find $(m/p)P \neq \infty$. Then s divides the order of E modulo any prime divisor of n.

If also $s > (n^{1/4} + 1)^2$, then n is prime.

PROOF Let q be a prime factor of n. The calculations on E modulo n, when reduced modulo q, show that s divides the order of P on E modulo q, just as in the proof of Theorem 11.17.

If also $s > (n^{1/4}+1)^2$, then the size of E modulo q must also be $> (n^{1/4}+1)^2$. But by Hasse's theorem 12.3, the size of E modulo q is $< (\sqrt{q}+1)^2$. Therefore, $(\sqrt{q} + 1)^2 > (n^{1/4} + 1)^2$, so $q > \sqrt{n}$. Since this is true for every prime factor q of n, n must be prime. ∎

The algorithm following Theorem 11.17 applied that theorem with $F = p$, a prime slightly larger than $\sqrt{n}$. We can use Theorem 12.5 in a similar way. To prove that n is prime, try to find an elliptic curve E modulo n and a point P on E whose order is $p > (n^{1/4} + 1)^2$. This computation shows that if p is prime, then n is prime. How do we know that p is prime? Apply the theorem recursively to produce a decreasing sequence of numbers, each of which is prime if the next smaller one is prime. The sequence ends when it reaches a number small enough to be proved prime by trial division or some other method. When the last number is proved prime, the primality of all the other numbers, including n, is demonstrated.

The one catch in the algorithm is determining the order of P on E. How do we find the number m in Theorem 12.5? It is supposed to be the order of the elliptic curve modulo n. If n really is prime, then m is within $2\sqrt{n}$ of $n + 1$ by Hasse's theorem. The formula in Theorem 12.2 is useful only for primes up to a few million.

There is a variation of Shanks' baby-step-giant-step algorithm which will find the order m of E modulo n assuming that n is prime. Basically, this algorithm tries to find a discrete logarithm m of ∞ in the Hasse interval $n + 1 - 2\sqrt{n} < m < n + 1 - 2\sqrt{n}$. The giant steps have size $\sqrt{2\sqrt{n}}$. The algorithm has complexity $O(n^{1/4})$ group operations and is effective for n up to about 10^{30}. See Algorithm 7.5.3 of Crandall and Pomerance [33] for details. There is a twist.

As we wish to apply the elliptic curve prime proving algorithm to n much larger than 10^{30}, we need a faster way of computing orders of elliptic curves. Schoof [101] found a beautiful algorithm for computing the order m of E modulo n which runs in $O((\log n)^k)$ for fixed k. His method determines $m \bmod q$ for many small primes q and deduces m via the Chinese remainder theorem, just as Sun Tsu counted his soldiers. Schoof's algorithm uses division polynomials and is too complicated to present here. See [101] or Algorithm 7.5.6 of [33] for details.

12.4 Exercises

1. Derive the formulas for adding points $(x_1, y_1) + (x_2, y_2)$ stated after Theorem 12.1.

2. Consider the curve $y^2 = x^3 - 7x + 15$. Add the points $(1, 3) + (2, 3)$. Add the points $(1, -3) + (2, 3)$. Double the point $(1, 3)$. Be sure to check that the given points and your answers all lie on the curve.

3. Consider the curve $y^2 \equiv x^3 + 4x + 4 \pmod{11}$. Add the points $(1, 3) + (2, 3)$. Add the points $(1, 8) + (2, 3)$. Double the point $(1, 3)$. Find the order of the point $(1, 8)$. Find the number of points on the elliptic curve. Be sure to check that the given points and your answers all lie on the curve.

4. Show that the points $P = (-3, 2)$, $Q = (-1, 4)$ and $R = (1, 2)$ are on the elliptic curve $y^2 = x^3 - 7x + 10$. Compute $P + Q$, $2R$, $3R$ and $4R$ on this curve.

5. The point $(0, 16)$ has finite order on the elliptic curve $y^2 = x^3 + 256$ over the rational numbers. Find its order.

6. Prove that if p is prime, $p \equiv 3 \pmod 4$, and $b = 0$, then the elliptic curve $E_{a,0}$ modulo p has exactly $p + 1$ points.

7. Find formulas for adding points in homogeneous coordinates $[X, Y, Z]$.

8. Let g be a quadratic nonresidue modulo the prime p. Let m and n be the orders of the two elliptic curves $y^2 = x^3 + ax + b$ and $y^2 = x^3 + g^2 ax + g^3 b$ modulo p. (The second curve is called the "twist" of the first curve.) Prove that $m + n = 2p + 2$.

9. Let $P \neq \infty$ be a point on an elliptic curve over the real numbers. Give a geometric condition (something involving tangent lines, points of inflection, etc.) that is equivalent to P being a point of order

   a. 2.

   b. 3.

   c. 4.

10. For the following values of p and B determine the fraction of the integers between $p + 1 - 2\sqrt{p}$ and $p + 1 + 2\sqrt{p}$ having no prime divisor greater than B:

    a. $p = 109$, $B = 5$.

    b. $p = 127$, $B = 17$.

Chapter 13

Subexponential Factoring Algorithms

This chapter deals with integer factoring algorithms whose complexity grows more slowly than an exponential function of $\log n$, the length of the input number n. (An exponential function of $\log n$ means a function of the form $\exp(c \log n) = e^{c \log n} = n^c$, for some constant $c > 0$. The complexity functions in this chapter have roughly the form $\exp(c\sqrt{\log n})$ for some constant $c > 0$, or even slower growth.) We have already mentioned one of these algorithms, the elliptic curve factoring method, which factors n in about $\exp(\sqrt{(\ln n) \ln \ln n})$ steps. The complexity of that algorithm is actually a slowly increasing function of the prime factor p it discovers. The complexity of the factoring algorithms described in this chapter depends only on n and not on the size of any prime factor of n. In this respect, they are similar to SQUFOF, whose complexity is $O(n^{1/4})$ steps.

The factoring algorithms in this chapter have several other similarities. They all factor many relatively small auxiliary numbers using the primes in a fixed set, called a factor base, and they use linear algebra over the field $\mathbf{F}_2$ with two elements to combine the factorizations of the auxiliary numbers to construct x and y with $x^2 \equiv y^2 \pmod{n}$, which gives the factorization of n.

13.1 Factoring with Continued Fractions

This factoring algorithm has been superseded by the algorithms in the following sections. We study it here because, historically, it was the first subexponential integer factoring algorithm and because the later algorithms build on the ideas invented to make this one work.

The following theorem has been known for a long time.

THEOREM 13.1 Factoring by congruent squares

If n is a composite positive integer, x and y are integers, and $x^2 \equiv y^2 \pmod{n}$, but $x \not\equiv \pm y \pmod{n}$, then $\gcd(x - y, n)$ and $\gcd(x + y, n)$ are proper factors of n.

PROOF   The congruence shows that n divides $x^2 - y^2 = (x - y)(x + y)$, while the incongruences imply that n does not divide either $x - y$ or $x + y$. Hence, at least one prime factor of n does not divide $x - y$ and so must divide $x + y$. Likewise, at least one prime factor of n divides $x - y$. Therefore both gcd's are > 1. Neither gcd can equal n because of the incongruences.   ∎

Although we will describe ways to find x and y with $x^2 \equiv y^2 \pmod{n}$, it is difficult to ensure that $x \not\equiv \pm y \pmod{n}$, so we ignore this condition. We can easily factor even numbers. We can also factor prime powers n by computing $n^{1/k}$ by Newton's method for $2 \leq k \leq \log_2 n$. Suppose now that n is odd and has $k > 1$ distinct prime factors. Suppose we can find x and y satisfying $x^2 \equiv y^2 \pmod{n}$. Then $x^2 \equiv y^2 \pmod{p^e}$ for each of the k distinct prime divisors p of n. The number y^2 is clearly a quadratic residue modulo p. By Theorem 7.16, the congruence $z^2 \equiv y^2 \pmod{p^e}$ has exactly two solutions z. Since y and $-y$ are clearly two solutions, $z \equiv \pm y \pmod{p^e}$. By the Chinese remainder theorem, given y, there are 2^k solutions x to $x^2 \equiv y^2 \pmod{n}$, one for each choice of the $\pm$ sign in each $x \equiv \pm y \pmod{p^e}$. The solutions with $x \equiv \pm y \pmod{n}$ are two of these 2^k solutions. Therefore, if x and y are chosen randomly subject to $x^2 \equiv y^2 \pmod{n}$, the probability that $x \not\equiv \pm y \pmod{n}$ is $(2^k - 2)/2^k = 1 - 2^{k-1}$. Since $k > 1$, the probability is at least $1/2$ that a random congruence $x^2 \equiv y^2 \pmod{n}$ will yield a factorization of n. We have proved:

THEOREM 13.2 Each congruence has at least a 1/2 chance of factoring

If n is an odd positive integer with at least two different prime factors, and if x and y are chosen randomly subject to $x^2 \equiv y^2 \pmod{n}$, then, with probability $\geq 1/2$, $\gcd(x - y, n)$ is a proper factor of n.

It would be futile to try to guess integers x and y satisfying $x^2 \equiv y^2 \pmod{n}$. If n is the product of two primes, then for each x only two values of y lead to a factorization. The continued fraction and quadratic sieve algorithms both find many "relations" $x_i^2 \equiv q_i \pmod{n}$, where $0 < |q_i| < n$. A subset of the q_i's is found whose product is a square, say, y^2. If we let x be the product of the corresponding x_i's, then we have $x^2 \equiv y^2 \pmod{n}$.

The idea of using relations $x_i^2 \equiv q_i \pmod{n}$ to help factor n actually goes back to Legendre [61] in 1830. He observed that if p is any prime factor of n, then each q_i must be a quadratic residue modulo p. For each i he would find a list of congruence classes modulo q_i or $4q_i$ containing all primes having q_i

as a quadratic residue, just as we did in Examples 7.2 and 7.3. He used the arithmetic progressions to limit his search for a factor of n by trial division, testing only primes in the possible congruence classes. Each q_i that was not a square times some other q_j eliminated half of the remaining eligible primes. This scheme works fine if you are using pencil and paper to factor a ten-digit integer. The sieves we mentioned in the discussion of Fermat's method are ideal hardware devices for finding primes that lie in specified congruence classes modulo various small primes q_i. But even though the fastest sieves today [128] process billions of candidate factors per second, they are still using trial division and cannot compete with the methods in this chapter for factoring numbers of cryptographic interest.

Seventy-five years ago, Kraitchik obtained relations by *ad hoc* means. He [60], page 201, tried to factor $n = 193541963777$ and found the relations

$$439935^2 \equiv 2^8 \cdot 7^2 \cdot 67 \pmod{n}$$
$$1609^2 \cdot 7^2 \cdot 67 \equiv 449490^2 \pmod{n}.$$

He multiplied the two congruences, canceled the $7^2 \cdot 67$ and got

$$(439935 \cdot 1609)^2 \equiv (2^4 \cdot 449490)^2 \pmod{n},$$

from which the factorization of n is easy to deduce.

Although, Kraitchik's methods were *ad hoc*, they do suggest a way to find a subset of a given set of relations $x_i^2 \equiv q_i \pmod{n}$ for which the product of the q_i is a square. If the numbers q_i have been factored completely, we can try to match the primes in the relations so that each appears an even number of times in the relations we select. Suppose $p_1, \ldots, p_k$ are all of the primes which appear in the factorizations of all the q_i. We can write the i-th relation as

$$x_i^2 \equiv q_i = p_1^{e_{i1}} p_2^{e_{i2}} \cdots p_k^{e_{ik}} \pmod{n},$$

where we allow $e_{ij} = 0$ if p_j does not divide q_i. Think of the list of exponents as a vector $v_i = (e_{i1}, e_{i2}, \ldots, e_{ik})$, and let v_i represent the i-th relation. These vectors are added when two relations are multiplied to form a new relation. Such a vector represents a relation with a square right side if and only if each component of the vector is even, because a positive integer is a square if and only if every prime that divides it does so an even number of times.

To select a subset of the q_i's whose product is a square, we can write a system of k congruences modulo 2, one for each prime p_j. Let $z_i = 1$ if the i-th relation is to be selected and $z_i = 0$ if it is not selected. Then the system of congruences is $\sum_i z_i e_{ij} \equiv 0 \pmod{2}$ for $1 \le j \le k$. If we have only a few relations and many primes, it is unlikely that there will be a nontrivial solution. But if there are more relations than primes, then we certainly will have a solution with some $z_i = 1$.

Since we are concerned only with the parity of the components of the exponent vectors, we may regard v_i as a vector of dimension k over the field $\mathbf{F}_2$ of integers modulo 2. Then the system of congruences becomes the homogeneous system of linear equations $\sum_i z_i e_{ij} = 0$ for $1 \le j \le k$. We may use techniques of linear algebra, such as Gaussian elimination, to solve this system.

With these preliminaries, we now head towards the continued fraction integer factoring algorithm. A **simple continued fraction** is an expression of the form

$$x = q_0 + \cfrac{1}{q_1 + \cfrac{1}{q_2 + \cfrac{1}{q_3 + \cdots + \cfrac{1}{q_k}}}},$$

which we will denote by $[q_0, q_1, q_2, q_3, \ldots, q_k]$. The numbers q_i, except for the last one, q_k, are required to be integers. Every real number x has a simple continued expansion which may be computed by this algorithm:

```
i = 0
q₀ = ⌊x⌋
x = x - q₀
while (x > 0) {
        i = i + 1
        qᵢ = ⌊1/x⌋
        x = x - qᵢ
        }
```

The algorithm terminates if and only if x is a rational number. If $x = a/b$, then the q_i are the quotients in the Euclidean algorithm for $\gcd(a, b)$.

Given a finite simple continued fraction $[q_0, q_1, q_2, \ldots, q_k]$, with all q_i integers, we can find its value A_k/B_k as a rational number by clearing the denominators starting from the end and working backwards. We can also find A_k/B_k by working forwards, using the formulas $A_{-1} = 1$, $B_{-1} = 0$, $A_0 = q_0$, $B_0 = 1$ and $A_i = q_i A_{i-1} + A_{i-2}$, $B_i = q_i B_{i-1} + B_{i-2}$ for $i = 1, 2, \ldots, k$.

Continued fractions have an important application in finding rational numbers a/b which closely approximate real numbers x. With A_k and B_k as above, one can prove that

$$\left| \frac{A_k}{B_k} - x \right| < \frac{1}{B_k B_{k+1}} < \frac{1}{B_k^2}$$

for every $k > 0$. See Theorem 7.11 of [78] for a proof. Also, if A and B are integers with $\gcd(A, B) = 1$ and

$$\left| \frac{A}{B} - x \right| < \frac{1}{2B^2}, \tag{13.1}$$

then there is a $k > 0$ for which $A = A_k$ and $B = B_k$. See Theorem 7.14 of [78] for a proof.

When n is not a perfect square, the simple continued fraction expansion of $\sqrt{n}$ is infinite because $\sqrt{n}$ is not a rational number. However, the expansion is periodic, that is, the q_i's repeat after a while. Usually, the length of the period is roughly $\sqrt{n}$. One would have to know $\sqrt{n}$ to very high precision to compute its simple continued fraction by the algorithm above. But, there is a simple iteration that computes the q_i's using only integer arithmetic. Two or three other integers are computed during the iteration. One of them is an integer Q_i which satisfies $A_i^2 - nB_i^2 = (-1)^i Q_i$ and $0 < Q_i < 2\sqrt{n}$. (The numbers $(-1)^i Q_i$ are the same as the a in SQUFOF, but that algorithm iterates the continued fraction expansion until a is a square.) If we regard the equation as a congruence modulo n, we have $A_i^2 \equiv (-1)^i Q_i \pmod{n}$. In other words, the continued fraction iteration produces a sequence $\{(-1)^i Q_i\}$ of quadratic residues modulo n whose absolute values are $< 2\sqrt{n}$! This is very small indeed, since the average quadratic residue between 1 and n is about $n/2$.

The continued fraction factoring algorithm, CFRAC, first implemented by Morrison and Brillhart [77], uses the fact that, since the Q_i are small, they are more likely to be smooth than numbers near $n/2$, say, because u will be only half as big in Theorem 4.9. The continued fraction expansion for $\sqrt{n}$ generates the sequences $\{Q_i\}$ and $\{A_i \bmod n\}$ and tries to factor each Q_i by trial division using the primes below some bound B, called the **factor base**. It saves the B-smooth Q_i's, together with the corresponding A_i, representing the relation $A_i^2 \equiv (-1)^i Q_i \pmod{n}$. When enough relations have been collected, Gaussian elimination is used to find linear dependencies (modulo 2) among the exponent vectors of the relations. (In linear algebra terminology, we find a basis for the null space of the linear system.) There are enough relations when there are more of them than primes in the factor base. Each linear dependency produces a congruence $x^2 \equiv y^2 \pmod{n}$ and a chance to factor n by Theorem 13.2.

Suppose the prime p divides Q_i. The equation $A_i^2 - nB_i^2 = (-1)^i Q_i$ shows that $(A_i/B_i)^2 \equiv n \pmod{p}$, so the Legendre symbol (n/p) is 1 (or 0 if $p|n$). The factor base should contain only primes p with $(n/p) = 1$, that is, about half of the primes up to B. One might think that the probability that Q_i is B-smooth would be lessened by having only about half of the possible primes available. But there is a heuristic argument (see Section 4.5.4 of Knuth [56]) that if $p < B$ does not divide n, then p divides Q_i with probability $2/(p+1)$ rather than the expected $1/p$. This higher chance of dividing Q_i compensates for the smaller number of useful primes $< B$ and leaves the estimate in Theorem 4.9 essentially unchanged. Assuming a couple of plausible hypotheses, Pomerance [87] proved that the time complexity of CFRAC is $L(n)^{\sqrt{2}}$, where $L(x)$ is defined in Theorem 12.4.

Here is a simple example of CFRAC.

Let us factor $n = 13290059$. The continued fraction expansion for $\sqrt{n}$ yields the relations below, and many others.

i	$A_i \bmod n$	$(-1)^i$	Q_i	Q_i factored
10	6700527	$+1$	1333	$31 \cdot 43$
23	1914221	-1	226	$2 \cdot 113$
26	11455708	$+1$	3286	$2 \cdot 31 \cdot 53$
31	1895246	-1	5650	$2 \cdot 5^2 \cdot 113$
40	3213960	$+1$	4558	$2 \cdot 43 \cdot 53$

Of course, a square cannot be negative. We handle this requirement by treating (-1) as another "prime" factor of $(-1)^i Q_i$. Each relation in the table above is represented by one row in the next table. Each row holds one exponent vector v_i modulo 2.

i	(-1)	2	5	31	43	53	113
10	0	0	0	1	1	0	0
23	1	1	0	0	0	0	1
26	0	1	0	1	0	1	0
31	1	1	0	0	0	0	1
40	0	1	0	0	1	1	0

By Gaussian elimination modulo 2, or otherwise, one sees that the rows with $i = 10, 26$ and 40 are linearly dependent, as are the rows with $i = 23$ and 31. The first dependency is

$$(6700527 \cdot 11455708 \cdot 3213960)^2 \equiv (2 \cdot 31 \cdot 43 \cdot 53)^2 \quad (\bmod\ 13290059)$$

or $141298^2 \equiv 141298^2 \pmod{13290059}$, which fails to factor n. The second dependency is

$$(1914221 \cdot 1895246)^2 \equiv (2 \cdot 5 \cdot 113)^2 \quad (\bmod\ 13290059)$$

or $12677605^2 \equiv 1130^2 \pmod{13290059}$, which gives the factors

$$\gcd(12677605 - 1130, 13290059) = \gcd(12676475, 13290059) = 4261 \quad \text{and}$$
$$\gcd(12677605 + 1130, 13290059) = \gcd(12678735, 13290059) = 3119.$$

Smith and Wagstaff [113] and [118] fabricated a special computer for factoring large integers by CFRAC. It had a 128-bit wide main processor with a bit-slice architecture to generate the A_i and Q_i, and sixteen simple remaindering units (the "Mod Squad") to factor sixteen Q_i's in parallel.

13.2   The Quadratic Sieve

The quadratic sieve factoring algorithm, QS, is quite similar to CFRAC. The difference is in the method of producing relations $x^2 \equiv q \pmod{n}$ with q

factored completely. CFRAC forms x and q from the continued fraction expansion of $\sqrt{n}$ and factors q by trial division, a slow process. The quadratic residues q in CFRAC are likely to be smooth because they are $< 2\sqrt{n}$.

QS produces x and q with a quadratic polynomial $q = f(x)$ and factors the q's with a **sieve**, a much faster process than trial division. The quadratic polynomial $f(x)$ is chosen so that the q's will be as small as possible. This means that most of them will be larger than $2\sqrt{n}$, but not too many times larger, so that they are almost as likely to be smooth as the q's in CFRAC.

Let $f(x) = x^2 - n$ and $s = \lceil \sqrt{n} \rceil$. Consider the numbers $f(s), f(s+1), f(s+2), \ldots$. Fermat's factoring method considered the same numbers and sought $f(s + i) = y^2$. As we saw in Theorem 10.1, this could take a long time. Suppose we could factor some of these numbers, not by trial division, but by a faster method which would find the B-smooth numbers quickly. If a prime p divides some $f(x) = x^2 - n$, then $x^2 \equiv n \pmod{p}$, so n is a quadratic residue modulo p (unless $p|n$). If there are K primes $p \le B$ with $(n/p) = +1$ and we can find $R > K$ B-smooth numbers $f(x)$, then we will have R relations involving K primes and linear algebra will give us at least $R - K$ congruences $x^2 \equiv y^2 \pmod{n}$, each of which has probability at least $1/2$ of factoring n, by Theorem 13.2. Typically, R is only $K + 10$ or $K + 20$.

How do we find the B-smooth numbers among $f(s), f(s+1), f(s+2), \ldots$? We sieve them by some primes $< B$. As with CFRAC, the factor base for QS consists of the primes $p < B$ for which the Legendre symbol $(n/p) = +1$. Write down the numbers $f(s + i)$ for i in an interval $a \le i < b$ of convenient length, say a few million. The first interval will have $a = s$. Subsequent intervals will begin with a equal to the endpoint b of the previous interval. For each prime $p < B$, divide out all factors of p from those $f(s + i)$ which p divides. For which i does p divide $f(s + i)$? Since $f(x) = x^2 - n$, p divides $f(x)$ precisely when $x^2 \equiv n \pmod{p}$. We know from Theorem 7.2 that the solutions x to this congruence lie in the union of two arithmetic progressions with common difference p, and we learned how to find the starting points of these two arithmetic progressions in Section 7.5. If the roots of $x^2 \equiv n \pmod{p}$ are x_1 and x_2, then the arithmetic progressions begin with the first numbers $\equiv x_1$ and $x_2 \pmod{p}$ which are $\ge a$. The prime factor p is removed from each $f(s + i)$ which it divides. There is no trial division. We divide only when we already know that the remainder will be 0.

The number of sieve operations for a prime p is about $\frac{2}{p}(b - a)$ because exactly two of every p numbers are divided by p. The complexity of the sieve is $\sum_{p<B, p \text{ prime}} \frac{2}{p}(b-a)$. It can be shown that this sum is $O((b-a)\ln \ln B)$. The amortized cost of sieving one i value is thus $\ln \ln B$. Trial division would have taken about $O((b - a)B/\ln B)$ steps to find the B-smooth numbers between a and b, or $B/\ln B$ steps per i value. The sieve saves much time.

If one replaces $f(s + i)$ and p by their logarithms, one can replace the slow division of large numbers with subtraction of small numbers. Initialize an array F[i] with the logarithm of $f(a + i)$. During the sieve, subtract $\log p$ from F[i] when p divides $f(a + i)$. Most implementations use scaled approximate logarithms which are integers between 0 and 255, so that fast integer byte arithmetic may be used. After the sieve, the array is scanned for small values of F[i]. We do not require that the byte value be zero because of the approximation of logarithms, because some primes may divide $f(a+i)$ to a higher power than the first, and because small primes are treated differently. Really small primes, $p < 100$, say, are replaced in the factor base by powers p^e and the sieve subtracts $\log p^e$ whenever p^e divides $f(a+i)$. This is done because the small primes contribute little to the factorization of $f(a+i)$, unless p^e divides $f(a+i)$, and sieving by them is expensive. When F[i] is smaller than a threshold T, we form the integer $f(a+i)$ and try to factor it. This factoring is facilitated by the fact that we already know the two roots of $f(x)$ modulo each p in the factor base. We need only compare $x \pmod{p}$ with these two numbers to determine whether p divides $f(x)$. The threshold is adjusted so that we will often succeed in factoring the number completely. Each success represents a relation, which is saved in a set as $a+i$ and perhaps the factorization of $f(a+i)$. When the set contains a few more than K entries, the sieving stops and linear algebra constructs congruences $x^2 \equiv y^2 \pmod{n}$ which will likely yield a factorization of n.

In the final step of the algorithm, x in $x^2 \equiv y^2 \pmod{n}$ is formed as the product modulo n of the x_i's on the left sides of the relations $x_i^2 \equiv q_i \pmod{n}$ which participate in the dependency. The number y^2 is the product of the q_i's in the same relations. Here is a good way to compute y. The complete prime factorization of each q_i is known, and only primes from the factor base appear in these factorizations. For each prime p in the factor base count the number of times it appears as a factor in any of the q_i's. This count must be an even number $2e$ because of the linear algebra. Multiply the prime powers p^e modulo n to find y.

The size K of the factor base is about $\frac{1}{2}\pi(B) \approx \frac{1}{2}B/\ln B$ and should be optimized to minimize the total work. We want to choose K small so that we will need fewer relations to complete the factorizations. But if we choose K too small, the B-smooth numbers $f(a+i)$ will be very rare and we will search for them forever. We must choose K large enough so that B-smooth numbers will appear at a steady pace.

In order to determine the complexity of QS, we must estimate the size of the numbers $f(s+i)$. If $0 \le i < M$, say, where M is much smaller than $\sqrt{n}$, we have

$$f(s+i) = (s+i)^2 - n = s^2 + 2si + i^2 - n \approx 2\sqrt{n}i < 2\sqrt{n}M.$$

This shows that the numbers we hope are B-smooth are only about M times larger than the corresponding numbers in CFRAC. We will estimate the probability that these numbers are smooth by assuming that they are about the size of $\sqrt{n}$. By Theorem 4.9, the probability that $f(s+i)$ is B-smooth is about u^{-u}, where $u = (\ln \sqrt{n})/\ln B = \frac{1}{2}(\ln n)/\ln B$.

We expect to have to try about u^u values of i to get one B-smooth $f(s+i)$. Therefore, we will need to try about $M = Ku^u$ values of i to get about K

relations. We need just a few more than K relations. Using the complexity of the sieve mentioned above, we see that the total work to factor n is about $W(B) = Ku^u \ln \ln B$. Write $L(x) = \exp(\sqrt{(\ln x) \ln \ln x})$. An analysis like that in the proof of Theorem 12.4 shows that the optimal smoothness bound B is about $(L(n))^{1/2}$ and that the total work using this B is about $W(B) = L(n)$. The total number of values of $f(s+i)$ sieved is about

$$M = L(n)/\ln \ln B = \exp\left(\sqrt{\frac{\ln n}{\ln \ln n}}\right).$$

This analysis ignores the time for the linear algebra needed to find the dependencies. Ordinary Gaussian elimination takes $O(K^3)$ steps, which is about $(L(n)^{3/2})$ and too slow in theory. In practice, Gaussian elimination is a fine method for finding the relations because the constant implied in $O(K^3)$ is tiny. One can pack 32 vector components into a 32-bit word and perform 32 subtractions modulo 2 with a single exclusive-or operation. Furthermore, the matrix of exponents is sparse because few primes divide any particular $f(s+i)$. One can use "structured" Gaussian elimination (see [79]) to preserve the sparseness as long as possible. Other sparse matrix methods, like the block Lanczos method (see [79] and [75]), which run in essentially $O(K^2)$ steps, can replace Gaussian elimination and preserve the theoretical estimate of $L(n)$ for the complexity of QS.

13.3 Variations of the Quadratic Sieve

The version of QS described in the previous section was close to the initial design of Pomerance [87] implemented by Gerver [46]. Several variations on this basic algorithm accelerate it in practice, although they do not improve the theoretical complexity below $O(L(n))$.

13.3.1 Large Primes

Recall that during the scan after the sieve, the value of $f(s+i)$ is factored by trial division for each i for which F[i] is less than a threshold T, and i is saved provided $f(s+i)$ was completely factored. These relations are called **fulls**. The size K of the factor base would have to be quite large for this to work well and the sieving process would take a long time. This problem was solved already in CFRAC. Morrison and Brillhart [77] proposed saving the relations that have at most one prime factor larger than the largest prime F in the factor base and smaller than some upper bound P. This technique has been used in every implementation of the quadratic sieve, even the first one [46]. These relations with one prime beyond F are called **partial relations**. The partial relations are stored and sorted in order of their large primes. Any two partial relations containing the same large prime can be multiplied, and the common large prime removed, to form a full relation. It takes no extra effort to find partial relations when $P \leq F^2$ since we know that the remaining cofactor is prime because the trial division has already searched for all possible prime divisors below its square root. By Theorem 2.4, we will begin to get partial relations having large primes, which have already appeared as soon as we have about $\sqrt{\pi(P)/2}$ relations. Many more duplicate large primes appear as the number of relations increases above this number. (There is a factor $1/2$ inside the square root because large primes p, like those in the factor base, must satisfy $(p/n) = 1$, and this equality holds for about half of all primes.)

Another variation, due to A. K. Lenstra and Manasse [65], saves relations that have at most two large primes less than P and greater than F. This method takes a small additional effort, since the cofactor remaining after trial division may have to be factored into the two large primes. Because the remaining cofactor could also be a single large prime, a probable prime test is performed to distinguish prime and composite cofactors so that factorization is attempted only for the composite ones. SQUFOF and ECM are good choices for factoring these numbers. Relations that contain two large primes are called **partial-partial relations**, or **pp**'s, and can be combined with partial and other partial-partial relations by a graph cycle-finding algorithm to form full relations [65].

Three large primes, or **ppp**'s, have also been used occasionally. See [68] for an example.

The quadratic residues factored in partial, partial-partial and ppp relations are k-semismooth numbers for $k = 1, 2, 3$, respectively. They are counted by the function $\psi_k(x, F, B)$, where $x \approx \sqrt{n}$. By Theorem 4.11, the probability is about $\rho_k((\ln n)/(2 \ln B), (\ln n)/(2 \ln F))$ that any particular $f(s+i)$ is semismooth. Zhang [131] analyzed the use of three large primes in [68].

13.3.2   Multiple Polynomials

As the size of x in the sieve increases, the probability of successfully factoring $f(x)$ decreases. It was proposed by Davis and Holdridge [34] and Montgomery (see [88]) to use many polynomials for shorter sieve intervals. The multiple polynomial quadratic sieve (MPQS) is significantly faster than the single polynomial version but requires expensive multi-precision and modular inverse operations. One must find the two zeros of $f(x)$ modulo p for each new polynomial. The algorithm spends much time calculating the new zeros for each polynomial when compared to the sieving time.

In QS, the relations $x^2 \equiv q \pmod{n}$ with q factored completely are produced as follows. Let the factor base consist of the first K small primes $p_1, \ldots, p_K$ for which n is a quadratic residue. To construct many suitable polynomials, choose pairs a, b of integers with $a = c^2$ for some integer c, $b^2 \equiv n \pmod{a}$ and $0 < b < a/2$. (Of course, $(a-b)^2 \equiv n \pmod{a}$, but it leads to an equivalent polynomial.) Then the quadratic polynomial

$$Q(t) = \frac{1}{a}[(at + b)^2 - n] = at^2 + 2bt + \frac{b^2 - n}{a}$$

will take integer values at every integer t. Since each polynomial of this form has discriminant n, the factor base is the same for each polynomial, namely the primes for which n is a quadratic residue. If a value of t is found for which the right hand side is factored completely, a relation $x^2 \equiv q \pmod{n}$ is produced, with $x \equiv (at + b)c^{-1} \pmod{n}$ and $q = Q(t) = \prod_{j=1}^{K} p_j^{f_j}$, as desired. No trial division by the primes in the factor base is necessary. A sieve factors millions of $Q(t)$'s at once. Let t_1 and t_2 be the two solutions of $(at + b)^2 \equiv n \pmod{p_i}$ in $0 \le t_1, t_2 < p_i$. This congruence has two solutions because n is a quadratic residue modulo p_i. Then all solutions of $Q(t) \equiv 0 \pmod{p_i}$ are $t_1 + kp_i$ and $t_2 + kp_i$ for $k \in \mathbf{Z}$. In most implementations, $Q(t)$ is represented by one byte $Q[t]$, initialized at 0, and $\log p_i$ is added to this byte to avoid division of $Q(t)$ by p_i, a slow operation. The two inner loops are

```
t = t_1
while t < upper_limit
      Q[t] = Q[t] + log p_i
      t = t + p
end
```

and a similar loop for the other root of the quadratic congruence. After the sieve completes, one harvests the relations $x^2 \equiv q \pmod{n}$ from those t for which $Q[t]$ exceeds a threshold T less than $\log Q(t)$. Only at this point is $Q(t)$ formed and factored, the latter operation being done by trial division with the primes in the factor base.

13.3.3   The Self-Initializing Quadratic Sieve

We have seen how to change polynomials easily. It is good to change polynomials because each new one gives us a new set of small numbers to try to factor. On the other hand, we have to solve the congruences $(at+b)^2 \equiv n \pmod{p_i}$ for each prime p_i in the factor base for each new polynomial, and this requires a lot of extended precision arithmetic, which may take as long as the sieving itself. The next version of the QS algorithm amortizes the root finding over many polynomials. It was invented independently by Peralta [81] and Alford and Pomerance [4], who respectively called it the hypercube multiple polynomial quadratic sieve and the self-initializing quadratic sieve. The algorithm uses polynomials with two coefficients, a and b, of the form $f(x) = (ax + b)^2 - n$. We omit the details of this polynomial construction, but in summary it is:

- a has s prime factors $q_1 \ldots q_s$, where q_i is in the factor base, $1 \le i \le s$.

- b is the sum of s values.

- $b^2 \equiv n \pmod{a}$. There exist 2^s solutions to this equation by Theorem 7.16, but only 2^{s-1} are of interest because the other half represent the negative values of the first 2^{s-1} values, and would yield duplicate relations.

- The 2^{s-1} values of b and corresponding zeros of $f(x) \pmod{p}$ are quickly computed with a Gray code using single precision addition or subtraction instead of the multi-precision multiplication and inversion needed in the original multiple polynomial QS.

The sieving and trial division process of the hypercube multiple polynomial QS is the same as with the single polynomial QS, except that the hypercube multiple polynomial QS does not sieve by the prime factors of a.

13.4   The Number Field Sieve

A good general reference for the Number Field Sieve, NFS, is the book [64] by Lenstra and Lenstra. The book [33] by Crandall and Pomerance has an excellent treatment of this algorithm.

In the quadratic sieve, we produced many relations $x_i^2 \equiv q_i \pmod{n}$ with q_i factored completely. When we had enough relations, we matched the prime factors of the q_i and selected a subset of them for which the product of the q_i was square. In this way, we found congruences $x^2 \equiv y^2 \pmod{n}$ which could factor n.

Let us now drop the requirement that the left side of a relation must be square. Let us seek relations $r_i \equiv q_i \pmod{n}$ in which both r_i and q_i have been factored completely. We could use linear algebra as in QS to match the prime factors of r_i and the prime factors of q_i and select a subset of the relations for which both the product of the r_i's and the product of the q_i's are square. This is fine idea, but unfortunately, no one has been able to make it work.

NFS tries to make the idea work by letting the numbers on one side of each relation be algebraic integers from an algebraic number field. The plan is to match the irreducible factors so that each occurs an even number of times and the product of the algebraic integers in the selected subset of the relations is a square in the algebraic number field.

The first difficulty of this approach is in writing a congruence modulo n with a noninteger on one side. We solve this problem by using a homomorphism h from the ring of integers of the algebraic number field to Z_n, the integers modulo n. Suppose we have many algebraic integers θ_i, each factored into irreducibles, and also every $h(\theta_i)$ factored into the product of primes. Then we may match the irreducibles and match the primes to choose a subset of the θ_i's whose product is a square γ^2 in the ring of algebraic integers and so that the product of the $h(\theta_i)$'s is a square y^2 in the integers. Let $x = h(\gamma)$, a residue class modulo n. Since homomorphisms preserve multiplication, we have

$$x^2 = (h(\gamma))^2 = h(\gamma^2) = h(\prod \theta_i) = \prod h(\theta_i) \equiv y^2 \pmod{n}.$$

The congruence $x^2 \equiv y^2 \pmod{n}$ may lead to a factorization of n by Theorem 13.2. In order for this theorem to apply, we assume that n is odd and has at least two different prime factors.

Now we choose the algebraic number field and construct the homomorphism. Let

$$f(x) = x^d + c_{d-1}x^{d-1} + \cdots + c_1 x + c_0$$

be an irreducible monic polynomial with integer coefficients and let α be a zero of f in $\mathbf{C}$. The algebraic number field will be $\mathbf{Q}(\alpha)$ and our ring will be the ring $\mathbf{Z}[\alpha]$ of all $\sum_{j=0}^{d-1} a_j \alpha^j$, where the a_j are integers. This ring is contained in the ring $\mathcal{I}$ of integers of $\mathbf{Q}(\alpha)$. We must also know an integer m for which $f(m) \equiv 0 \pmod{n}$. The homomorphism from $\mathbf{Z}[\alpha]$ to Z_n will be defined by $h(\alpha) = m \pmod{n}$. This implies that $h\left(\sum_{j=0}^{d-1} a_j \alpha^j\right) \equiv \sum_{j=0}^{d-1} a_j m^j \pmod{n}$.

The numbers θ will all have the simple form $a - b\alpha$. We will seek a set $\mathcal{S}$ of pairs (a, b) of integers such that

$$\prod_{(a,b)\in\mathcal{S}} (a - bm) \text{ is a square in } \mathbf{Z},$$

and

$$\prod_{(a,b)\in\mathcal{S}} (a - b\alpha) \text{ is a square in } \mathbf{Z}[\alpha]. \tag{13.2}$$

Let the integer y be a square root of the first product. Let $\gamma \in \mathbf{Z}[\alpha]$ be a square root of the second product. We have $h(\gamma^2) \equiv y^2 \pmod{n}$, since $h(a - b\alpha) \equiv a - bm \pmod{n}$. Let $x = h(\gamma)$. Then $x^2 \equiv y^2 \pmod{n}$, which will factor n with probability at least $1/2$, by Theorem 13.2.

In addition to being irreducible and having a known zero m modulo n, we want the polynomial $f(x)$ to have "small" coefficients compared to n. There are several ways one might satisfy all these conditions. In practical applications, one should choose the degree d of $f(x)$ to be 4 for n near 10^{100}, 5 for n near 10^{150} and 6 for n near 10^{200}.

The requirements on $f(x)$ are easily met in the Special Number Field Sieve, SNFS, which factors numbers of the form $n = r^e - s$, where r and $|s|$ are small positive integers. While numbers of this special form are not likely to be cryptographic keys, their factorizations arise in many problems in mathematics and have been studied extensively. The numbers in [18], which have this form, are often used to test new factoring algorithms. Let k be the least positive integer for which $kd \geq e$. Let $t = sr^{kd-e}$. Let $f(x)$ be the polynomial $x^d - t$. Let $m = r^k$. Then $f(m) = r^{kd} - sr^{kd-e} = r^{kd-e}n \equiv 0 \pmod{n}$. See [42] for some other interesting ways to choose the polynomial.

In the general case, called the General Number Field Sieve, GNFS, one standard approach to finding a good polynomial (of degree 5, say) to factor n is to let $m = \lfloor n^{1/5} \rfloor$ and write $n = \sum_{i=0}^{5} d_i m^i$ in base m. The digits d_i will be in the interval $0 \leq d_i < m$, which is small compared to n. Then let the polynomial be $f(x) = \sum_{i=0}^{5} d_i x^i$.

With the choice of $f(x)$ for either the SNFS or the GNFS, we assume f is irreducible. If f is not irreducible, then we can factor n immediately. If $f(x) = g(x)h(x)$ in $\mathbf{Z}[x]$, then the integer factorization $n = g(m)h(m)$ gives a nontrivial factorization of n. See Brillhart et al. [16] for details.

We will have two sieves, one for $a - bm$ and one for $a - b\alpha$. The sieve on $a - bm$ is simple. Choose a bound M for a and b. Note that if a and b are replaced by their negatives, so are $a - bm$ and $a - b\alpha$, and no new relation is produced. We eliminate duplicate relations by requiring that $b > 0$. Also, if both a and b are multiplied by the same integer $g > 1$, we get a relation which is just a multiple of the first one, and which does not provide additional help in forming congruent squares. We avoid these useless relations by requiring that $\gcd(a, b) = 1$. Then for each fixed $0 < b < M$ we try to factor the numbers $a - bm$ for $-M < a < M$ by a sieve much like that of Eratosthenes. During the scan after the sieve, we ignore otherwise good a's with $\gcd(a, b) > 1$.

The goal of the sieve on the numbers $a - b\alpha$ is to allow us to choose a set $\mathcal{S}$ of pairs (a, b) so that the product in Equation (13.2) is a square. Rather than try to factor the algebraic integers $a - b\alpha$, let us work with their norms. The norm function is multiplicative and the norm of an algebraic integer is an integer. The norm of a square γ^2 is a square because $N(\gamma^2) = (N(\gamma))^2$. Thus, if the product in Equation (13.2) is a square, then its norm is a square, and its norm is the product of all $N(a - b\alpha)$ with $(a, b) \in \mathcal{S}$. Since the norms are rational integers, rather than algebraic integers, it is easy to match their prime factors to form squares. Furthermore, the norm of $a - b\alpha$ is a polynomial in a and b and therefore something we know how to factor with a sieve.

Let the complex numbers $\alpha_1, \ldots, \alpha_d$ be all of the zeros of the minimal polynomial $f(x)$ of α. These numbers are the conjugates of α. The conjugates of $a - b\alpha$ are $a - b\alpha_1, \ldots, a - b\alpha_d$, and so

$$N(a - b\alpha) = \prod_{i=1}^{d} (a - b\alpha_i) = b^d \prod_{i=1}^{d} (a/b - \alpha_i) = b^d f(a/b),$$

because $f(x) = (x - \alpha_1) \cdots (x - \alpha_d)$. If we define

$$F(x, y) = y^d f(x/y) = x^d + c_{d-1} x^{d-1} y + \cdots + c_0 y^d,$$

then $N(a - b\alpha) = F(a, b)$.

We can perform the second sieve this way: For each $0 < b < M$, sieve the polynomial $F(a, b)$ for $-M < a < M$ and find smooth values of $N(a - b\alpha)$. We want $a - bm$ to be simultaneously smooth, too. For each fixed b we sieve both $a - bm$ and $F(a, b)$ and save the pairs (a, b) for which both of these integers are smooth and also $\gcd(a, b) = 1$. The two sieves might have different factor bases corresponding to different smoothness bounds. The exponent vectors will have one entry for each prime in each factor base. When we have found many relations, linear algebra will construct sets of pairs (a, b) for which the product of $a - bm$ is a square and the product of the norms of $a - b\alpha$ is a square.

The product of the norms of $a - b\alpha$ is the norm of the product of the $a - b\alpha$. Will this product be the square of a number in $\mathbf{Z}[\alpha]$ as we require? Not in general. One problem is that the norm function does not distinguish among associates. For example, $3 + 2i$ and $3 - 2i$ are associates in the Gaussian integers $\mathbf{Z}[i]$. They have the same norm, 13, and

$$N((3 - 2i)(3 + 2i)) = N(3 - 2i)N(3 + 2i) = 13^2$$

is a square although $(3 - 2i)(3 + 2i)$ is not the square of a Gaussian integer. This problem is easy to solve using data already computed to perform the sieve. For each prime p in the factor base, let $R(p)$ denote the set of all $0 \le r < p$ with $f(r) \equiv 0 \pmod{p}$. In the case of the Gaussian integers, the polynomial is $f(x) = x^2 + 1$, so $R(2) = \{1\}$, $R(7)$ is empty and $R(13) = \{5, 8\}$. If $\gcd(a, b) = 1$, then p divides $F(a, b)$ if and only if $a \equiv br \pmod{p}$ for some r in $R(p)$. For the Gaussian integers, $F(a, b) = a^2 + b^2$. We have

$$N(3 + 2i) = F(3, -2) = F(-3, 2) = 13 = F(3, 2) = F(-3, -2) = N(3 - 2i),$$

and $3 = 2 \cdot 8 \pmod{13}$ while $-3 = 2 \cdot 5 \pmod{13}$. For another example,

$$N(7 - 4i) = F(7, 4) = 65 = F(8, 1) = N(8 - i),$$

and $7 \equiv 4 \cdot 5 \pmod{13}$, showing that $3 + 2i$ divides $7 - 4i$, while $8 \equiv 1 \cdot 8 \pmod{13}$, showing that $3 - 2i$ divides $8 - i$. We can remember this information in the exponent vectors. Use one entry for each pair p, r, where p is a prime in the factor base and r is in $R(p)$. Suppose p divides $F(a, b)$ and $\gcd(a, b) = 1$. If $a \not\equiv br \pmod{p}$, then the exponent vector will have entry 0 for the pair p, r. But if $a \equiv br \pmod{p}$, then the entry for the pair p, r in the exponent vector will be the exponent on p in the prime factorization of $F(a, b)$. Note that the sets $R(p)$ should already be computed during sieve setup. Fix b and sieve $F(a, b)$ as a polynomial in the single variable a. The a for which a given prime p divides $F(a, b)$ are the a in the residue class $a \equiv br \pmod{p}$ for each r in $R(p)$.

So far, we have solved the problem of the norm function not distinguishing among associates. The units cause further problems, as do the possible lack of unique factorization in $\mathbf{Z}[\alpha]$ and the fact that $\mathbf{Z}[\alpha]$ might be a proper subset of the ring $\mathcal{I}$ of integers in $\mathbf{Q}(\alpha)$. All of these conditions may cause the product of the $a - b\alpha$ in Formula 13.2 to fail to be the square of a number in $\mathbf{Z}[\alpha]$ even though the product of the norms of $a - b\alpha$ is a square integer. Here is one trick that solves many of these problems. Remember that if a is a square, then the

Legendre symbol $(a/q) = +1$ for any prime q not dividing a. The converse statement is false, of course, but if a is an integer such that $(a/q) = +1$ for many primes q, then a is likely to be square. Recall that when we multiply positive and negative integers to form a square, we include the sign as one more column, for the "prime" -1. To solve the problem with the square norm not guaranteeing a square, we add a few more entries to the exponent vectors. Choose several primes q not in the factor base. For each of these q's, find a solution s to $f(s) \equiv 0 \pmod{q}$ with $f'(s) \not\equiv 0 \pmod{q}$. For each pair (a, b) and each prime q, evaluate the Legendre symbol $((a + bs)/q)$, put its value (0 for $+1$, 1 for -1) in the exponent vector and extend the linear algebra to ensure that

$$\prod_{(a,b) \in \mathcal{S}} \left(\frac{a + bs}{q} \right) = 1$$

for every q. The resulting sets $\mathcal{S}$ will very likely produce squares in Formula 13.2.

The final step of the NFS is to find the square roots and then compute $\gcd(x - y, n)$. The square root x in $\mathbf{Z}$ may be found just as for QS. But it is much harder to compute the square root γ of the product in Formula 13.2. See Buhler et al. [19], Couveignes [32] and Montgomery [74] for ways of finding this square root.

Now let us consider why NFS is a fast factoring algorithm. The answer to this question involves the proper choice of the parameters. In QS, we factor many numbers near $\sqrt{n}$ until we can find a subset of them whose product is square. This size estimate leads to the complexity $L(n)$ for QS. We will show that in NFS the numbers we try to factor are much smaller than $\sqrt{n}$, so that it is easier to find enough smooth ones to produce a subset of them whose product is square.

In the polynomial constructions for SNFS and GNFS above, the number m was chosen to be near $n^{1/d}$. Let us assume m has this approximate size. Suppose the absolute values of the coefficients c_i of $f(x)$ and $F(x, y)$ are also bounded by $n^{1/d}$. Suppose we sieve the rectangle $0 < b < M$, $-M < a < M$. Then $|a - bm| < 2n^{1/d}M$ and $|F(a, b)| < (d + 1)n^{1/d}M^d$. Requiring that *both* $a - bm$ and $F(a, b)$ be smooth for the same pair (a, b) is essentially the same as requiring that their product be smooth for (a, b). Hence we seek smooth numbers bounded above by $2(d + 1)n^{2/d}M^{d+1}$. Comparing with QS, we see that if d and M are fixed, then NFS should be faster than QS for large n when $d > 4$. Indeed, if d is fixed, the complexity of NFS is essentially $L(n)^{\sqrt{4/d}}$. A more careful analysis (see Section 6.2.3 of [33]) shows that if one lets d increase slowly, so that

$$d \approx \left(\frac{3 \ln n}{\ln \ln n} \right)^{1/3},$$

then the complexity of the number field sieve is

$$\exp \left(c(\log n)^{1/3} (\log \log n)^{2/3} \right)$$

for some constant $c > 0$. The constant c is a bit smaller for SNFS than for GNFS because the coefficients are smaller.

There are several variations on the basic NFS algorithm. One can use large primes, as in QS, on either side of the relations. The multiple polynomials, which work so well in QS, do not work in a practical way for NFS because each new polynomial defines a new number field and has different root sets $R(p)$ for each prime p. See Section 6.2.7 of [33] for other variations.

13.5   *Exercises*

1. Factor Kraitchik's number 193541963777.

2. Devise an algorithm to solve this problem in polynomial time. The input is a composite integer n, not a power, and a proper factor a of n. The output consists of two relatively prime integers c, d satisfying $n = cd$ and $1 < c < n$.

3. Given a positive integer n, not a power, and an integer d, let $m = \lfloor n^{1/d} \rfloor$. Write $n = \sum_{i=0}^{d} d_i m^i$ with $0 \le d_i < d$. Define a polynomial $f(x) = \sum_{i=0}^{d} d_i x^i$. Prove that if d is fixed and n is large enough, then $f(x)$ is monic.

Chapter 14

Computing Discrete Logarithms

Many cryptosystems could be broken if we could compute discrete logarithms quickly, that is, if we could solve the equation $a^x = b$ in a large finite field. For convenience of computation, usually the finite field is either the integers modulo a prime p or the field with 2^n elements.

The discrete logarithm algorithms considered here and in the first two sections apply in any group. In particular, they are about the best one can do in elliptic curve groups. The **discrete logarithm problem for elliptic curves** is to find an integer x for which $Q = xP$, where P and Q are two given points on an elliptic curve E modulo p. It is also given that such an integer x exists, perhaps because the elliptic curve group is cyclic and P generates it.

The algorithms in the final two sections depend on the notion of smoothness and solve the discrete logarithm problem only in the group R_p, the integers modulo a prime p, where one can define smooth numbers. The index calculus and other fast algorithms for discrete logarithms are much faster than the methods of Shanks and Pollard. Hence, the group R_p must be much larger than an elliptic curve group to achieve the same security. A rough rule of thumb is that R_p with a 1024-bit prime p is about as safe as an elliptic curve modulo a 128-bit prime.

Consider first the exponential congruence $a^x \equiv b \bmod p$. By analogy to ordinary logarithms, we may write $x = \text{Log}_a b$ when p is understood from the context. These discrete logarithms enjoy many properties of ordinary logarithms, such as $\text{Log}_a bc = \text{Log}_a b + \text{Log}_a c$, except that the arithmetic with logarithms must be done modulo $p - 1$ because $a^{p-1} \equiv 1 \bmod p$. This is explained in Theorem 6.19. Neglecting powers of $\log p$, the congruence may be solved in $O(p)$ time and $O(1)$ space by raising a to successive powers modulo p and comparing each with b. It may also be solved in $O(1)$ time and $O(p)$ space by looking up x in a precomputed table of pairs $(x, a^x \bmod p)$, sorted by the second coordinate. The next section explains an intermediate method which takes essentially $O(\sqrt{p})$ time and $O(\sqrt{p})$ space.

14.1 Shanks' Baby-Step-Giant-Step Method

This algorithm was described for general groups in Section 9.3.

Shanks' baby-step-giant-step algorithm solves the congruence $a^x \equiv b \bmod p$ in $O(\sqrt{p}\log p)$ time and $O(\sqrt{p})$ space as follows. Let $m = \lceil \sqrt{p-1} \rceil$. Compute and sort the m ordered pairs $(j, a^{mj} \bmod p)$, for j from 0 to $m-1$, by the second coordinate. Compute and sort the m ordered pairs $(i, ba^{-i} \bmod p)$, for i from 0 to $m-1$, by the second coordinate. Find a pair (j, y) in the first list and a pair (i, y) in the second list. This search will succeed because every integer between 0 and $p-1$ can be written as a two-digit number ji in base m. Finally, $x = mj + i \bmod p - 1$.

14.2 Pollard's Methods

Pollard [86] invented two methods for finding discrete logarithms analogous to his rho method for factoring integers. Like Shanks' baby-step-giant-step algorithm, these algorithms work in any group and have complexity $O(\sqrt{p})$, where p is the group order. However, their space requirements are tiny.

14.2.1 The Rho Method for Discrete Logarithms

We will describe the rho method for solving the congruence $a^x \equiv b \bmod p$, where p is prime, although it works in any group. The method is quite similar to his rho method for factoring, which is described in Section 10.2 and which the reader should review before continuing.

We are given a prime $p > 3$, a primitive root g modulo p and an element h of R_p, the group of nonzero integers modulo p. We seek the x modulo $p - 1$ for which $g^x \equiv h \pmod{p}$. The answer x may be written $x = \text{Log}_g h$. We use Theorem 6.19 often in the following.

Define three sequences $\{x_i\}, \{a_i\}, \{b_i\}$ by $x_0 = 1$, $a_0 = b_0 = 0$ and
if $0 < x_i < p/3$, then $x_{i+1} = hx_i \bmod p$,
$\qquad a_{i+1} = 1 + a_i \bmod p - 1$ and $b_{i+1} = b_i \bmod p - 1$,
if $p/3 < x_i < 2p/3$, then $x_{i+1} = x_i^2 \bmod p$,
$\qquad a_{i+1} = 2a_i \bmod p - 1$ and $b_{i+1} = 2b_i \bmod p - 1$, and
if $2p/3 < x_i < p$, then $x_{i+1} = gx_i \bmod p$,
$\qquad a_{i+1} = a_i \bmod p - 1$ and $b_{i+1} = 1 + b_i \bmod p - 1$.

A simple induction argument shows that $x_i \equiv h^{a_i} g^{b_i} \pmod{p}$.

The mapping $x_i \to x_{i+1}$ is a random mapping from R_p to itself. By Theorem 2.4, after about $\sqrt{p}$ iterations of the mapping there will be a repeated value $x_i = x_j$. As in the Pollard rho factoring method, we can use the Floyd cycle-finding algorithm to find two repeated values by computing two iterates of the mapping in the same loop, with one instance running twice as fast as the other. This gives us a subscript e with $x_{2e} = x_e$.

Now we have a congruence $h^{a_{2e}} g^{b_{2e}} \equiv h^{a_e} g^{b_e} \pmod{p}$. As we can easily find inverses modulo p, this leads at once to a congruence $h^m \equiv g^n \pmod{p}$, where $m \equiv a_e - a_{2e} \pmod{p-1}$ and $n \equiv b_{2e} - b_e \pmod{p-1}$. Using Theorem 6.19, we can rewrite this as

$$mx \equiv m\mathrm{Log}_g h \equiv n \pmod{p-1}. \tag{14.1}$$

Let $d = \gcd(m, p-1)$. We know that Congruence (14.1) must have a solution because g is a primitive root modulo p and p does not divide h. By Theorem 5.6, $d|n$, and by Theorem 5.7, Congruence (14.1) has d solutions, one of which is the answer x we seek. One can show that d is usually small, say, $d = 1$ or 2, so we can try all d solutions to Congruence (14.1) and find x.

Example 14.1

Let $p = 999959$, $g = 7$ and $h = 3$. Find $x = \mathrm{Log}_g h$.

At $e = 1174$ we have $x_e = x_{2e} = 11400$, $m = 310686$ and $n = 764000$. Congruence (14.1) becomes $310686x \equiv 764000 \pmod{999958}$. The extended Euclidean algorithm gives

$$2 = \gcd(310686, 999958) = 148845 \cdot 310686 - 46246 \cdot 999958,$$

and we find that $3^2 \equiv 7^{356324} \pmod{p}$ and $3 \equiv \pm 7^{178162}$. Since 3 is a quadratic residue modulo p and -1 is not, the plus sign is correct and $x = \mathrm{Log}_g h = 178162$.

In the setting of an elliptic curve group E, we are given two points P and Q, are told that $Q = xP$ for some integer x, and must find x. The group is partitioned into three pieces of roughly equal size. The random mapping of $E \to E$ takes a point X into $X + P$, $X + X$ or $X + Q$, according to which piece of the group contains X. The initial value of the variable point X is the identity ∞. The a_i and b_i are defined just as above. A repeated point yields an equation $mQ = nP$, which means that $mx \equiv n \pmod{N}$, where N is the order of P in E. Since we know that $Q = xP$, this congruence must have a solution.

14.2.2   The Lambda Method for Discrete Logarithms

We describe Pollard's lambda method in the general setting of groups. This method is also called the kangaroo method, since it employs two kangaroos to hop around in the group.

Let G be a finite cyclic group with generator g and let h be an element of G. We seek the least positive integer x so that $h = g^x$. Suppose we know that x lies in the interval $a \le x < b$. Pollard [86] defined two kangaroos, a tame one $\mathcal{T}$ starting at $t_0 = g^b$ (the upper end point of the interval) and a wild one $\mathcal{W}$

starting at $w_0 = h$ (an unknown point in the interval). Define $d_0(\mathcal{T}) = b$, the initial distance of $\mathcal{T}$ from the origin. Let $d_0(\mathcal{W}) = 0$, the initial distance of $\mathcal{W}$ from h. Let $S = \{g^{s_1}, \ldots, g^{s_k}\}$ be a set of jumps. Let G be partitioned into k pieces and for each $a \in G$, let $f(g)$, with $1 \le f(g) \le k$, be the number of the piece to which g belongs. The exponents s_i should be positive and small compared $b - a$. Pollard suggested that $s_i = 2^i$ might be good choices. The reader should experiment with various choices. Think of the s_i as the lengths of the hops of the kangaroos.

Now let the two kangaroos hop around in the group G. The tame one $\mathcal{T}$ hops from t_i to $t_{i+1} = t_i g^{s_{f(t_i)}}$ for $i \ge 0$. Keep track of $\mathcal{T}$'s distance from the origin by computing $d_{i+1}(\mathcal{T}) = d_i(\mathcal{T}) + s_{f(t_i)}$ for $i \ge 0$. It follows that $t_i = g^{d_i(\mathcal{T})}$ for $i \ge 0$. After a while $\mathcal{T}$ stops and sets a trap at its final location, say t_m. Then the wild kangaroo hops along the path from w_i to $w_{i+1} = w_i g^{s_{f(w_i)}}$ for $i \ge 0$. Keep track of $\mathcal{W}$'s distance from the unknown starting position (the discrete logarithm of h) by computing $d_{i+1}(\mathcal{W}) = d_i(\mathcal{W}) + s_{f(w_i)}$ for $i \ge 0$. Then $w_i = g^{d_i(\mathcal{W})}$ for $i \ge 0$.

After each hop, we check to see whether $\mathcal{W}$ has fallen into the trap by testing whether $w_i = t_m$. With a good choice of the parameters s_i, it is highly likely that eventually $w_n = t_m$ for some n. Then we have $x = d_m(\mathcal{T}) - d_n(\mathcal{W})$.

If we find that $d_n(\mathcal{W}) > d_m(\mathcal{T})$, then $\mathcal{W}$ has passed the trap. In this case, we start a new wild kangaroo at $w_0 = hg^z$ for some small integer $z > 0$ and hope it falls into the trap.

If the two kangaroos ever land on the same spot $(w_i = t_j)$, then their paths will coincide from that point on and $\mathcal{W}$ will be trapped. If you draw their paths going upwards, the paths will form the Greek letter lambda: λ. This is the reason for the name.

The most important property of the jumps sizes s_i is their average. Van Oorschot and Wiener [116] have shown that if the mean value of the s_i is about $\frac{1}{2}\sqrt{b - a}$ and if $\mathcal{T}$ makes about $0.7\sqrt{b - a}$ hops before setting the trap, the running time will be minimal. With these choices, $\mathcal{W}$ will hop about $2.7\sqrt{b - a}$ times before getting trapped, which happens three-fourths of the time, or passing the trap. The space requirement is about $O(\log(b - a))$.

14.3   Discrete Logarithms via Index Calculus

There are faster ways to solve $a^x \equiv b \pmod{p}$ using methods similar to the two integer factoring algorithms QS and NFS. Here is the analogue for QS. It is called the **index calculus method**. Choose a factor base of primes $p_1, \ldots, p_k$, usually all primes $\le B$. Perform the following precomputation which depends on a and p but not on b. For many random values of x, try to factor $a^x \bmod p$ using the primes in the factor base. Use trial division or a more powerful method such as Pollard's rho method or the elliptic curve method. The complexity will be subexponential regardless of the factoring algorithm used. Save at least $k + 20$ of the factored residues:

$$a^{x_j} \equiv \prod_{i=1}^{k} p_i^{e_{ij}} \pmod{p} \text{ for } 1 \leq j \leq k + 20,$$

or equivalently

$$x_j \equiv \sum_{i=1}^{k} e_{ij} \mathrm{Log}_a p_i \pmod{p - 1} \text{ for } 1 \leq j \leq k + 20.$$

Use linear algebra to solve for the $\mathrm{Log}_a p_i$. This is not as simple as it sounds because $p - 1$ is composite for prime $p > 3$. Linear algebra is much easier over a field. Solve the system of congruences modulo each prime q dividing $p - 1$. Use Hensel's lemma to lift the solutions to solutions modulo the highest power q^e of q dividing $p - 1$. Finally, combine the prime power solutions with the Chinese remainder theorem.

When b is given, perform the following main computation to find $\mathrm{Log}_a b$. Try many random values for s until one is found for which $ba^s \bmod p$ can be factored using only the primes in the factor base. Write it as

$$ba^s \equiv \prod_{i=1}^{k} p_i^{c_i} \pmod{p}$$

or

$$(\mathrm{Log}_a b) + s \equiv \sum_{i=1}^{k} c_i \mathrm{Log}_a p_i \pmod{p - 1}.$$

Substitute the values of $\mathrm{Log}_a p_i$ found in the precomputation to get $\mathrm{Log}_a b$. Using arguments like those for the running time of the elliptic curve and quadratic sieve factoring algorithms, one can prove that if a fast factoring algorithm like the elliptic curve method is used, the precomputation takes time

$$\exp\left(\sqrt{2 \log p \log \log p}\right),$$

while the main computation takes time

$$\exp\left(\sqrt{\log p \log \log p}\right).$$

See Section 6.4 of [33] for more details.

14.4 Other Fast Methods for the Group R_m

There is an algorithm similar to the index calculus method for solving congruences of the form $a^x \equiv b \pmod{p}$ which is analogous to NFS factoring algorithm and runs faster than the index calculus method for large p. See Gordon [49] for a method with time complexity

$$\exp\left(c(\log p)^{1/3}(\log\log p)^{2/3}\right)$$

for some constant $c > 0$.

The Pohlig-Hellman cipher, which we will describe in Chapter 16, could be broken if one could solve the discrete logarithm problem in R_p quickly. In the paper [83] in which this cipher was published, Pohlig and Hellman give an algorithm for computing discrete logarithms modulo a prime p when $p - 1$ is B-smooth and B is small enough so that one can perform $O(B)$ operations. Suppose $p - 1 = \prod_i q_i^{e_i}$ is the prime factorization of $p - 1$. For each i we will find y_i so that $x \equiv y_i \pmod{q_i^{e_i}}$. Then we will find the common solution to these congruences by the Chinese remainder theorem.

Let q^e be one of the prime power factors. Let us find y so that $x \equiv y \pmod{q^e}$. As $0 \leq y < q^e$, we may write y as an e-digit number in base q:

$$y = y_0 + y_1 q + y_2 q^2 + \cdots y_{e-1} q^{e-1} \text{ where } 0 \leq y_i \leq q - 1.$$

We will find $y_0, y_1, \ldots, y_{e-1}$ in that order to compute y. We have

$$(x \bmod q^e)\left(\frac{p-1}{q}\right) = y\left(\frac{p-1}{q}\right) = y_0\left(\frac{p-1}{q}\right) + (p-1)(y_1 + y_2 q + \cdots)$$

$$= y_0\left(\frac{p-1}{q}\right) + (p-1)Y,$$

for some integer Y. Raise both sides of $b \equiv a^x \pmod{p}$ to the power $(p-1)/q$ and get

$$b^{(p-1)/q} \equiv a^{x(p-1)/q} \equiv a^{y_0(p-1)/q}(a^{p-1})^Y \equiv a^{y_0(p-1)/q} \pmod{p}.$$

We used Fermat's little theorem to get the last congruence. To obtain y_0, form the powers $(a^{(p-1)/q})^n \bmod p$ for $n = 0, 1, \ldots, q - 1$ until one of them is congruent to $b^{(p-1)/q}$. Then y_0 is the exponent n that worked.

If q^2 divides $p - 1$, let $b_1 \equiv ba^{-y_0} \equiv a^{q(y_1 + y_2 q + \cdots)} \pmod{p}$. Raise both sides to the power $(p - 1)/q^2$ and get $b_1^{(p-1)/q^2} \equiv a^{y_1(p-1)/q} \pmod{p}$, where we have used Fermat's little theorem again. To obtain y_1, form the powers $(a^{(p-1)/q})^n \bmod p$ for $n = 0, 1, \ldots, q - 1$ until one of them is congruent to $b_1^{(p-1)/q^2}$. Then y_1 is the exponent n that worked.

If $e > 2$, repeat this process until all of $y_0, y_1, \ldots, y_{e-1}$ have been computed. Then $y = y_0 + y_1 q + y_2 q^2 + \cdots y_{e-1} q^{e-1}$.

Example 14.2

Solve $2^x \equiv 15 \pmod{19}$.

This example is trivial, but the method we use to solve the congruence would work if 19 were replaced by a large prime p such that the largest prime factor q of $p - 1$ was small enough so that we could do q operations.

We have $19 - 1 = 2 \cdot 3^2$. It is easy to find $x \bmod 2$. We have $(19 - 1)/2 = 9$ and

$$(-1)^x \equiv 18^x \equiv (2^x)^9 \equiv 15^9 \equiv 18 \equiv (-1) \pmod{19},$$

so $x \equiv y_0 \equiv 1 \pmod 2$.

Now we find $x \bmod 3$. We have $(19 - 1)/3 = 6$ and

$$7^x \equiv (2^6)^x \equiv 15^6 \equiv 11 \pmod{19}.$$

We try the powers of 7 (mod 19): $7^0 \equiv 1$, $7^1 \equiv 7$, $7^2 \equiv 11 \pmod{19}$, and so $x \equiv y_0 \equiv 2 \pmod 3$. To compute y_1, we let $b_1 \equiv 15 \cdot 2^{-2} \equiv 181 \pmod{19}$. Raising both sides to the power $(19 - 1)/3^2 = 2$, we see

$$1 \equiv 18^2 \equiv (2^6)^{y_1} \equiv 7^{y_1} \pmod{19},$$

so $y_1 = 0$ and $x \equiv y \equiv 2 + 0 \cdot 3 = 2 \pmod 9$. Now apply the Chinese remainder theorem to the pair of congruences $x \equiv 1 \pmod 2$ and $x \equiv 2 \pmod 9$ to obtain $x \equiv 11 \pmod{18}$.

There is one other case in which it is easy to solve a discrete logarithm problem. If m is B-smooth, where B is not too large, then one can solve $a^x \equiv b \pmod m$. Roughly speaking, solve the congruence modulo each prime divisor of m and combine the solutions with the Chinese remainder theorem. For example, suppose $m = pq$ is the product of two primes small enough so that we can solve the congruences $a^{x_1} \equiv b \pmod p$ and $a^{x_2} \equiv b \pmod q$. Then we know that $x \equiv x_1 \pmod{p-1}$ and $x \equiv x_2 \pmod{q-1}$. The Chinese remainder theorem does not apply directly because $p - 1$ and $q - 1$ are not relatively prime. However, if there is a solution x, then the two congruences must be compatible, and one can solve them with the methods of Exercise 17 of Section 5.4.

Example 14.3

Solve $2^x \equiv 35 \pmod{1003}$.

Note that $1003 = 17 \cdot 59$. We solve $2^{x_1} \equiv 35 \equiv 1 \pmod{17}$ and find $x_1 \equiv 8 \pmod{16}$. Then we solve $2^{x_2} \equiv 35 \pmod{59}$ and find $x_2 \equiv 24 \pmod{58}$. The $\gcd(16, 58) = 2$ divides $(24 - 8) = 16$, so the two congruences are compatible and we find $x \equiv 24 \pmod{464}$ since $\operatorname{lcm}(16, 58) = 464$.

There is a method of Coppersmith [30] for solving equations of the form $a^x = b$ in the field with 2^n elements that is practical for n up to about 1000.

Empirically, it is about as difficult to solve $a^x = b$ in the field with p^n elements as it is to factor a general number about as large as p^n.

14.5   *Exercises*

1. Use Shanks' baby-step-giant-step method to solve the discrete logarithm problem $2^x \equiv 82 \pmod{107}$.

2. Use Pollard's rho method to solve the discrete logarithm problem $5^x \equiv 20 \pmod{103}$.

3. Use Pollard's lambda method to solve the discrete logarithm problem $2^x \equiv 39 \pmod{101}$.

4. Solve the discrete logarithm problem $10^x \equiv 83 \pmod{97}$ by the index calculus method, using the following information. The factor base consists of the three primes 2, 3, 5. The precomputation, which depends on 10 and 97, but not on 83, generated many random exponents y and tried to factor $(10^y \bmod 97)$ using just the primes in the factor base. It produced these congruences:

$$10^1 \equiv 10 \equiv 2 \cdot 5 \quad (\bmod\ 97)$$
$$10^2 \equiv 3 \quad\quad\quad\ (\bmod\ 97)$$
$$10^{13} \equiv 15 \equiv 3 \cdot 5 \quad (\bmod\ 97).$$

   The main computation generated many random z and tried to factor $(83 \cdot 10^z \bmod 97)$ using just the primes in the factor base. After a while, it found the congruence

$$83 \cdot 10^{93} \equiv 6 \equiv 2 \cdot 3 \ (\bmod\ 97).$$

   Restate these congruences in terms of discrete logarithms modulo 97. Solve these congruences (modulo $96 = \phi(97)$) for the discrete logarithm x of 83. Do not perform any exponentiation modulo 97, except to check your answer after you find it.

5. Solve the discrete logarithm problem $3^x \equiv 282 \pmod{391}$.

6. Devise a probabilistic algorithm to solve the following discrete logarithm problem in expected time $O(\sqrt{n/m})$ group operations. The input consists of a cyclic group G of order n generated by g, integers m and a with $2 \le m < n$ and $0 \le a < m$, and an element h of G. It is given that there exists an integer $x \equiv a \pmod{m}$ such that $h = g^x$ and $0 < x \le n$, but x is unknown. The output is x.

Chapter 15

Random Number Generation

This chapter is the last one in Part I because it nearly fits into Part II. Random numbers have many uses. They are used in simulation. In cryptography they are used in stream ciphers and for choosing a secret key.

Some desirable properties a sequence of random numbers might have are:

1. The sequence looks random—it passes statistical tests of randomness.

2. The sequence is unpredictable: knowing the algorithm and previous bits, one cannot guess the next bit(s), but the sequence can be reproduced. Such sequences of random numbers might be used as key streams for stream ciphers.

3. The sequence cannot be reliably reproduced: If you run the random number generator (RNG) twice with the same input (as closely as possible), you get two different random sequences. The sequence cannot be compressed. Sequences of this sort might be used to select a secret key, like a large prime.

Most computer libraries provide a simple random number generator called a **linear congruential generator**. It works this way. Fix a multiplier a, an increment b, a modulus m and a seed x_0. Define x_i for $i \geq 1$ by $x_i = (ax_{i-1} + b) \bmod m$. The random numbers x_i are periodic and the period is always $\leq m$ because each x_i depends only on x_{i-1}, and all x_i are in the interval $0 \leq x_i < m$. One example with maximum period uses $a = 9301$, $b = 49297$ and $m = 233280$. Linear congruential generators pass some statistical tests, are fine for simulation and are efficient. However, they are worthless for cryptography because their linearity makes them easy to break.

Other generators for reproducible random numbers are described in the next two sections. They have been used for cryptography.

15.1   *Linear Feedback Shift Registers*

A **linear feedback shift register** LFSR is a device that generates a pseudorandom bit stream. It consists of an n-bit shift register and an exclusive-or gate. Let the vector $R = (r_0, r_1, \ldots, r_{n-1})$ hold the bits in the the shift register, with r_0 the bit at the right end. At each clock cycle, the bits in the register shift one position to the right. The bit r_0 is shifted out the right end and used. The output bit of the exclusive-or gate is shifted into the bit r_{n-1} at the left end of the register.

The inputs to the exclusive-or gate are several bits selected (tapped) from fixed bit positions in the register. Let the vector $T = (t_1, t_2, \ldots, t_n)$ specify the tapped bit positions: $t_i = 1$ means "bit r_{n-i} was selected" and $t_i = 0$ means "bit r_{n-i} was not selected." Let M^t denote the transpose of the matrix M. The output of the exclusive-or gate may be regarded as the scalar product

$$ TR^t = \left(\sum_{i=1}^{n} t_i r_{n-i} \right) \bmod 2 = t_1 r_{n-1} \oplus t_2 r_{n-2} \oplus \cdots \oplus t_n r_0. $$

In this section, the sum $\sum$ means exclusive-or $\oplus$, the sum modulo 2.

Define r_j for $j \geq n$ by $r_j = (\sum_{i=1}^{n} t_i r_{j-i}) \bmod 2$. Then $\{r_j\}$ for $j \geq 0$ is the sequence of pseudorandom bits generated by the LFSR. This bit stream is sometimes used as the key of a stream cipher. If the plaintext bit stream is $\{m_j\}$ for $j \geq 0$, then the ciphertext bit stream is $\{c_i\}$, defined for $j \geq 0$ by $c_j = m_j \oplus r_j$.

Example 15.1

Show the operation of the LFSR with $n = 3$, $T = (1, 0, 1)$ and initial $R = (1, 1, 1)$ and find the period of its output stream.

We have $r_0 = 1$, $r_1 = 1$, $r_2 = 1$, $r_3 = 1 \cdot 1 \oplus 0 \cdot 1 \oplus 1 \cdot 1 = 0$, etc. The column r_0 of this table shows the output stream $\{r_j\}$.

j	r_2	r_1	r_0
0	1	1	1
1	0	1	1
2	1	0	1
3	0	1	0
4	0	0	1
5	1	0	0
6	1	1	0
7	1	1	1

The period is seven.

The bit stream $\{r_j\}$ must be periodic because eventually the n bits in the shift register will be duplicated, and the bits r_j will repeat from that point. There are 2^n possible contents of the shift register; so, the period certainly cannot exceed that number. However, if the register contains all zero bits, then every r_j will be zero since the exclusive-or of zeros is zero. Thus, the period cannot be more than $2^n - 1$. If the period has this maximal value, then every bit pattern, other than all zeros, will appear as the content of the shift register sometime during each period. One can prove that for every $n \geq 1$ there is at least one tap vector T which achieves the maximal period $2^n - 1$.

Let $r(x) = \sum_{i=0}^{n-1} r_i x^i$ in $\mathbf{F}_2[x]$. The degree of $r(x)$ is, hopefully, defined and less than n. A useful tool for studying the period of an LFSR is the **generating function** $G(x) = \sum_{j=0}^{\infty} r_j x^j$. Since the coefficients r_j are 0 or 1, the sum converges at least for $|x| < 1$. We prove first that $G(x)$ is the ratio of two polynomials. We have

$$G(x) = \sum_{j=0}^{\infty} r_j x^j = \sum_{j=0}^{n-1} r_j x^j + \sum_{j=n}^{\infty} r_j x^j = r(x) + \sum_{j=n}^{\infty} r_j x^j.$$

In the last sum, replace r_j by $r_j = \sum_{i=1}^{n} t_i r_{j-i}$ and interchange the order of summation.

$$G(x) = r(x) + \sum_{i=1}^{n} \sum_{j=n}^{\infty} t_i r_{j-i} x^j = r(x) + \sum_{i=1}^{n} t_i x^i \sum_{j=n}^{\infty} r_{j-i} x^{j-i} =$$

$$= r(x) + \sum_{i=1}^{n} t_i x^i \left(\sum_{j=i}^{\infty} r_{j-i} x^{j-i} - \sum_{j=i}^{n-1} r_{j-i} x^{j-i} \right) =$$

$$= r(x) + \sum_{i=1}^{n} t_i x^i \left(G(x) - \sum_{j=i}^{n-1} r_{j-i} x^{j-i} \right).$$

Solving for $G(x)$ gives $G(x) = s(x)/t(x)$, where $t(x) = 1 - \sum_{i=1}^{n} t_i x^i$ and

$$s(x) = r(x) - \sum_{i=1}^{n} t_i x^i \sum_{j=i}^{n-1} r_{j-i} x^{j-i}.$$

The polynomial $t(x)$ must have degree n in $\mathbf{F}_2[x]$ because if $t_n = 0$, then the last bit position r_0 would serve merely to delay the output of a bit and would not participate in generating them. The degree of $s(x)$ is less than n since $x^i x^{j-i} = x^j$ and $1 \leq j \leq n-1$. The polynomial $t(x)$, which may also be written $t(x) = 1 + \sum_{i=1}^{n} t_i x^i$ since $-1 = +1$ in $\mathbf{F}_2$, is called the **characteristic polynomial** of the tap sequence. We will assume that $t(x)$ is irreducible. One can show via the partial fraction decomposition of $s(x)/t(x)$ that if $t(x)$ were not irreducible, then the LFSR could not have maximal period. See Theorem 2.3 of Golomb [48] for a proof. Since $t(x)$ is irreducible and the degree of $s(x)$ is less than the degree of $t(x)$ we must have $\gcd(s(x), t(x)) = 1$ and so no common factors can be canceled in the ratio $s(x)/t(x)$.

Now let p be the minimum period of the LFSR. Then p is the smallest positive integer such that $r_{p+j} = r_j$ for every $j \geq 0$. Then

$$G(x) = \sum_{j=0}^{\infty} r_j x^j = \sum_{j=0}^{p-1} r_j x^j + \sum_{j=0}^{\infty} r_{p+j} x^{p+j} = \sum_{j=0}^{p-1} r_j x^j + x^p G(x).$$

Therefore, $G(x)(1 - x^p) = \sum_{j=0}^{p-1} r_j x^j$. Recall that $G(x) = s(x)/t(x)$. Hence, $(s(x)/t(x))(1 - x^p) = \sum_{j=0}^{p-1} r_j x^j$. The left side must be a polynomial because the right side is one. Since $\gcd(s(x), t(x)) = 1$, $t(x)$ must divide $1 - x^p = x^p + 1$ in $\mathbf{F}_2[x]$.

Conversely, it is not hard to show that if $t(x)$ divides $x^q + 1$ for some positive integer q, then q is a multiple of the period p of the LFSR. See Theorem 2.4 of Golumb [48] for a proof. Therefore, p is the smallest positive integer for which $t(x)$ divides the polynomial $x^p + 1$. The condition that $t(x)$ be irreducible is necessary but not sufficient for the LFSR to have maximal period. A sufficient condition is that $t(x)$ be primitive.

DEFINITION 15.1   *A polynomial $t(x)$ of degree n in $\mathbf{F}_2[x]$ is called* **primitive** *if it is irreducible, it divides $x^{2^n - 1} + 1$, but it does not divide $x^d + 1$ for any divisor d of $2^n - 1$.*

One can prove that, for every $n \geq 1$, there is a primitive tap polynomial of degree n. Primitive trinomials $x^n + x^a + 1$ are especially popular for use as the tap polynomial of an LFSR because exclusive-or gates with only two inputs are much cheaper than those with more than two inputs.

Unfortunately, LFSR's do not produce cryptographically strong random sequences. If n is a few thousand and $t(x)$ is primitive, the period of the bit stream is $2^n - 1$, which is the maximum possible. This huge period gives the cipher the appearance of security, but the linearity makes it easy to break.

Recall that $R = (r_0, r_1, \ldots, r_{n-1})$ is the contents of the LFSR. Let $R' = (r'_0, r'_1, \ldots, r'_{n-1})$ be the contents of the register after the shift. Then $r'_i = r_{i+1}$ for $0 \leq i < n - 1$ and $r'_{n-1} = TR^t$. In other words, $R'^t \equiv HR^t \bmod 2$, where H is the $n \times n$ matrix with T as its first row, 1's just below the main diagonal and 0's elsewhere.

$$H = \begin{bmatrix} t_1 & t_2 & \ldots & t_{n-2} & t_{n-1} & t_n \\ 1 & 0 & \ldots & 0 & 0 & 0 \\ 0 & 1 & \ldots & 0 & 0 & 0 \\ \vdots & \vdots & \ddots & \vdots & \vdots & \vdots \\ 0 & 0 & \ldots & 1 & 0 & 0 \\ 0 & 0 & \ldots & 0 & 1 & 0 \end{bmatrix}$$

Suppose $2n$ consecutive key bits, $r_0, \ldots, r_{2n-1}$, are known. Let X and Y

be the $n \times n$ matrices

$$X = \begin{bmatrix} r_{n-1} & r_n & \cdots & r_{2n-2} \\ r_{n-2} & r_{n-1} & \cdots & r_{2n-3} \\ \vdots & \vdots & \ddots & \vdots \\ r_0 & r_1 & \cdots & r_{n-1} \end{bmatrix} \quad \text{and} \quad Y = \begin{bmatrix} r_n & r_{n+1} & \cdots & r_{2n-1} \\ r_{n-1} & r_n & \cdots & r_{2n-2} \\ \vdots & \vdots & \ddots & \vdots \\ r_1 & r_2 & \cdots & r_n \end{bmatrix}.$$

From $R'^t \equiv HR^t \bmod 2$ it follows that $Y \equiv HX \bmod 2$, so H may be computed from $H \equiv YX^{-1} \bmod 2$. The inverse matrix $X^{-1} \bmod 2$ is easy to compute by Gaussian elimination for n up to at least 10^4. The tap vector T is the first row of H and the initial contents R of the shift register are $(r_{n-1}, \ldots, r_0)$.

See Golumb [48] and Ding, Xiao and Shan [41] for more information about linear feedback shift registers and variations of them. Some variations use several LFSR's connected by some nonlinear muddle. These are not good sources of cryptographically secure random numbers either. All can be broken with linear algebra.

15.2 A Quadratic Residue Random Number Generator

Blum, Blum and Shub [11] invented a random bit generator called the BBS generator. It chooses two Blum primes, that is, primes $p \equiv q \equiv 3 \pmod{4}$. Let $n = pq$ and let s be relatively prime to n. Define $x_0 = s^2 \bmod n$ and $x_i = x_{i-1}^2 \bmod n$ for $i > 0$. The i-th pseudorandom bit is the low-order bit b_i of x_i.

A simple induction shows that $x_i = x_0^{2^i} \bmod n$. It follows from Theorem 6.15 that

$$x_i = x_0^{(2^i \bmod \phi(n))} \bmod n = x_0^{(2^i \bmod ((p-1)(q-1)))} \bmod n.$$

We can compute $2^i \bmod ((p-1)(q-1))$ with only $O(\log i)$ multiplications modulo $(p-1)(q-1)$. Hence, we can compute x_i with only $O((\log i)(\log n)^2)$ bit operations, rather than the $O(i(\log n)^2)$ bit operations it would take using the definition. If we used the BBS generator to form a key stream to encipher a random-access file, we could use this property to decipher the file from any starting point without forming the key stream from its beginning.

The number -1 is a quadratic nonresidue modulo every Blum prime by Part 5 of Theorem 7.5. Hence, r is a quadratic residue modulo p if and only if $-r$ is a quadratic nonresidue modulo p. The same statement holds with p replaced by q. By Theorem 7.18, every quadratic residue r modulo n has exactly four square roots x modulo n. The square roots of r modulo p are $\pm r^{(p-1)/4} \pmod{p}$ by Theorem 7.13, and exactly one of these two numbers is a quadratic residue. The same is true when p is replaced by q. The four square roots x of r modulo n are constructed by using the Chinese remainder theorem to solve $x \equiv$ one square root modulo p and $x \equiv$ one square root modulo q. Now x is a quadratic residue modulo $n = pq$ if and only if it is a quadratic residue modulo p and a quadratic residue modulo q. Therefore, exactly one of the square roots of a quadratic residue r modulo n is itself a quadratic residue. Since every x_i in the BBS generator is clearly a quadratic residue modulo n, we have a way to compute x_{i-1} from x_i. It is the unique square root of x_i that is a quadratic residue. This shows that if we know the factorization of n, then we can compute the sequence x_i backwards.

Conversely, suppose we can compute x_{i-1} somehow for any given x_i. Then we can factor n. Just pick a t for which the Jacobi symbol $(t/n) = -1$. This t must be a quadratic nonresidue modulo n. Let $x_i = t^2 \bmod n$. Compute x_{i-1} somehow. Then $x_{i-1}^2 \equiv x_i \equiv t^2 \pmod{n}$ and $\gcd(t + x_{i-1}, n) = p$ or q by Theorem 13.1, so n has been factored.

Suppose n is made public, but p and q are kept secret. Then anyone can use the BBS generator with that modulus n to compute the sequence $\{x_i\}$ forwards. By what was just proved, one can compute the sequence $\{x_i\}$ backwards if and only if they know the factorization of n.

There are $(p-1)/2$ quadratic residues modulo p and $(q-1)/2$ quadratic residues modulo q. Therefore, by Theorem 5.9, there are $(p-1)(q-1)/4$ quadratic residues modulo n. The mapping $x_i \to x_{i+1}$ is a permutation of the set of quadratic residues modulo n. Blum, Blum and Shub [11] show how to ensure that the period of the sequence $\{x_i\}$ is long.

They also prove this result. If the factorization of n is unknown, then the key stream $\{b_i\}$ is unpredictable in a strong sense. Given k consecutive key bits $b_j, b_{j+1}, \ldots, b_{j+k-1}$, one cannot guess the bits b_{j+k} or b_{j-1} with probability more than 0.5. Although the algorithm is slow, one can accelerate it somewhat by using the low-order $\log_2 n$ bits of x_i rather than just the low-order bit.

15.3 Hash Functions

A **weak hash function** is a function h of a message M of arbitrary length that produces a **message digest** or **hash value** $h(M)$, which is a bit string of fixed length, say, m bits, such that:

1. Given M, it is easy to compute $h(M)$,
2. Given h_0, it is hard to compute any M for which $h(M) = h_0$, and
3. Given M, it is hard to find $M' \neq M$ for which $h(M') = h(M)$.

A **strong hash function** is a weak hash function that also satisfies:

4. It is hard to find *any* two messages $M' \neq M$ for which $h(M') = h(M)$.

A common use for hash functions is authentication of messages. Property 3 provides the authentication. Suppose a long message M is not secret, but the sender wants the recipient to be sure that M was not changed by an active wiretapper during transmission. Then M would be sent along with a shorter, signed message containing $h(M)$. The recipient would check the signature, compute $h(M)$ from M, and compare this message digest with the signed one. If they agreed, he would know that M was the same message that was sent. Usually, $h(M)$ would be either enciphered or transmitted separately from M.

We will use hash functions in the next section as an aid to generating random numbers. They are also used in many protocols, such as signing contracts digitally. If the hash function has Property 4, then only $h(M)$ need be signed, where M is a long text like a contract.

Property 4 protects M against birthday attacks. Let m be the length in bits of the message digest $h(M)$. Property 4 says that m is large enough so that one party to a contract cannot compute $h(M)$ for $2^{m/2}$ messages M, which would be needed to mount a birthday attack.

A **one-way function** is a function f which can be computed easily but which has the property that given any y in the range of f it is infeasible to compute any x with $f(x) = y$. One example is a sparse polynomial of high degree modulo a large prime. See Purdy [91].

Most hash functions are built from a one-way function f which takes one argument of length b bits and one of length m bits and produces a value of length m bits. The whole message M is broken into blocks M_i of length b bits each. One computes $h_i = f(M_i, h_{i-1})$ with some standard initial value h_0 of length m bits. The hash value or message digest is the final h_i.

Some examples of hash functions are SNEFRU, N-Hash, MD4, MD5 and SHA. MD5 produces a 128-bit message digest, while SHA's message digest has 160 bits, and so is even more resistant to birthday attacks.

Both MD5 and SHA begin by padding the message M with a 1 and as many 0's as needed to make the total length $\equiv 448 = 512 - 64 \pmod{512}$. The last 64 bits hold the length of the message (in bits) before padding (modulo 2^{64}). This makes the message length a multiple of 512 bits. Call the 512-bit blocks $M_0, \ldots, M_{L-1}$.

MD5 defines h_{-1} to be a 128-bit constant stored in four 32-bit words. SHA defines h_{-1} to be a 160-bit constant stored in five 32-bit words. Both compute $h_i = f(M_i, h_{i-1})$ for $0 \le i < L$, where the one-way function f is easy to compute using addition, shift and Boolean operations on 32-bit integers. See Schneier [100] or Stallings [114] for more about MD5 and SHA.

15.4 Generating Truly Random Numbers

In designing secure random number generators, it is best to assume your adversary has a copy of your key-generating program, any master key in it, and knows the time-of-day, process number, machine name, network address, etc., of your program and its machine.

Use as many of the following sources of randomness as possible: Use the computer's clocks: UNIX gives seconds since January 1, 1970 (a whole number plus a fractional part to the microsecond). Also set an alarm and increment a counter rapidly until interrupted. Then use the low-order bits of the counter.

Use any available special hardware to produce random bits: a Geiger counter with a speck of plutonium, a capacitor to charge, an unstable oscillator, thermal noise, radio static, /dev/audio with no mike attached, the disk position or time to read one block. Use random system values such as CPU load and arrival time of network packets. If there is a user present, have the user provide randomness by typing on the keyboard, moving the mouse or speaking into the mike. Hash together, with SHA, say, anything with at least some randomness.

If the random bits you generate by using the techniques above appear to be biased, you should make them less biased in one of the following ways.

1. Exclusive-or several such bits together. Say the bit is 0 with probability $1/2 + e$ and 1 with probability $1/2 - e$, for some $0 \le e < 1/2$. Then the exclusive-or of two such bits is 0 with probability $(1/2+e)^2+(1/2-e)^2 = 1/2 + 2e^2$. Then the exclusive-or of four such bits is 0 with probability $1/2 + 8e^4$. In the limit when many such bits are exclusive-or'ed, the probability that the exclusive-or will be 0 will converge to $1/2$.

2. Use the biased "random" bits in pairs: If the two bits in a pair are the same, skip; else output the first bit.

Neither of these techniques works if adjacent bits are correlated.

Use at least two independent sources of random bits. The random numbers for generating session keys should come from the timing of the users' keystrokes. Private keys may be encrypted by a passphrase, a character string remembered and typed by the user. SHA produces a 160-bit hash of the passphrase.

15.5   *Exercises*

1. A linear congruential generator with $m = 65537$ produces the three consecutive x_i values 10413, 9953, 14267. Find a and b.

2. Let $M = 10001011$ and $C = 11110011$ be corresponding bit streams in a known-plaintext attack, where the key was generated by a four-bit LFSR. Find the matrix H and the tap sequence T.

3. The definition of primitive polynomial is redundant. Prove that if $t(x)$ is an irreducible polynomial of degree $n > 1$, then $t(x)$ divides $x^{2^n-1}+1$. Hint: Apply Lagrange's theorem to x in $\mathbf{F}_{2^n} = \mathbf{F}_2[x]/(t(x))$.

4. Let $n > 1$ be an integer. Prove that set of all quadratic residues modulo n is a group under multiplication modulo n.

5. Design two programs for choosing truly random numbers on your computer, one which accepts randomness input by a user and one which does not.

Part II

The Cryptographic Algorithms

Chapter 16

Private Key Ciphers

In this part of the book we describe many cryptographic functions and algorithms that use number theory.

This chapter describes several private key encryption functions that use some number theory. There are many other private key encryption functions that use little or no number theory, such as the Digital Encryption Standard, DES, and the International Data Encryption Algorithm, IDEA.

Also called symmetric ciphers, private key ciphers feature very fast enciphering and deciphering. They are used to transmit lots of data securely between two people who have previously agreed on a common secret key, or to encipher the private files of one person. Each encryption function has a parameter that determines its secrecy, that is, the difficulty of breaking it by trying all possible keys.

The plaintext input to each cipher must be broken into blocks of fixed length and the characters encoded as numbers, for example, their ASCII codes. These numbers are concatenated into one large number M that represents one block. We assume it is trivial to encode the characters into M and to decode M back into characters. The descriptions of the ciphers that follow tell how to encipher M to form the ciphertext C, a number about the same size as M, and how to decipher C to recover M.

16.1 Rijndael, the Advanced Encryption Standard

Rijndael, the new Advanced Encryption Standard, AES, was invented by Joan Daemen and Vincent Rijmen in Belgium. The name Rijndael is pronounced like "Rain Doll" and not like "Region Deal."

Rijndael is a block cipher. The block size and key length can be chosen independently to be 128, 192 or 256 bits. It has 10, 12 or 14 steps called **rounds**, depending on the block and key lengths. It was designed to be simple, to be resistant against all known attacks and to have fast and compact code on many platforms. Each round is composed of four basic steps called **layers**, which operate either on eight-bit bytes or 32-bit words. We begin by describing the arithmetic operations for these types of data.

16.1.1   Byte Arithmetic in Rijndael

A byte $b_7 b_6 \ldots b_1 b_0$ is considered to be a polynomial of degree 7 in $\mathbf{F}_2[x]$, that is the coefficients are in $\{0, 1\}$:

$$b(x) = b_7 x^7 + b_6 x^6 + \cdots + b_1 x + b_0.$$

For example, the byte $\texttt{0xB7} = 1011\ 0111$ is the polynomial

$$x^7 + x^5 + x^4 + x^2 + x + 1.$$

Bytes are added as polynomials in $\mathbf{F}_2[x]$, which is the same as combining them with exclusive-or ($\oplus$). Addition is associative and commutative. The identity element is $\texttt{0x00} = 0$. Every byte is its own additive inverse, since $x \oplus x = 0$.

Example 16.1

We have $\texttt{0xB7} \oplus \texttt{0xA5} = 1011\ 0111 \oplus 1010\ 0101 = 0001\ 0010 = \texttt{0x12}$.

Bytes are multiplied as polynomials modulo $m(x) = x^8 + x^4 + x^3 + x + 1 = \texttt{0x11B}$. Multiplication is associative and commutative. The identity element is $\texttt{0x01} = 1$. Every nonzero polynomial (byte) has a unique inverse with respect to this multiplication. The inverse may be computed by the extended Euclidean algorithm for greatest common divisor of the polynomial with $m(x)$. This multiplication is denoted $\bullet$. Thus, Rijndael treats bytes as elements of the field $\mathbf{F}_{2^8}$, discussed in Example 9.1.

Example 16.2

Multiply the bytes $\texttt{0xB7} \bullet \texttt{0xA5} = 1011\ 0111 \bullet 1010\ 0101$.
Multiplying them as polynomials, we have

$$
\begin{aligned}
(x^7 + x^5 + x^4 + x^2 + x + 1)\ (x^7 + x^5 + x^2 + 1) = \\
= x^{14} + x^{12} + x^{11} + x^9 + x^8 + x^7 + \\
x^{12} + x^{10} + x^9 + x^7 + x^6 + x^5 + \\
x^9 + x^7 + x^6 + x^4 + x^3 + x^2 + \\
x^7 + x^5 + x^4 + x^2 + x + 1 \\
= x^{14} + x^{11} + x^{10} + x^9 + x^8 + x^3 + x + 1.
\end{aligned}
$$

To reduce this polynomial modulo $m(x)$ we replace x^8 by $x^4 + x^3 + x + 1$.

$$x^{14} + x^{11} + x^{10} + x^9 + x^8 + x^3 + x + 1 \equiv$$

$$(x^{10} + x^9 + x^7 + x^6) + x^{11} + x^{10} + x^9 + x^8 + x^3 + x + 1$$
$$\equiv x^{11} + x^8 + x^7 + x^6 + x^3 + x + 1$$
$$\equiv (x^7 + x^6 + x^4 + x^3) + x^8 + x^7 + x^6 + x^3 + x + 1$$
$$\equiv x^8 + x^4 + x + 1$$
$$\equiv (x^4 + x^3 + x + 1) + x^4 + x + 1$$
$$\equiv x^3 \pmod{m(x)}.$$

The polynomial x^3 is the byte 0x08, and this is the product.

Multiplication of $b(x)$ by $x = $ 0x02 is a left shift of one bit position, followed by an exclusive-or with $m(x)$ if and only if the bit shifted out was 1. Therefore, multiplication of two polynomials may be performed by up to eight left shifts and conditional exclusive-ors. Let xtime(z) denote a left shift of the byte z by one bit position, followed by an exclusive-or with $m(x) = $0x11B if the bit shifted out of z was a 1 bit. In pseudocode, xtime(z) is

```
function xtime(z)
z = 2z
if (z ≥ 256) { z = z⊕0x11B }
return z
```

Example 16.3

Multiply the bytes 0xB7 • 0xA5 = 1011 0111 • 1010 0101.

We begin by multiplying 0xB7 by x^i for $0 \le i \le 7$, that is, computing 0xB7 •z, where z is a byte having exactly one 1 bit. Of course, 0xB7 • 0x01 = 0xB7.

$$0xB7 • 0x02 = \text{xtime}(B7) = 0x75$$
$$0xB7 • 0x04 = \text{xtime}(75) = 0xEA$$
$$0xB7 • 0x08 = \text{xtime}(EA) = 0xCF$$
$$0xB7 • 0x10 = \text{xtime}(CF) = 0x85$$
$$0xB7 • 0x20 = \text{xtime}(85) = 0x11$$
$$0xB7 • 0x40 = \text{xtime}(11) = 0x22$$
$$0xB7 • 0x80 = \text{xtime}(22) = 0x44.$$

Now we exclusive-or the needed bytes to form the product. Since

$$0xA5 = 10100101 = 0x80 \oplus 0x20 \oplus 0x04 \oplus 0x01,$$

we have

$$0xB7 • 0xA5 = 0xB7 • (0x80 \oplus 0x20 \oplus 0x04 \oplus 0x01)$$
$$= 0xB7 • 0x80 \oplus 0xB7 • 0x20 \oplus 0xB7 • 0x04 \oplus 0xB7 • 0x01$$
$$= 0xB7 \oplus 0xEA \oplus 0x11 \oplus 0x44 = 0x08,$$

which is the same product we found in the previous example.

16.1.2   Word Arithmetic in Rijndael

Here is how Rijndael operates on thirty-two bit words.

Thirty-two bit words are regarded as four bytes, which are the coefficients of a polynomial of degree three with coefficients in $\mathbf{F}_{2^8}$, that is, cubic polynomials in $\mathbf{F}_{2^8}[x]$.

Addition of two 32-bit words is simple: Add them as polynomials. This is the same operation as exclusive-or'ing the coefficients and the same as exclusive-or'ing the two 32-bit words.

Multiplication of two 32-bit words is done by multiplying the polynomials modulo $M(x) = x^4 + 1$. This multiplication is denoted $\otimes$. If

$$a(x) = a_3 x^3 + a_2 x^2 + a_1 x + a_0$$

and

$$b(x) = b_3 x^3 + b_2 x^2 + b_1 x + b_0,$$

then

$$d(x) = a(x) \otimes b(x) = d_3 x^3 + d_2 x^2 + d_1 x + d_0$$

may be computed by

$$d_0 = a_0 \bullet b_0 \oplus a_3 \bullet b_1 \oplus a_2 \bullet b_2 \oplus a_1 \bullet b_3$$

$$d_1 = a_1 \bullet b_0 \oplus a_0 \bullet b_1 \oplus a_3 \bullet b_2 \oplus a_2 \bullet b_3$$

$$d_2 = a_2 \bullet b_0 \oplus a_1 \bullet b_1 \oplus a_0 \bullet b_2 \oplus a_3 \bullet b_3$$

$$d_3 = a_3 \bullet b_0 \oplus a_2 \bullet b_1 \oplus a_1 \bullet b_2 \oplus a_0 \bullet b_3$$

The reason this works is that multiplication of a cubic polynomial by x modulo $M(x)$ is equivalent to a circular left shift of the bytes of a 32-bit word:

$$xa(x) = x(a_3 x^3 + a_2 x^2 + a_1 x + a_0) =$$
$$= a_3 x^4 + a_2 x^3 + a_1 x^2 + a_0 x \equiv$$
$$\equiv a_2 x^3 + a_1 x^2 + a_0 x + a_3 \pmod{x^4 + 1}.$$

Example 16.4

Multiply 0xB7A5662F $\otimes$ 0x03010102 modulo $M(x) = x^4 + 1$.

We use the formulas above with $a_0 = $ 0x2F, $a_1 = $ 0x66, $a_2 = $ 0xA5, $a_3 = $ 0xB7, $b_0 = $ 0x02, $b_1 = $ 0x01, $b_2 = $ 0x01 and $b_3 = $ 0x03. In the formula for d_0 we have

$$a_0 \bullet b_0 = \text{0x2F} \bullet \text{0x02} = \text{0x5E}$$

$$a_3 \bullet b_1 = \text{0xB7} \bullet \text{0x01} = \text{0xB7}$$

$$a_2 \bullet b_2 = \text{0xA5} \bullet \text{0x01} = \text{0xA5}$$

$$a_1 \bullet b_3 = \text{0x66} \bullet \text{0x03} = \text{0xAA}$$

and so

$$d_0 = a_0 \bullet b_0 \oplus a_3 \bullet b_1 \oplus a_2 \bullet b_2 \oplus a_1 \bullet b_3$$
$$= \text{0x5E} \oplus \text{0xB7} \oplus \text{0xA5} \oplus \text{0xAA} = \text{0xE6}.$$

Similarly,

$$d_1 = \text{0xCC} \oplus \text{0x2F} \oplus \text{0xB7} \oplus \text{0xF4} = \text{0xA0}$$
$$d_2 = \text{0x51} \oplus \text{0x66} \oplus \text{0x2F} \oplus \text{0xC2} = \text{0xDA}$$
$$d_3 = \text{0x75} \oplus \text{0xA5} \oplus \text{0x66} \oplus \text{0x71} = \text{0xC7}.$$

Finally, $\text{0xB7A5662F} \otimes \text{0x03010102} = \text{0xC7DAA0E6}$.

16.1.3 The Structure of Rijndael

Rijndael has 10, 12 or 14 rounds, depending on the block and key lengths. The block length and key length can be chosen independently to be 128, 192 or 256 bits. Let Nb be the length of the block in 32-bit words (Nb = 4, 6 or 8). Let Nk be the length of the key in 32-bit words (Nk = 4, 6 or 8). Let Nr be the number of rounds. Then Nr = 14 if either Nb or Nk = 8. Otherwise, Nr = 12 if either Nb or Nk = 6. Finally, Nr = 10 if both Nb and Nk = 4.

Different parts of the Rijndael cipher operate on the intermediate result, called the **State**. The State is a rectangular array of bytes with four rows and Nb columns. The key begins as a rectangular array of bytes with four rows and Nk columns. The key is expanded and placed in an array W[Nb*(Nr+1)] of 32-bit words. We will describe the key expansion in the next section.

Each round of Rijndael consists of four different transformations or layers, expressed here in pseudo C code.

```
Round(State, RoundKey)
{
ByteSub(State);
ShiftRow(State);
MixColumn(State);
AddRoundKey(State, RoundKey);
}
```

The FinalRound omits the MixColumn. The plaintext is the initial State. The final State is the ciphertext. The complete Rijndael cipher consists of:

```
Rijndael(State, CipherKey)
{
KeyExpansion(CipherKey, ExpandedKey);
AddRoundKey(State, ExpandedKey);
For (i=1; i<Nr; i++) Round(State, ExpandedKey + Nb*i);
FinalRound(State, ExpandedKey + Nb*Nr);
}
```

Before we describe these transformations, we define a simple substitution cipher S on bytes. If a is a byte, compute $S(a)$ as follows.

1. First, if $a \neq 0$, take the multiplicative inverse of a in $\mathbf{F}_{2^8}$, that is, the inverse with respect to the $\bullet$ multiplication. Map $a = 0$ to itself. Label the bits of the resulting byte $x_7 x_6 x_5 x_4 x_3 x_2 x_1 x_0$.

2. Apply the affine transformation (over $\mathbf{F}_2$)

$$
\begin{bmatrix} y_0 \\ y_1 \\ y_2 \\ y_3 \\ y_4 \\ y_5 \\ y_6 \\ y_7 \end{bmatrix}
=
\begin{bmatrix}
1 & 0 & 0 & 0 & 1 & 1 & 1 & 1 \\
1 & 1 & 0 & 0 & 0 & 1 & 1 & 1 \\
1 & 1 & 1 & 0 & 0 & 0 & 1 & 1 \\
1 & 1 & 1 & 1 & 0 & 0 & 0 & 1 \\
1 & 1 & 1 & 1 & 1 & 0 & 0 & 0 \\
0 & 1 & 1 & 1 & 1 & 1 & 0 & 0 \\
0 & 0 & 1 & 1 & 1 & 1 & 1 & 0 \\
0 & 0 & 0 & 1 & 1 & 1 & 1 & 1
\end{bmatrix}
\begin{bmatrix} x_0 \\ x_1 \\ x_2 \\ x_3 \\ x_4 \\ x_5 \\ x_6 \\ x_7 \end{bmatrix}
+
\begin{bmatrix} 1 \\ 1 \\ 0 \\ 0 \\ 0 \\ 1 \\ 1 \\ 0 \end{bmatrix}.
$$

Then $S(a)$ is the byte $y_7 y_6 y_5 y_4 y_3 y_2 y_1 y_0$. Each of these two steps performs a permutation of bytes. There is an inverse function S^{-1} so that $S^{-1}(S(a)) = a$ for every byte a. In most implementations of Rijndael, the function S and its inverse are precomputed, so that they may be evaluated by table look-up during encryption.

Now we describe the round transformations.

ByteSub(State) transforms each byte a in the State by replacing it with $S(a)$.

ShiftRow(State) is a circular left shift of the rows in the State by various byte offsets which depend on Nb and on the row. The shift offsets are specified in this table.

Shift offsets for different rows and block lengths.

Nb	Row0	Row1	Row2	Row3
4	0	1	2	3
6	0	1	2	3
8	0	1	3	4

In MixColumn(State) the columns of the State are considered to be cubic polynomials with coefficients in $\mathbf{F}_{2^8}$ and each is multiplied ($\otimes$) modulo $M(x) = x^4 + 1$ with the fixed polynomial

$$c(x) = \mathtt{0x03}x^3 + \mathtt{0x01}x^2 + \mathtt{0x01}x + \mathtt{0x02}.$$

The polynomial $M(x) = x^4 + 1$ is not irreducible in $\mathbf{F}_{2^8}[x]$, so not all cubic polynomials are invertible modulo $M(x)$. However, the polynomial $c(x)$ is relatively prime to $x^4 + 1$ and therefore invertible. The inverse operation to MixColumn(State) is multiplication of each column by

$$d(x) = \mathtt{0x0B}x^3 + \mathtt{0x0D}x^2 + \mathtt{0x09}x + \mathtt{0x0E}.$$

AddRoundKey(State, RoundKey) is simply a byte-by-byte exclusive-or of State with RoundKey. The generation of RoundKey is described in the next section.

16.1.4 The Key Schedule of Rijndael

Here we tell how the key is expanded and the round keys are produced.

Recall that the key begins as a rectangular array of bytes with four rows and Nk columns. The key is expanded and placed in an array W[Nb*(Nr+1)] of 32-bit words.

The first Nk words of the array W are the key. Each subsequent word is the exclusive-or of the previous word and the word Nk words back in the array, except that words whose subscript is a multiple of Nk have the previous word transformed before the exclusive-or. When $Nk \leq 6$, the key expansion is described in this pseudo C code.

```
For (i = Nk; i < Nb*(Nr + 1); i++) {
temp = W[i - 1];
if (i % Nk == 0) temp = T(temp);
W[i] = W[i - Nk] xor temp;
}
```

Here xor is the exclusive-or operation $\oplus$ and $T(w)$ is a transformation of a word w described as follows. First, the bytes of w are rotated left one byte position. Next, S is applied to each of the four bytes. Finally, the high order byte is exclusive-or'ed with a byte representing the element $x^{(i/Nk)-1}$, where i is the loop variable i in the pseudocode.

The RoundKey used in the i-th AddRoundKey transformation consists of the Nb consecutive words of the key array W beginning with W[Nb*i].

16.1.5 Summary of Rijndael

Ideas from finite field theory are used to give a concise and elegant description of the substitution and transposition of bits and bytes that define Rijndael. It does not have the linearity that is the weakness of linear feedback shift registers. See Chapter 5 of Trappe and Washington [115] for another presentation of Rijndael.

Rijndael is by far the fastest cipher described in this chapter, which is one reason it was chosen as the new AES to replace DES. Implementations of the cipher in hardware can encipher and decipher at disk transfer speeds.

We did not describe the deciphering function for Rijndael, which is slightly slower and more complicated than the enciphering function. One reason it takes longer is that multiplication ($\otimes$) by $d(x)$ is slower than multiplication by $c(x)$. The interested reader can either derive the deciphering function or look it up on the Web.

16.2 The Pohlig-Hellman Cipher

The Pohlig-Hellman cipher is an example of an **exponentiation cipher**, one which uses exponentiation modulo a large number as its encryption function. Here is how an exponentiation cipher functions.

Choose a large integer n for modulus. Encode plaintext as blocks in $0 \leq M < n$. Encipher M as $C = E(M) = M^e \bmod n$. Decipher C as $M = D(C) = C^d \bmod n$.

This works, that is, $D(E(M)) = M$ for all M in $0 \leq M < n$, provided that $ed \equiv 1 \pmod{\phi(n)}$ since $M^{\phi(n)} \equiv 1 \pmod{n}$, by Euler's theorem. (Proof: Write $ed = t\phi(n) + 1$ for some integer t.) This implies that e and d are relatively prime to $\phi(n)$.

The **Pohlig-Hellman cipher** is *not* a public-key cipher. It is a symmetric cipher which is used in one of the two following ways.

Let $n = p = $ prime. Then $\phi(p) = p - 1$ and $ed \equiv 1 \pmod{p-1}$.

Method 1: Keep all of p, e, d secret. All three are the "key." There is just one user or one pair of users.

Method 2: Let p be public and keep e and d secret. The key is the pair (e, d). Each user has a secret pair to safeguard her personal secrets. Each pair of users who wish to communicate choose a common key pair.

Since it may take a while to generate a large prime, Method 2 is more commonly used than Method 1. Furthermore, Method 2 has interesting mathematical properties which foster its use in special ways discussed later (Massey-Omura, mental poker).

Here is the cryptanalysis. For a known-plaintext attack on Method 2, one is given a prime p, C and M, and must find an exponent e so that $C \equiv M^e \pmod{p}$, or equivalently, d so that $M \equiv C^d \pmod{p}$. These problems are instances of the discrete logarithm problem, in which one is given positive integers a, b and m, and must find x so that $a^x \equiv b \pmod{m}$. This is a well known difficult problem in number theory, and ways to solve it are discussed in Chapter 14.

16.3 Elliptic Curve Pohlig-Hellman

This cipher works just like the Pohlig-Hellman cipher except that the multiplicative group R_p of integers modulo p is replaced by an elliptic curve.

Let p be a large prime and let E be an elliptic curve modulo p that has order N, that is, E has N points including the identity ∞.

We will explain shortly how a plaintext block M might be embedded into the x-coordinate of a point P on E. Assume this has been done.

A point P on E is enciphered by adding it to itself e times, using fast multiplication; the ciphertext point is $Q = eP$. The latter is deciphered by multiplying by d: $P = dQ$.

In order for the deciphering to return to P, the multipliers e and d must satisfy $ed \equiv 1 \pmod{N}$, because $NP = \infty$ by Lagrange's theorem, and so $cP = P$ for any $c \equiv 1 \pmod{N}$. This implies that e (and d) must be chosen relatively prime to N. Of course, one should choose a random e in $1 < e < N$, relatively prime to N, and then compute d by the extended Euclidean algorithm.

By Hasse's theorem, N is approximately p. But this approximation is not good enough. We must know N exactly in order to choose e and d. In typical cryptographic applications, N and p must be large enough so only Schoof's algorithm is fast enough to compute N. Schoof's algorithm is complicated. If you have a program for it, then you are free to choose any elliptic curve E for the analogue of the Pohlig-Hellman cipher. Otherwise, you must choose an elliptic curve whose order has been published.

In a known-plaintext attack on this system, one is given E, p, N (which are public anyway), P and Q, and one must find e with $Q = eP$ or, equivalently, d with $P = dQ$. Either problem is the discrete logarithm problem on the elliptic curve E, whose solution is discussed in Chapter 14.

Now we deal with the matter of embedding plaintext into points. There are two methods in common use. Both embed a plaintext M in $0 < M < p$ into the x-coordinate of a point $P = (x, y)$ on a given elliptic curve E.

The first method is probabilistic and may fail to embed M with a positive probability. The overall encryption function must handle this failure gracefully. It may

1. skip M,

2. change M in some way, or

3. ask for human assistance in changing M.

In any case, it is easy to make the probability of failure minuscule. Let us reserve k bits of the x-coordinate for a small integer. Then the blocks M must be k bits shorter, that is, $0 < M < p/2^k$ rather than $0 < M < p$. The probability of failure will be only 1 chance in 2^{2^k}. This is less than one chance in a billion if $k = 5$. The x-coordinate will be $x = 2^k M + i$, where i is a k-bit integer $0 \leq i < 2^k$. When P is recovered during deciphering, M is extracted from x by $M = \lfloor x/2^k \rfloor$, which may be done quickly with a right shift of x by k bits. Let the elliptic curve have equation $y^2 \equiv x^3 + ax + b \pmod{p}$. Choose i by this algorithm:

```
for (i = 0 to 2^k - 1) {
        x = 2^k M + i
        if (((x^3 + ax + b)/p) = +1) { return i }
        }
return "Failure:  could not choose i"
```

The algorithm returns the first $i < 2^k$, if any, for which the Legendre symbol $((x^3 + ax + b)/p) = +1$. Since the Legendre symbol (r/p) is $+1$ for

$(p-1)/2$ values of r modulo p, and since, for each i, the value $x^3 + ax + b$ is more or less random modulo p, the probability that all 2^k choices for i yield $((x^3 + ax + b)/p) \neq +1$ is about 2^{-2^k}, as claimed.

Once we have i with $((x^3 + ax + b)/p) = +1$, where $x = 2^k M + i$, we find a square root y of x modulo p by the methods of Chapter 7 and let $P = (x, y)$. Then P lies on E.

The second method of embedding plaintext into points is deterministic but only works for special primes p and elliptic curves E. Plenty of primes and elliptic curves satisfy the requirements.

Assume that $p \equiv 3 \pmod 4$, that is, p is a Blum prime. For such primes p, -1 is a quadratic nonresidue, so $(-1/p) = -1$. Let $b = 0$ in the congruence defining E, so that E is $y^2 \equiv x^3 + ax \pmod p$.

Plaintext M is restricted to $0 < M < p/2$. Thus, one bit of possible plaintext storage space is lost. Given M, form $t = M^3 + aM \bmod p$. Since $(-1/p) = -1$, exactly one of t and $-t$ is a quadratic residue modulo p by Theorem 7.5. If $(t/p) = +1$, let $x = M$. If $(t/p) = -1$, let $x = p - M$. Then $((x^3 + ax)/p) = +1$ and we can find y with $y^2 \equiv x^3 + ax \pmod p$ by Theorem 7.13. Let $P = (x, y)$. When P is recovered during deciphering, look at x. If $x < p/2$, then $M = x$. If $x > p/2$, then $M = p - x$.

16.4   *Exercises*

1. This question concerns the 8-bit and 32-bit arithmetic operations used in Rijndael.

   a. Add the bytes 0xB3 $\oplus$ 0x95.

   b. Multiply the bytes 0xB4 $\bullet$ 0x4F.

   c. Find the inverse with respect to the $\bullet$ multiplication of the byte 0xB3.

   d. Multiply the 32-bit numbers (regarded as cubic polynomials over $\mathbf{F}_{2^8}$) 0x21A68490 $\otimes$ 0x03010102 (modulo $M(x)$, of course), just as Rijndael's MixColumn() procedure would multiply them. Your answer should be a 32-bit number, given as hexadecimal digits. (The low-order byte is the constant term of the cubic polynomial.)

   e. Show that $d(x)$ is the inverse of $c(x)$ by multiplying them modulo $M(x) = x^4 + 1$.

2. We mentioned that the polynomial $M(x) = x^4 + 1$ is not irreducible in $\mathbf{F}_{2^8}[x]$. Factor it.

3. The Pohlig-Hellman cipher with prime modulus $p = 2591$ and enciphering exponent $e = 13$ was used to encipher a secret message. Two-letter blocks were used. Note that the largest block would be 2525 (meaning ZZ) and this is less than p. Decipher the cipher text 1213 0902 0539 1208 1234 1103 1374.

Chapter 17

Public Key Ciphers

This chapter introduces several public key ciphers. Public-key ciphers are generally slower than private key ciphers. They are used for short communications, like a private key for a longer attached ciphertext. In contrast to private key ciphers, they do not require the exchange or establishment of a secret key before communication begins.

Another advantage of public key ciphers is that fewer keys are needed when many users wish to communicate with each other. If n people communicate with public key cryptography, then there are n public keys, one per person. If the same n people wished to use private key cryptography, then each would have to manage $n-1$ keys, one for each other user, and there would be a total of $\binom{n}{2} = n(n-1)/2$ keys.

17.1 Rivest-Shamir-Adleman

The Rivest-Shamir-Adleman public key cipher, RSA, [97] is another exponentiation cipher. See Section 16.2 for exponentiation ciphers.

Each user of RSA chooses two large primes p and q. She lets $n = pq$. She chooses a random e in $1 < e < n-1$ with $\gcd(e, \phi(n)) = 1$. Now $\phi(n) = \phi(pq) = (p-1)(q-1)$, so she calculates d so that $ed \equiv 1 \pmod{(p-1)(q-1)}$. She makes n and e public, but keeps d secret. The factors p and q are not needed after e and d are computed, but in any case should not be revealed. The encryption function is $E(M) = M^e \bmod n$, which anyone can compute, since n and e are public. The deciphering function is $D(C) = C^d \bmod n$.

Since n is public and one can easily compute d from e and the factors of n, a direct approach to breaking RSA is to factor n. Using the best currently-known methods, this is about as hard as solving a discrete logarithm problem with the same sized modulus. For a modulus n of 300 decimal digits, this is too hard for current algorithms and computers.

It may be possible to break RSA without factoring n. However, it is probably true that if you know an algorithm that can decipher any ciphertext C

with positive probability, then you can use the algorithm to factor n, but no one has ever proved this statement. If the hypothetical deciphering algorithm raises C to a fixed power modulo n, then, with high probability, you can factor n.

17.2   Massey-Omura

One can change the Pohlig-Hellman private-key cipher in Section 16.2 slightly to create a public-key cipher. This was done [70] by Massey and Omura. Their system is not used much because it is inefficient. (But the elliptic curve version is used.)

Consider a Pohlig-Hellman cipher with common prime p. This was called Method 2 in the previous chapter. Suppose users A and B have encryption functions E_A and E_B and decryption functions D_A and D_B. (So $E_A(M) = M^{e_A} \bmod p$ and $D_A(C) = C^{d_A} \bmod p$, where $e_A d_A \equiv 1 \pmod{p-1}$, etc.) Since the encryption and decryption functions all consist of exponentiation modulo a fixed modulus, they all *commute*, that is, they may be done in any order and give the same result. For example, $E_A(D_B(x)) = D_B(E_A(x))$ for every x because both are just $x^{e_A d_B} \equiv x^{d_B e_A} \pmod{p}$.

How do A and B use this property as a public-key cipher? The "public key" is the common prime modulus p. The private keys are *all* of the exponents (unlike RSA). If Alice wants to send a message $0 < M < p$ to Bob, she first sends $E_A(M)$ to Bob. Bob replies by sending $E_B(E_A(M))$ to Alice. Then Alice sends $D_A(E_B(E_A(M))) = E_B(D_A(E_A(M))) = E_B(M)$ to Bob. Bob deciphers the message by applying D_B to $E_B(M)$.

Note that this requires three messages to pass between Alice and Bob. This means that they must communicate in close to real time.

An eavesdropper would see the messages $r = E_A(M)$, $s = E_B(E_A(M))$ and $t = D_A(E_B(E_A(M))) = E_B(D_A(E_A(M))) = E_B(M)$ pass between Alice and Bob. If the eavesdropper could solve the discrete logarithm problem modulo p, then he could read M in either of two ways. First, $s \equiv r^{e_B} \pmod{p}$. He knows s, r and p. If he can solve for e_B, then he can compute d_B by the extended Euclidean algorithm. Then he can compute $M = t^{d_B} \bmod p$. The other way to read M is to use the congruence $s \equiv t^{e_A} \pmod{p}$ to find e_A, by solving a different discrete logarithm problem. Then compute d_A from e_A by the extended Eulcidean algorithm and find $M = r^{d_A} \pmod{p}$. It is likely that the two discrete logarithm problems are equally hard. But it is possible that one of the bases r, t might have a much smaller order modulo p than the other, and so produce an easier discrete logarithm problem. In a direct attack on this communication, one should try to solve both problems together. The two discrete logarithm problems are intertwined. It would be interesting to find an attack on both of them together, using information from each congruence to facilitate the solution of the other, so that the total effort is easier than solving either one separately.

17.3   Elliptic Curve Massey-Omura

Consider an elliptic curve Pohlig-Hellman cipher with elliptic curve E having N points modulo a prime p. Suppose users A and B have encryption functions E_A and E_B and decryption functions D_A and D_B. (So $E_A(P) = e_A P$ on E. $D_A(Q) = d_A Q$ on E, where $e_A d_A \equiv 1 \pmod{N}$, etc.) Since the encryption and decryption functions are all multiplication of integers times points on E, they all *commute*, that is, they may be done in any order and give the same result. For example, $E_A(D_B(P)) = D_B(E_A(P))$ for every P because both are just $e_A d_B P = d_B e_A P$.

How do A and B use this property as a public-key cipher? The "public key" consists of E, N, and p. The private keys are *all* of the multipliers e_A, d_A, etc. If Alice wants to send a message $0 < M < p$ to Bob, she first embeds it in a point P of E, as explained in the previous chapter. She sends $E_A(P)$ to Bob. Bob replies by sending $E_B(E_A(P))$ to Alice. Then Alice sends $D_A(E_B(E_A(P))) = E_B(D_A(E_A(P))) = E_B(P)$ to Bob. Bob deciphers the message by applying D_B to $E_B(P)$.

Note that this requires three messages to pass between Alice and Bob. This means that they must communicate in close to real time.

An eavesdropper would see the messages $R = E_A(P)$, $S = E_B(E_A(P))$ and $T = D_A(E_B(E_A(P))) = E_B(D_A(E_A(P))) = E_B(P)$ pass between Alice and Bob. If the eavesdropper could solve the discrete logarithm problem for points of E, then he could read P in either of two ways. First, $S = e_B R$. He knows S, R, E, N, and p. If he can solve for e_B, then he can compute d_B by the extended Euclidean algorithm. Then he can compute $P = d_B T$. The other way to read P is to use the equation $S = e_A T$ to find e_A, by solving a different discrete logarithm problem on E. Then compute d_A from e_A by the extended Eulcidean algorithm and find $P = d_A R$. It is likely that the two discrete logarithm problems are equally hard. But it is possible that one of the points R, T might have a much smaller order on E than the other, and so produce an easier discrete logarithm problem. In a direct attack on this communication, one should try to solve both problems together. The two discrete logarithm problems are intertwined. It would be interesting to find an attack on both of them together, using information from each congruence to facilitate the solution of the other, so that the total effort is easier than solving either one separately.

17.4   ElGamal

The **ElGamal public key cryptosystem** is defined as follows: Fix a large prime p which is public. Also public is a primitive root g modulo p in $1 < g < p$. Each user A who wishes to participate in this public-key cryptosystem chooses a secret a_A in $0 < a_A < p-1$ and publishes $b_A = g^{a_A} \bmod p$. When a user B wants to send a secret message M in $0 < M < p$ to A, she chooses a random k in $0 < k < p - 1$ and sends to A the pair

$$C = (g^k \bmod p, (Mb_A^k) \bmod p).$$

The plaintext M is enciphered by multiplying it by b_A^k in the second component of C. Note that $b_A^k \equiv (g^{a_A})^k \equiv g^{a_A k} \pmod{p}$. The first component of C provides a hint for deciphering M from the second component of C, but one which is useful only to A. Only A knows the secret key a_A, so only A can compute $(g^k)^{a_A} \equiv g^{a_A k} \pmod{p}$. If the multiplicative inverse of this number is multiplied times the second component, one recovers M:

$$\left(g^{a_A k}\right)^{-1} \left(Mb_A^k\right) \equiv \left(g^{a_A k}\right)^{-1} \left(Mg^{a_A k}\right) \equiv M \pmod{p}.$$

An eavesdropper who could solve the discrete logarithm problem modulo p could compute M from C and public data without knowing a_A as follows. The first component of C is $h = g^k \bmod p$. This number and $T = (Mb_A^k) \bmod p$ are observed by the eavesdropper. The eavesdropper knows p and g because these numbers are public. He can also obtain A's public key b_A from A's directory, just as B did. He would solve the discrete logarithm problem $g^k \equiv h \pmod{p}$ for k and then compute

$$T \left(b_A^k\right)^{-1} \equiv \left(Mb_A^k\right) \left(b_A^k\right)^{-1} \equiv M \pmod{p}.$$

17.5   Elliptic Curve ElGamal

There is an elliptic curve analogue to the ElGamal public key cryptosystem defined as follows: Fix an elliptic curve E modulo p and a point P_0 of large order on E. All of this data is public. Each user A who wishes to participate in this public-key cryptosystem chooses a secret a_A in $0 < a_A < p$ and publishes $P_A = a_A P_0$ on E. When a user B wants to send a secret message M to A, she first embeds M into a point P of E (explained in the previous chapter). She chooses a random k in $0 < k < p$ and sends to A the pair $C = (kP_0, kP_A + P)$.

The plaintext P is enciphered by adding the point kP_A in the second component of C. Note that $kP_A = k(a_A P_0) = (ka_A)P_0$. The first component of C provides a hint for deciphering P from the second component of C, but one which is useful only to A. Only A knows the secret key a_A, so only A can compute $a_A(kP_0) = (ka_A)P_0$. If this point is subtracted from the second component, one recovers P: $kP_A + P - (ka_A)P_0 = (ka_A)P_0 + P - (ka_A)P_0 = P$.

An eavesdropper who could solve the discrete logarithm problem on E could compute P from C and public data without knowing a_A as follows. The first component of C is $P_1 = kP_0$. This and $T = kP_A + P$ are observed by the eavesdropper. The eavesdropper knows p, E and P_0 because this information is public. He can also obtain A's public key P_A from A's directory, just as B did. He would solve the elliptic curve discrete logarithm problem $kP_0 = P_1$ for k and then compute $T - kP_A = (kP_A + P) - kP_A = P$.

17.6   Rabin-Williams

This is a public key cipher invented by Rabin [92]. Each user chooses two large Blum primes p and q, that is, $p \equiv q \equiv 3 \pmod 4$. The user publishes the product $n = pq$ as her public key. The factors p and q are her private key. Someone who wishes to send a plaintext M in $0 < M < n$ to the user encrypts M as $C = M^2 \bmod n$. The user, knowing the factors of n, can compute the four square roots of C modulo n by the methods of Chapter 7. If the original M was written in English, then, with high probability, only one of the four square roots of C will make sense, and this one is M.

If M were a binary string, then a standard header must be prepended to M before enciphering to allow the recipient to tell which square root is M. For example, one might use a two-bit number to indicate which square root is M: "00" means "the smallest one," "01" means "the second smallest one," etc. The enciphering function would have to compute all four square roots to determine this two-bit number.

Here is a slightly faster and more elegant way to fashion the two bits. During deciphering, the square roots of M^2 are first computed modulo p and modulo q, and then they are combined with the Chinese remainder theorem in the four possible ways. Let the bits indicate whether (i) $M \bmod p < p/2$ and whether (ii) $M \bmod q < q/2$. These bits are easy to compute during enciphering and they prevent unnecessary work during deciphering.

Aside from a lack of elegance, the downside of using two bits to distinguish a square root is that they provide a modicum of information about M to a cryptanalyst.

Recall that the RSA public key system is probably equivalent to factoring its modulus, but no one has proved this statement. This equivalence can be proved for the Rabin cipher.

THEOREM 17.1 Breaking Rabin's cipher is equivalent to factoring n
Breaking the Rabin cipher is equivalent to factoring its modulus n.

PROOF   Clearly, anyone who can factor n can decipher any message the same way the intended recipient can decipher it.

Breaking the Rabin cipher means having an algorithm that will decipher any message M in a reasonable time. In the first version of the cipher above, the algorithm would have to return all four square roots of C so that the human user could decide which one was meaningful. One could factor n by squaring an arbitrary x modulo n, using the algorithm to find the four square roots of x^2 modulo n, and picking one of them, say, $y \not\equiv \pm x \pmod n$. Then $\gcd(x + y, n) = p$ or q, by Theorem 13.1, and n has been factored.

In the versions of the cipher with two extra bits, square an x modulo n and use the algorithm one, two or three times, with different two-bit numbers, until it gives you a $y \not\equiv \pm x \pmod n$. Then factor n as before. ∎

Williams [122] improved Rabin's cipher by eliminating the ambiguity in deciphering without adding a two-bit number. Williams' cipher also has the property that breaking it is provably equivalent to factoring the modulus. In his scheme, the user chooses large primes $p \equiv 3 \pmod 8$ and $q \equiv 7 \pmod 8$. Let $n = pq$. Then $n \equiv 5 \pmod 8$. Let $d = ((p-1)(q-1)/4 + 1)/2$, which is an integer. The public key is n. The secret key is d. The primes p and q are not needed after d is computed. They may be discarded, and certainly should not be revealed. Let (r/n) denote the Jacobi symbol.

The set of allowed plaintext M is not all of $0 < M < n$. Let $\mathcal{M}$ be the set of all positive integers M such that $2(2M+1) < n$ when $((2M+1)/n) = -1$ and $4(2M+1) < n$ when $((2M+1)/n) = +1$. Only M in $\mathcal{M}$ are allowed to be plaintext. This set includes all M in $0 < M < n/8 - 1$ and some larger M.

Williams defines the following functions for encryption and decryption.

For $M \in \mathcal{M}$, let

$$E_1(M) = \begin{cases} 4(2M+1), & \text{when } ((2M+1)/n) = 1, \\ 2(2M+1), & \text{when } ((2M+1)/n) = -1. \end{cases}$$

We have $(2/n) = -1$ because $n \equiv 5 \pmod 8$, so $(E_1(M)/n) = 1$ for every M. It is possible for $((2M+1)/n) = 0$, but this event is so unlikely that we ignore it. Note that the Jacobi symbols are easy to compute.

Define $E_2(N) = N^2 \bmod n$. This is Rabin's enciphering function.

Define $D_2(C) = C^d \bmod n$.

Finally, let

$$D_1(L) = \begin{cases} (L/4 - 1)/2, & \text{when } L \equiv 0 \pmod 4, \\ ((n-L)/4 - 1)/2, & \text{when } L \equiv 1 \pmod 4, \\ (L/2 - 1)/2, & \text{when } L \equiv 2 \pmod 4, \\ ((n-L)/2 - 1)/2, & \text{when } L \equiv 3 \pmod 4. \end{cases}$$

THEOREM 17.2 Williams' version of Rabin's cipher works
If $M \in \mathcal{M}$, then $D_1(D_2(E_2(E_1(M)))) = M$.

If we define $E(M) = E_2(E_1(M))$ and $D(C) = D_1(D_2(C))$, then the theorem says that for every $M \in \mathcal{M}$, we have $D(E(M)) = M$. The enciphering function E is easy to compute by anyone who knows the public key n. The deciphering function D is easy to compute by anyone who knows the public key n and the secret key d. The proof of Theorem 17.2 requires one lemma.

LEMMA 17.1
If $n = pq$, where p and q are distinct Blum primes, and $(M/n) = 1$, then

$$M^{(p-1)(q-1)/4} \equiv \pm 1 \pmod n.$$

PROOF   Since $(M/(pq)) = 1$, we have $(M/p) = (M/q)$, by definition of the Jacobi symbol.

Suppose first that $(M/p) = (M/q) = 1$. By Euler's criterion, $M^{(p-1)/2} \equiv 1 \pmod{p}$ and $M^{(q-1)/2} \equiv 1 \pmod{q}$. Hence, $M^{(p-1)(q-1)/4} \equiv 1 \pmod{p}$ and $M^{(p-1)(q-1)/4} \equiv 1 \pmod{q}$ and so $M^{(p-1)(q-1)/4} \equiv 1 \pmod{pq}$.

Suppose now that $(M/p) = (M/q) = -1$. By Euler's criterion, $M^{(p-1)/2} \equiv -1 \pmod{p}$ and $M^{(q-1)/2} \equiv -1 \pmod{q}$. Since $(p-1)/2$ and $(q-1)/2$ are odd numbers (because $p \equiv q \equiv 3 \pmod 4$), we have $M^{(p-1)(q-1)/4} \equiv -1 \pmod{p}$ and $M^{(p-1)(q-1)/4} \equiv -1 \pmod{q}$ and so $M^{(p-1)(q-1)/4} \equiv -1 \pmod{pq}$.   ∎

Now we prove Theorem 17.2.

PROOF   Let $M \in \mathcal{M}$. Let $N = E_1(M)$. Then N is even and $0 < N < n$. We have $(N/n) = 1$ because $(2/n) = -1$.

Let $L = D_2(E_2(N))$. Then

$$L \equiv (N^2)^d \equiv N^{2d} \equiv N^{(p-1)(q-1)/4+1} \equiv \pm N \pmod n$$

by Lemma 17.1. Also, $0 < L < n$. Therefore, since N is always even, if L is even, then $L = N$, while if L is odd, then $L = n - N$.

If $L \equiv 0 \pmod 4$, then $2M + 1 = N/4$ and so $M = (L/4 - 1)/2 = D_1(L)$. We leave the other three cases of L modulo 4 to the reader.   ∎

THEOREM 17.3 *Breaking Williams' cipher is equivalent to factoring n*
If there is an efficient algorithm A such that for every C of the form $C = E(M)$ for some $M \in \mathcal{M}$, A can compute M given C, then there is an efficient algorithm for factoring the modulus n.

For a proof, see Williams [122].

17.7   *Exercises*

1. Alice uses $n = 2581$ and $e_A = 107$ for her public RSA key. How would Bob encipher $M = 1619$ to send to Alice? Decipher the ciphertext $C = 1674$, which Alice received from Chuck.

2. Alice uses the "double RSA" cipher. She makes public a modulus n, which is the product of two secret primes, and *two* public encryption exponents, e_1 and e_2. She tells people to encipher messages M in $0 < M < n$ to her by computing $C_1 = M^{e_1} \bmod n$ and then $C = C_1^{e_2} \bmod n$ and sending just C to her.

   a. Tell how Alice deciphers C, using her knowledge of the secret prime factors of n.

b. Is there an easy way to factor n, given e_1 and e_2?

c. Is the "double RSA" cipher more secure, less secure or just as secure as the regular RSA cipher with the same modulus n but only one encryption exponent?

d. Chuck got Alice's instructions confused, and enciphered a message M for Alice using e_1 and e_2 in the reverse order. What happened when Alice, unaware of Chuck's error, tried to decipher the ciphertext using her usual procedure? Did she get M or nonsense? If nonsense, could she recover M anyway?

3. Alice and Bob use the Massey-Omura cipher with common modulus $p = 2591$. Alice's secret enciphering exponent is $e_A = 107$; Bob's is $e_B = 257$. Compute the deciphering exponents and show the numbers passed between them when Alice sends Bob the plaintext $M = 1234$.

4. Alice and Bob use the elliptic curve Massey-Omura cipher with the elliptic curve $y^2 \equiv x^3 + 1441x + 611 \pmod{2591}$. Alice's secret enciphering multiplier is $e_A = 107$; Bob's is $e_B = 257$.

a. Find the number of points on the elliptic curve.

b. Compute the deciphering multipliers d_A and d_B.

c. Show the numbers in the messages passed between them when Alice sends Bob the plaintext $P = (1619, 2103)$.

5. A simple version of the ElGamal cipher uses the public common modulus $p = 97$ and the primitive root $g = 5$. Alice participates in this ElGamal system and uses $e_A = 37$ as her secret key and $b_A = g^{e_A} \bmod p = 56$ as her public key. How would Bob encipher $M = 82$ to send to Alice if he chose $k = 75$ for the random number? Show how Alice would decipher the ciphertext $(7, 84)$, which she received from Chuck.

6. Alice uses the Rabin-Williams cipher with public modulus $n = 11021$.

a. What ciphertext would Bob send to Alice if the plaintext is $M = 678$?

b. Factor n via Fermat's difference of squares method.

c. Find Alice's deciphering exponent d.

d. Decipher the ciphertext $C = 6525$, which Alice received from Chuck.

7. Finish the proof of Theorem 17.2 by showing that $D_1(L) = M$ for the three remaining cases of L modulo 4.

Chapter 18

Signature Algorithms

This chapter defines several signature algorithms. These are methods of "signing" messages to show their authenticity.

18.1 Rivest-Shamir-Adleman Signatures

RSA has no direct authentication: Anyone can send any message to you and claim it came from anyone. However, one can sign RSA [97] messages as follows.

Use the same notation for enciphering and deciphering functions as we did for Massey-Omura: E_A, D_B, etc. Alice can sign (and encipher) a message M to Bob by sending $C = E_B(D_A(M))$ to Bob. Bob can decipher C by applying D_B to it (to get $D_A(M)$) and then check the signature by applying E_A to the latter.

Note that Bob's cipher algorithms do not commute with Alice's because the modulus is different. Thus the order in which Bob applies the operations to C matters: Bob must use D_B first and then E_A second.

There is another problem caused by the different moduli. The functions D_A and E_A perform arithmetic modulo Alice's modulus n_A while E_B and D_B perform arithmetic modulo Bob's modulus n_B. This works fine if $n_A < n_B$, but part of the message will be lost if $n_A > n_B$.

There are three ways to solve this problem:

1. Re-block the message after D_A is applied.

2. Enforce an arbitrary threshold T and let every RSA user A have two complete sets of RSA keys, one with $n_{A_1} < T$ and one with $n_{A_2} > T$. The keys with the smaller modulus n_{A_1} are used for signing messages from A and the keys with the larger modulus n_{A_2} are used to encipher messages going to A.

3. A more elegant solution is for Alice to sign (and encipher) a message M to Bob by sending $C = E_B(D_A(M))$ to Bob when $n_A < n_B$, and by sending $C = D_A(E_B(M))$ to Bob when $n_A > n_B$. In either case, Bob undoes these operations in reverse order.

In the third method, what if Alice later denies sending M, and Bob goes to an independent judge to prove that M bears Alice's signature? In the first case $(n_A < n_B)$, Bob gives the judge M and $X = D_B(C)$, the judge computes $M' = E_A(X)$ and tests whether $M' = M$. If so, the judge rules that Alice signed M. In the second case $(n_A > n_B)$, Bob gives the judge M and C, the judge computes $X' = E_B(M)$ and $X' = E_A(C)$ and tests whether $X' = X$. If so, the judge rules that Alice signed M.

There is a trick that speeds RSA signature generation by a factor of four. Suppose the modulus is $n = pq$, where the primes p and q have about the same length. Let b be the number of bits in n, so that the length of p and q is about $b/2$ bits. If the decryption exponent is d, the plaintext M is signed as $D(M) = M^d \bmod n$. According to Theorems 6.2 and 3.5, this fast exponentiation takes about cb^3 bit operations, for some constant $c > 0$. The trick replaces this fast exponentiation by two fast exponentiations with b replaced by $b/2$. Let $M_p = M \bmod p$, $M_q = M \bmod q$, $d_p = d \bmod (p-1)$ and $d_q = d \bmod (q-1)$. The length of each of these four numbers is about $b/2$ bits. Compute $S_p = M_p^{d_p} \bmod p$ and $S_q = M_q^{d_q} \bmod q$ by fast exponentiation. Each of these exponentiations takes about $c(b/2)^3 = (c/8)b^3$ bit operations. Now the signature $D(M) \equiv S_p \pmod{p}$ and $D(M) \equiv S_q \pmod{q}$, so $D(M)$ can be computed from S_p and S_q by the Chinese remainder theorem. In the application of the Chinese remainder theorem, the inverses $p^{-1} \bmod q$ and $q^{-1} \bmod p$ may be precomputed. The result is that $D(M) = (aS_p + bS_q) \bmod n$ where a and b are precomputed constants. The total number of bit operations is essentially $2(c/8)b^3 = cb^3/4$, which is one-fourth as many as for computing $D(M) = M^d \bmod n$ directly. The same trick can be used in deciphering RSA messages, too, of course. But it can't be used to accelerate RSA encryption because p and q must be kept secret.

18.2 ElGamal Signatures

In addition to encryption, as explained in Section 17.4, one can sign messages with the ElGamal scheme. The security depends on the difficulty of the discrete logarithm problem.

All users have a common large prime p and a primitive root g for p. These numbers are public. Each user chooses a secret x in $1 < x < p-2$ and publishes $y = g^x \bmod p$. Thus, the public key is p, g and y, while the private key is x. We explained in Section 17.4 how to encipher a message to the user with public key y.

This user can sign a plaintext M as follows. Choose a random k in $1 <$

$k < p - 1$ and relatively prime to $p - 1$, and let $a = g^k \bmod p$. Then solve the congruence $kb \equiv M - xa \pmod{p-1}$ for b by Theorem 5.7. This congruence can be solved since $\gcd(k, p-1) = 1$. The signature for M is the pair (a, b).

The recipient verifies the signature by checking whether $y^a a^b \equiv g^M \bmod p$. Since $y \equiv g^x \pmod p$ and $a \equiv g^k \pmod p$, this is equivalent to $g^{xa} g^{kb} \equiv g^M \pmod p$. By Theorem 6.3, this will hold provided $xa + kb \equiv M \pmod{p-1}$, which is equivalent to the congruence defining b. Hence the signature verification will succeed if the signature was constructed according to the rules above. It is reasonable to call (a, b) a "signature" because it is hard to find a and b satisfying $y^a a^b \equiv g^M \bmod p$ without knowing x.

The random number k must not be revealed since it would allow one to compute the secret key x from the congruence $xa + kb \equiv M \pmod{p-1}$.

Example 18.1

Suppose $p = 19$, $g = 2$ and $x = 7$. Then $y = g^x \bmod p = 14$. To sign a message $M = 14$, the user would choose a random $k = 13$, say, and let $a = g^k \bmod p = 3$. Solve for b in $13b = kb \equiv M - xa \equiv 14 - 7 \cdot 3 \equiv 11 \pmod{p-1}$ to get $b = 5$. The signature for $M = 14$ is the pair $(3, 5)$. To verify the signature, one checks the congruence

$$6 \equiv 2^{14} \equiv g^M \equiv a^b y^a \equiv 3^5 14^3 \equiv 15 \cdot 8 \equiv 6 \pmod{19}.$$

18.3 Rabin-Williams Signatures

Recall the discussion of the Rabin-Williams cipher in Section 17.6. We will use the same notation in this section.

Theorem 17.2 says that if $M \in \mathcal{M}$, then $D_1(D_2(E_2(E_1(M)))) = M$. Since both of the functions E_2 and D_2 are exponentiations modulo n, they commute: $E_2(D_2(C)) = C^{2d} \bmod n = D_2(E_2(C))$. Therefore, from Theorem 17.2, we obtain the corollary that if $M \in \mathcal{M}$, then $D_1(E_2(D_2(E_1(M)))) = M$.

This corollary can be used to produce signatures as follows.

Suppose Alice uses $E_A = E_{2A} E_{1A}$ and $D_A = D_{1A} D_{2A}$ as her enciphering and deciphering functions. Let Bob use the corresponding functions $E_B = E_{2B} E_{1B}$ and $D_B = D_{1B} D_{2B}$. If Alice wishes to sign and encipher a message M to Bob, she computes the signature $S = D_{2A}(E_{1A}(M))$ and sends $C = E_B(S)$ to Bob. The mail header tells Bob that this is a signed message from Alice.

Bob deciphers C by computing $L = D_B(C)$. He finishes deciphering it and checks Alice's signature by computing $D_{1A}(E_{2A}(L)) = M$. Since only Alice knows D_{2A}, only she could have signed it. Only Alice could compute a ciphertext which would decipher through E_{2A} into a meaningful message. The reason this technique works is the corollary mentioned above.

As with RSA signatures, there may be a problem with the relative size of the moduli used in E_{2A} and E_{2B}. The signature S might have to be reblocked if it is too big for E_B. Reblocking can be avoided if a threshold T is enforced, and each user has two sets of Rabin-Williams enciphering and deciphering algorithms, one with a modulus below T and one with a modulus above T. Because of the limited message space $\mathcal{M}$ of the Rabin-Williams cipher, the second modulus should exceed $8T + 1$.

18.4   The Digital Signature Algorithm

The Digital Signature Standard, DSS, uses the Digital Signature Algorithm, DSA, to sign the output of hash functions. Compare this with signing a hash function with RSA.

DSA is a variation of signature schemes of ElGamal and Schnorr.

Here is the notation for DSA.

Let L be a multiple of 64 in the range $512 \leq L \leq 1024$, p be a prime of L bits, that is $2^{L-1} < p < 2^L$ and q be a 160-bit prime which divides $p - 1$. Let h be a primitive root modulo p in the interval $1 < h < p - 1$ and $g = h^{(p-1)/q} \bmod p$. Then g has order q modulo p.

DSA assumes that discrete logarithms modulo p are hard to compute.

Several people will use p, q, g as a global public key. Each user of the DSA chooses a secret private key x in $1 < x < q$ and publishes a public key $y = g^x \bmod p$. Each time a user wants to sign a message M, she chooses a secret random number k in $1 < k < q$ and computes SHA of M, called $h(M)$ below.

Alice signs message M with the pair r, s, where $r = (g^k \bmod p) \bmod q$, and $s = [k^{-1}(h(M) + xr)] \bmod q$.

If Bob receives the message M' with signature r', s' from Alice, he verifies her signature by computing $w = (s')^{-1} \bmod q$, $u_1 = [h(M')w] \bmod q$, $u_2 = (r')w \bmod q$, $v = [(g^{u_1}y^{u_2}) \bmod p] \bmod q$, and making the test, "Does $v = r'$?" If this equality holds, then Bob accepts that $M' = M$ is a message actually sent to him by Alice. Note that y is Alice's public key.

Why does the DSA work? That is, assuming that the message and signature are received correctly (so $M' = M$, $r' = r$ and $s' = s$), why should $v = r$? The following three lemmas and theorem prove that this equality should hold. Note that r doesn't even depend on M.

LEMMA 18.1

If $a \equiv b \pmod{q}$, then $g^a \equiv g^b \pmod{p}$.

PROOF   Write $a = b + qt$, where t is an integer. Then

$$g^a = g^{b+qt} = g^b g^{qt} \equiv g^b \cdot 1 = g^b \pmod{p}$$

because $g^q \equiv (h^{(p-1)/q})^q = h^{p-1} \equiv 1 \pmod{p}$, by Fermat's little theorem.

∎

LEMMA 18.2
With the notation above, $y^{((rw) \bmod q)} \equiv g^{((xrw) \bmod q)} \pmod{p}$.

PROOF By definition, $y = g^x \bmod p$, so

$$y^{((rw) \bmod q)} \equiv g^{x((rw) \bmod q)} \pmod{p}.$$

Lemma 18.2 follows from Lemma 18.1 and the fact that

$$x((rw) \bmod q) \equiv ((xrw) \bmod q) \pmod{q}.$$

∎

LEMMA 18.3
With the notation above, $((h(M) + xr)w) \bmod q = k$.

PROOF By definition, $w = s^{-1} \bmod q$ and $s = [k^{-1}(h(M) + xr)] \bmod q$. Therefore, $1 \equiv ws \equiv wk^{-1}(h(M) + xr) \bmod q \pmod{q}$. Since q is prime and q does not divide k, $k \equiv w(h(M) + xr) \pmod{q}$. The lemma follows because $1 < k < q$. ∎

THEOREM 18.1 The Digital Signature Algorithm works
If M is unchanged and really came from Alice, then $v = r$.

PROOF Using the definition of v and then those of u_1 and u_2, we find

$$v = ((g^{u_1} y^{u_2}) \bmod p) \bmod q$$

$$v = (g^{(h(M)w) \bmod q} \cdot y^{(rw) \bmod q} \bmod p) \bmod q.$$

Lemma 18.2 allows us to replace y by g:

$$v = (g^{(h(M)w) \bmod q} \cdot g^{(xrw) \bmod q} \bmod p) \bmod q$$

$$v = (g^{(h(M)w) \bmod q + (xrw) \bmod q} \bmod p) \bmod q.$$

Lemma 18.1 lets us combine terms in the exponent:

$$v = (g^{(h(M)w + xrw) \bmod q} \bmod p) \bmod q$$

$$v = (g^{((h(M) + xr)w) \bmod q} \bmod p) \bmod q.$$

Now use Lemma 18.3 and the definition of r:

$$v = (g^k \bmod p) \bmod q = r.$$

∎

18.5 *Exercises*

1. Show that the trick for speeding RSA signature generation produces the correct signature, and determine the constants a and b in the formula $D(M) = (aS_p + bS_q) \bmod n$.

2. Can the trick for speeding RSA signature generation be modified slightly to accelerate Rabin-Williams signature generation? Explain your answer.

3. Consider the following simple signature algorithm which is like DSA except that it does not require a secret random number.

   The public elements are a prime q and a primitive root g for q. There is a private key x in $1 < x < q$ and a public key $y = g^x \bmod q$.

   To sign a message M, compute $h = h(M)$ for some hash function h. We require that $\gcd(h, q-1) = 1$. If this is not so, then append the hash to the message and compute a new hash. Continue this process until a hash h is computed which is relatively prime to $q-1$. Then compute z satisfying $zh \equiv x \pmod{(q-1)}$. The signature for M is $s = g^z \bmod q$. The signature is verified by checking whether $s^h \equiv y \pmod q$.

   a. Show that the latter congruence will hold, provided the signature is valid.

   b. Show that the scheme is unacceptable by describing a simple technique for forging a user's signature on an arbitrary message.

4. Consider the following signature algorithm. The public elements are a prime q and a primitive root g for q. Everyone knows and uses the same q and g. Alice has a secret key x in $1 < x < q$ and a public key $y = g^{-x} \bmod q$. Let h be a standard hash function. The length in bits of the prime q is greater than the length of the output of h.

   To sign a message M, Alice generates a random integer k in $1 < k < q-1$ and computes $g^k \bmod q$. Then she computes $c = h(g^k \bmod q; M)$, where ";" means concatenation of bit strings. Finally, Alice computes $t = (k + cx) \bmod (q-1)$. She sends the pair (t, c) as the signature of M.

   When Bob receives M and the alleged signature (t, c), he obtains Alice's public key y from a secure site and computes $a = g^t y^c \bmod q$ and tests whether $h(a; M) = c$. If these are equal, then Bob accepts the signature; otherwise he rejects it.

   a. Show that equality will hold, provided the signature is valid.

   b. Compare the efficiency of this signature algorithm with that of DSA.

5. Design an elliptic curve variation of the Digital Signature Algorithm.

Chapter 19

Key Exchange Algorithms

This chapter discusses key exchange algorithms, which are protocols for two users, Alice and Bob, to agree on a common key or to learn each other's keys using a communication channel, like the Internet, which may have eavesdroppers or even malicious users who masquerade as others.

19.1 Key Exchange Using a Trusted Server

Here we assume there is a trusted server, Tracy, who helps the other parties choose a common key K to a symmetric cipher E_K.

In the first two protocols, Tracy shares a secret key with Alice and a different secret key with Bob. A message enciphered with the symmetric cipher E_A can be read only by Alice and Tracy. Likewise, the symmetric cipher E_B is used only for secret communication between Tracy and Bob. These secret keys were chosen when Tracy met with each party before the protocol begins. These functions are used only for key distribution and not to send messages between Alice and Bob.

In these protocols, A represents Alice's name and B represents Bob's name.

The first protocol of this type is called **Wide-mouthed Frog**. It appeared in [21] and is about as simple as such a protocol can be.

1. Alice generates a current time stamp t_A and a common secret key K. She sends the message $A, E_A(t_A, B, K)$ to Tracy.

2. Tracy knows that the message came from Alice because she sees Alice's name A in plaintext. She deciphers it with E_A and finds Bob's name B. She checks that the time stamp t_A is current to ensure that it is not being replayed by a malicious user. She makes a new current time stamp t_B and sends the message $E_B(t_B, A, K)$ to Bob.

3. Bob receives the message, deciphers it, checks that t_B is current, sees Alice's name and begins communicating with her using the secret key K.

No one could pretend to be Alice in Step 1 because they would not know E_A. No one could pretend to be Tracy in Step 2 because they would not know E_A. No one could pretend to be Bob in Step 3 because they would not know E_B. The messages could not be replayed later because the enciphered time stamps in them are generated and checked. No eavesdropper could learn K because it is enciphered in transit. Tracy could betray Alice and/or Bob in many ways, but they trust her. The most likely attack would be on Alice's random key generator. If it were badly designed, an eavesdropper might be able to guess K.

The next protocol is called **Yahalom**. It also appeared in [21]. This protocol and the previous one are mentioned in [22], which presents a way of analyzing the security of protocols like these using rules of inference as in mathematical logic.

1. Alice generates a random number, r_A, called a **nonce** because it is used just for this occasion. She sends the message A, r_A to Bob.

2. Bob receives the message, generates his own random number, r_B, and sends Tracy the message $B, E_B(A, r_A, r_B)$.

3. Tracy receives Bob's message, deciphers it, generates a random secret key K for Bob and Alice to use, and sends Alice the pair of enciphered messages $E_A(B, K, r_A, r_B), E_B(A, K)$.

4. Alice deciphers the first message and checks that r_A is the same nonce she created in Step 1. If it is, she sends Bob the message $E_B(A, K)$ and the message $E_K(r_B)$, which is encrypted with the session key K.

5. Bob decrypts $E_B(A, K)$, obtains K, decrypts $E_K(r_B)$ and checks that r_B is the same number he created in Step 2. Then he begins communicating with Alice using the secret key K.

Although Alice sends her nonce in plaintext in Step 1, it is enciphered in Steps 2 and 3. An eavesdropper could discover r_A, but would gain nothing from this knowledge because the eavesdropper could not forge the messages in Steps 2 and 3. Carol could pretend to be Alice in Step 1. If Carol managed to intercept the messages Tracy sent to Alice in Step 3, she would not be able to decrypt them, and so she could not perform Step 4. If Carol or someone else recorded and replayed messages from the protocol, they would not be accepted as genuine because the nonces would be wrong. Note how the nonces here play the role of the time stamps in the Wide-mouthed Frog protocol. No eavesdropper could learn K because it is enciphered in transit. An interesting feature of this protocol is that, although Alice initiates it, only Bob contacts Tracy.

The next two key exchange protocols use public-key cryptography. Here E_A, E_B and E_T are the public encryption functions of Alice, Bob and Tracy, respectively. Let K_A, K_B and K_T be the respective public keys. Likewise, D_A, D_B and D_T are the private decryption functions of Alice, Bob and Tracy, respectively. They are used also for signatures. Tracy maintains a database containing everyone's public keys, obtained securely before the protocol begins. Everyone knows Tracy's public key K_T, so everyone can verify Tracy's signature. Alice and Bob may learn each other's public keys during the protocol, but they communicate later using a symmetric cipher with a random key K created during the protocol. Symmetric ciphers are much faster than public-key ciphers.

Here is the key exchange protocol of Denning and Sacco [37].

1. Alice tells Tracy her identity and Bob's in the message A, B.

2. Tracy sends Alice Bob's public key and Alice's own public key, both signed. That is, she sends Alice the message $D_T(B, K_B), D_T(A, K_A)$.

3. Alice verifies the signatures. She chooses a random session key K and a current time stamp t_A. She signs these two numbers and enciphers them with Bob's public key. She sends this message, $E_B(D_A(K, t_A))$, to Bob together with the two messages she received from Tracy.

4. Bob verifies the signatures on the messages that came from Tracy via Alice. He uses his private key to decipher the message that originated with Alice and checks her signature using her public key, which he extracts from $D_T(A, K_A)$. If the time stamp t_A is still valid, he begins communicating with Alice using the symmetric cipher with key K.

An eavesdropper could learn from Step 1 that Alice wanted to communicate secretly with Bob. The eavesdropper could learn Alice and Bob's public keys from Step 2. But the eavesdropper could not decipher $E_B(D_A(K, t_A))$ because he would not know D_B. Hence the eavesdropper could not discover K or decipher the rest of the communication between Alice and Bob. There would be no point to replaying $E_B(D_A(K, t_A))$ later because its time stamp would be valid only for a short time.

Here is another key exchange protocol, due to Woo and Lam [129] and [130]. It uses nonces instead of time stamps.

1. Alice sends the message A, B to Tracy.

2. Tracy signs Bob's public key and sends it to Alice as the message $D_T(K_B)$.

3. Alice verifies Tracy's signature on the message. She chooses a nonce r_A and sends it with her name to Bob, enciphered with his public key: $E_B(A, r_A)$.

4. Bob sends Tracy Alice's name, his name and Alice's nonce enciphered with Tracy's public key: $A, B, E_T(r_A)$.

5. Tracy chooses a random secret key K for Alice and Bob to use in the symmetric cipher. Tracy sends Bob two messages. The first is $D_T(K_A)$, Alice's public key, signed by Tracy. The second is $E_B(D_T(r_A, K, A, B))$, which contains Alice's nonce r_A, Alice's name, and Bob's name, all signed by Tracy and enciphered with Bob's public key.

6. Bob deciphers the second message using D_B and verifies the signatures on both messages. Then he chooses a nonce r_B and sends Alice the signed second message from Step 5 and the new nonce, all enciphered with Alice's public key; that is, he sends Alice the message $E_A(D_T(r_A, K, A, B), r_B)$.

7. Alice deciphers the message using D_A. She verifies Tracy's signature and checks that r_A is the same nonce she chose in Step 3. Then she sends Bob his nonce enciphered with the session key K, $E_K(r_B)$.

8. Bob deciphers the message and checks that r_B is the same nonce he chose in Step 6.

An eavesdropper could learn from Step 1 that Alice wanted to communicate secretly with Bob. The eavesdropper could learn Alice's public key in Step 2 and Bob's public key in Step 5. But the eavesdropper could not decipher any of the enciphered messages, and so could not see the session key K or either nonce. Hence the eavesdropper could not decipher the rest of the communication between Alice and Bob. There would be no point to replaying any enciphered message because of the nonces in them.

19.2 The Diffie-Hellman Key Exchange

This protocol allows two users to choose a common secret key, for a symmetric cipher like DES or Rijndael, say, while communicating over an insecure channel, without the aid of a trusted third party.

The two users agree on a common large prime p and a constant value g, probably a primitive root, which may be publicly known and available to everyone. The algorithm is most secure when the order of g modulo p is large.

Alice secretly chooses a random x_A in $0 < x_A < p - 1$ and computes $y_A = g^{x_A} \bmod p$. Bob secretly chooses a random x_B in $0 < x_B < p - 1$ and computes $y_B = g^{x_B} \bmod p$.

Alice sends y_A to Bob. Bob sends y_B to Alice. An eavesdropper, knowing p and g, and seeing y_A and y_B, cannot compute x_A or x_B from this data unless he can solve the discrete logarithm problem quickly.

Alice computes $K_A = y_B^{x_A} \bmod p$. Bob computes $K_B = y_A^{x_B} \bmod p$. Then

$$K_A \equiv y_B^{x_A} \equiv (g^{x_B})^{x_A} \equiv g^{x_A \cdot x_B} \equiv (g^{x_A})^{x_B} \equiv y_A^{x_B} \equiv K_B \pmod{p}$$

and $0 < K_A, K_B < p$, so $K_A = K_B$.

Alice and Bob choose certain agreed-upon bits from K_A to use as their key for a single-key cipher like DES or Rijndael.

Although this protocol provides secure communication between Alice and whomever is at the other end of the communication line, it does not prove that Bob is the other party. To guarantee that Bob is at the other end, they would have to use a signature system like RSA or one of the protocols in the previous section.

There is an elliptic curve variation of the algorithm in which the group R_p is replaced by an elliptic curve. In it, Alice and Bob agree on an elliptic curve $E = E_{a,b}$ modulo a prime p and a point P of high order on E, perhaps a generator of the group. Let N be the order of the group. The group E and the point P need not be secret and Alice and Bob do not need to know N exactly. By Hasse's theorem, N is approximately p, and that approximation is good enough.

Alice secretly chooses a random x_A in $0 < x_A < N$ and computes $P_A = x_A P$ on E. Bob secretly chooses a random x_B in $0 < x_B < N$ and computes $P_B = x_B P$ on E.

Alice sends P_A to Bob. Bob sends P_B to Alice. An eavesdropper, knowing E and P, and seeing P_A and P_B, cannot compute x_A or x_B from this data unless he can solve the discrete logarithm problem for elliptic curves quickly.

Alice computes $K_A = x_A P_B$ on E. Bob computes $K_B = x_B P_A$ on E. Then

$$K_A = (x_A \cdot x_B)P = K_B.$$

Alice and Bob choose certain agreed-upon bits from K_A to use as their key for a private key cipher like DES or Rijndael.

Since the discrete logarithm problem is harder to solve for an elliptic curve than for the multiplicative group of integers modulo p, the modulus of the elliptic curve may be chosen smaller than the prime p for R_p.

The protocol clearly generalizes to any large group.

19.3   The X.509 Key Exchange

X.509 is a directory authentication service that solves the following problems without the aid of a trusted third party who communicates with you during the protocol.

How do you get the public key of someone to whom you wish to send mail? How do you know it is valid and not a forgery? How can you and another user agree on a private key to use to communicate over an insecure network?

ITU-T recommendation X.509 defines a framework for provision of authentication services. Each user has a public key certificate issued by a trusted certification authority CA. The signature of the certificate consists of the hash codes of its other fields, signed by the CA's private key.

The certificates form a tree-structured hierarchy.

Each certificate contains fields for Version, Serial number, Algorithm for signature, Name of issuer (CA), Period of validity, Subject name, Subject public key information, and the Signature of the CA, and perhaps other fields depending on the version.

Use `finger` or `ftp` or a web browser to obtain the certificate of a user to whom you wish to send mail via public key cryptography.

Use the "Issuer" field in the certificate to find the certificate for the CA, etc., to the root (whom everyone trusts) or up to some CA in the chain from you to the root.

Note that:

- Certificates need not be specially protected since they are unforgeable.

- Any user with access to the public key of the CA can recover the user public key that was certified.

- No one other than the CA can modify the certificate without the change being detected.

- CA's may certify each other, to make it easier for users to reach a CA they trust when obtaining the certificate of a new user.

- To revoke a certificate (for example, if the user's key was compromised), the CA of that key puts it on a public list, with its serial number and revocation date.

X.509 also includes three alternative authentication procedures.

Let us use the notation $X\{M\}$ to mean "X signs M," that is, M followed by the signed hash code of M. The notation "$A \rightarrow B$:" means "A sends the following message to B." The t_A is a time stamp, giving the date and time the message was sent, r_A is a nonce, that is, a random number generated and used just this one time, $sessionkey_{AB}$ is the key to a single-key cipher A and B will use to communicate for a while, and $sgnData$ is the signature of the message digest of the other fields. Consider the following three messages.

1. $A \rightarrow B$: $A\{t_A, r_A, B, sgnData, sessionkey_{AB}\}$

2. $B \rightarrow A$: $A\{t_B, r_B, A, r_A, sgnData, sessionkey_{AB}\}$

3. $A \rightarrow B$: $A\{r_B\}$.

Either Message 1, or Messages 1 and 2, or all three messages may be used.

Message 1 establishes the identity of A, that the message was generated by A, that the message was intended for B, and that the message has not been changed or sent more than once.

Message 2 establishes the identity of B, that the reply was generated by B, that the reply was intended for A, and that the reply has not been changed or sent more than once.

The purpose of the third message, if it is used, is to obviate the need to check time stamps. It is used when synchronized clocks are not available. It works because both nonces are echoed, so *they* can be checked to detect replay attacks.

See Chapter 14 of Stallings [114] for more about X.509.

19.4 Exercises

1. There is a flaw in the key exchange protocol of Denning and Sacco. Suppose Alice and Bob have a brief secret conversation, so that the time stamp t_A is still valid when they finish. Explain how Bob could pretend to be Alice and communicate with Carol so that Carol thought she was talking to Alice. Repair this flaw with a minor change to one message.

2. Alice and Bob communicate securely every day. Eve knows the parameters p and g of the Diffie-Hellman algorithm they use to choose their daily Rijndael keys. The prime modulus p is so large that Eve cannot solve discrete logarithm problems modulo p. She has recorded all messages passing between them, including every y_A, y_B and the Rijndael-enciphered traffic. She notices that Alice's random number generator must be defective, since Alice sends the same y_A every day. However, Bob uses a new y_B every day.

   Eve tells Bob that she is madly in love with him, is jealous of his daily secret conversations with Alice, and has recorded today's ciphertext. Bob recalls that today's conversation with Alice was pretty boring and, in a moment of weakness, tells Eve today's symmetric key K_B (the whole K_B, not just the part used for the Rijndael key) to let her read today's conversation and put her jealousy to rest.

   Given this information, how many days of recorded ciphertext can Eve decipher?

3. There is a flaw in the three-way authentication procedure for X.509. In simplest form, the protocol is:

$$A \rightarrow B: \quad A\{t_A, r_A, B\}$$
$$B \rightarrow A: \quad B\{t_B, r_B, A, r_A\}$$
$$A \rightarrow B: \quad A\{r_B\}.$$

   The X.509 rules state that checking the time stamps t_A and t_B is optional for three-way authentication. The time stamps are set to 0 when they are not used.

   a. Explain how an active wiretapper C can impersonate A to B. Assume that C is in the hierarchy, can capture messages passing between A and B, and can cause A to initiate authentication with C.

b. One solution to this problem is to use time stamps. Suggest another solution, one not using time stamps.

4. Design a variation of the Diffie-Hellman key exchange protocol that will allow Alice, Bob and Chuck to choose a single common Rijndael key securely for their mutual communication. There is a public large prime p and a primitive root g. The three participants choose secret keys a, b, c, respectively. The common key should be some bits of $g^{abc} \bmod p$. It should not be possible for an eavesdropper, who sees all messages passing among the three parties, but cannot change them, to deduce the common key.

Chapter 20

Simple Protocols

This chapter presents a few simple protocols. Some of them make sense; others may seem strange because they are used as components of larger protocols in the next chapter.

20.1 Bit Commitment

Alice wants to commit to a "bit" (it can be a string of bits) now so that she can't change it later, but only Alice knows it for now. She wants to tell Bob something now that he can remember, and which is connected to the bit. Later, when Alice reveals the bit, she will tell Bob the bit and more information which he will see is connected to the bit and what Alice told him earlier. At that time Bob can check something and be convinced that the bit Alice has revealed must have been the one she had in mind earlier. He will know that she could not have changed it.

Alice commits to b by generating two random strings R_1, R_2.

She creates a message, (R_1, R_2, b), and computes a hash value of it.

She reveals, that is, tells Bob, the hash value and R_1.

When the time comes to reveal b, Alice shows the message (R_1, R_2, b).

Bob checks that R_1 is the same as it was earlier and verifies the hash value.

Why is R_2 needed? Because if b were a short string (literally one bit, say), then Bob could guess it from the hash value.

20.2 Mental Poker

In the card game of poker, each player is dealt five of the 52 cards. Each player can see his hand, but not any other player's hand. Players bet based on their hands. The "best" hand wins. In some variations, some cards are revealed and some cards may be replaced by cards not yet dealt.

The "e-mail" or "mental" poker protocol requires a fair deal with these properties. Players see their own hands, but not other hands. The hands are disjoint. All hands are equally likely. A player can "draw" (replace) selected cards. A player can reveal individual cards one at a time without revealing other cards. All players can check at the end of the game that there was no cheating.

We use a variation of the Pohlig-Hellman cipher to implement mental poker.

Assume there are two players, Alice and Bob. (There are similar protocols for three or more players.)

The players jointly choose a large prime p as modulus. Each secretly chooses e_A, d_A, e_B, d_B, as in the Pohlig-Hellman cipher. Note that every e and d is relatively prime to $p - 1$ because we must have $e_A d_A \equiv 1 \pmod{p - 1}$, etc. Define $E_A(M) = M^{e_A} \bmod p$, etc. Recall that these functions commute: $E_A(E_B(M)) = E_B(E_A(M))$ for every M, etc. Let $M_1, \ldots, M_{52}$ be the encoded deck (more if there is a joker).

1. Bob enciphers the cards as $C_i = E_B(M_i)$ for $i = 1, \ldots, 52$. Bob sorts the C_i as numbers and sends them to Alice. The sorting operation shuffles the deck so that Alice cannot tell which C_i represents which M_i.

2. Alice selects five cards C_i at random and sends then to Bob, who decrypts them as his hand.

3. Alice selects five more random cards, say $C_1, \ldots, C_5$ (her hand) and enciphers them as $C_i' = E_A(C_i)$ for $i = 1, \ldots, 5$. She sends them to Bob.

4. Bob deciphers the C_i'. They are still enciphered with E_A after he applies D_B to undo E_B. He sends the $D_B(C_i')$ back to Alice.

5. Alice deciphers the five cards and uses them as her hand. They bet and play poker. A card is "revealed" by sending both the plain and cipher text of it to the other player.

6. At the end of the hand, Alice and Bob exchange their keys e_A, etc., and check everything that happened.

Unfortunately, one can cheat in mental poker because the cipher functions E_A, E_B, etc., preserve quadratic residues.

THEOREM 20.1 Quadratic residues are preserved
Let $0 < a < n$, $\gcd(a, n) = 1$, and $\gcd(e, \phi(n)) = 1$. Then a is a quadratic residue modulo n if and only if a^e is a quadratic residue modulo n.

PROOF   Let d be the inverse of e modulo $\phi(n)$: $ed \equiv 1 \pmod{\phi(n)}$. If a is a quadratic residue modulo n, then there exists an x so that $x^2 \equiv a \pmod{n}$. Let $y = x^e \bmod n$. Then

$$y^2 \equiv (x^e)^2 \equiv (x^2)^e \equiv a^e \pmod{n}.$$

This shows that a^e is a quadratic residue modulo n. Conversely, if a^e is a quadratic residue modulo n with $y^2 \equiv a^e \pmod{n}$, then $(y^d)^2 \equiv a^{ed} \equiv a \pmod{n}$, so a is a quadratic residue modulo n. ∎

Alice can use this theorem to cheat: Perhaps most high cards are quadratic residue and most low cards are quadratic nonresidues. It is like playing with a deck in which most high cards are "marked." This attack can be foiled by (a) appending extra bits to each M_i or (b) multiplying some of the M_i by a fixed quadratic nonresidue in order to make all cards be quadratic residues or all cards be quadratic nonresidues.

In order for Alice to cheat and in order to foil the attack, one must be able to distinguish between quadratic residues and quadratic nonresidues modulo p (at least for prime p) quickly. This is easy. See Theorem 7.11.

20.3 Oblivious Transfer

An application of finding square roots modulo n is the Rabin-Blum oblivious transfer or coin tossing protocol. In it, Alice reveals a secret to Bob with probability $1/2$.

In the oblivious transfer version, Alice doesn't know whether Bob got the secret or not (and this outcome must be acceptable to both participants).

In the coin tossing version, Bob tells Alice whether he got the secret. He wins the coin toss if he did get it; loses otherwise.

Alice's secret is the factorization of a number $n = pq$ which is the product of two large Blum primes $p \equiv q \equiv 3 \pmod{4}$.

1. Alice sends n to Bob.

2. Bob picks a random x in $\sqrt{n} < x < n$ with $\gcd(x, n) = 1$. Bob computes $a = x^2 \bmod n$ and sends a to Alice.

3. Knowing p and q, Alice computes the four solutions to $x^2 \equiv a \pmod{n}$. They are x, $n - x$, y and $n - y$, for some y. These are just four numbers to Alice. She doesn't know which ones are x and $n - x$. She chooses one of the four numbers at random and sends it to Bob.

4. If Bob receives x or $n - x$, he learns nothing. But, if Bob receives y or $n - y$, he can factor n by computing $\gcd(x + y, n) = p$ or q.

Why can Bob factor n if he gets y or $n - y$? He can do so because of Theorem 13.1.

Here is another way to do the same protocol:

Alice will send Bob one of two messages. Bob will receive one. Alice won't know which one.

1. Alice generates two RSA public/private key pairs. She sends both public keys to Bob.

2. Bob chooses a Rijndael key K. He chooses one of Alice's public RSA keys and enciphers K with it. He sends the encrypted key to Alice without telling her which of her public keys he used to encipher it.

3. Alice decrypts Bob's key twice, using both of her private RSA keys. In one case, she gets K. In the other case, she gets garbage that looks like a Rijndael key. She can't tell which is which.

4. Alice encrypts the two messages with Rijndael, one using K and the other using the garbage key. She sends both ciphertexts to Bob.

5. Bob tries to decrypt the two ciphertexts using K. He can read one message; the other is gibberish.

Alice doesn't know which message Bob can read. Alice could cheat unless we used the next step.

6. Alice gives Bob both of her private RSA keys so that he can verify that she did not cheat. After all, she could have encrypted the same message with both keys in Step 4. Then she would know which message Bob received.

20.4   Zero-knowledge Proofs

A zero-knowledge proof is a dialog between two people, the Prover (Alice) and the Verifier (Bob), in which the Prover convinces the Verifier that she knows a certain secret, but without revealing to the Verifier (or to an eavesdropper) any part of the secret. After the protocol concludes, neither the Verifier nor an eavesdropper could masquerade as the Prover to convince someone else that they know the secret.

There are many different forms of zero-knowledge proof. We describe a zero-knowledge proof protocol that uses square roots modulo n. One application of these protocols is in identification. If Alice can show that she can compute arbitrary square roots modulo n, a number whose factorization is known only to her, then she can convince someone at the other end of an Internet connection that she really is Alice, by Theorem 13.2.

This protocol is closely related to the oblivious transfer protocol. The difference is that Alice wants to convince Bob that she knows the factors of $n = pq$, but does not want to reveal the factors to Bob.

Alice (the Prover) convinces Bob (the Verifier) that she knows the prime factorization of a large composite number n, but does not give Bob any hint that would help him find the factors of n. In terms of entropy, this means that if M is a message that tells the factors of n, and S is the set of messages exchanged by Alice and Bob during the zero-knowledge proof, then $H(M|S) = H(M)$. Thus, Bob learns nothing about the factorization of n during the protocol that he could not have deduced on his own without Alice's help.

Roughly speaking, Bob gives Alice some quadratic residues modulo n and Alice replies with their square roots. The difficulty with this simple approach is that when Alice replies to Bob with a square root, there is a 50% chance that she will reveal the factorization of n to Bob, as in the first oblivious transfer protocol. It is known that computing square roots modulo n is polynomial-time equivalent to factoring n.

Here is a good way to do the zero-knowledge proof protocol:

Alice knows n, p and q. Bob knows n but not p or q.

1. Alice chooses a in $\sqrt{n} < a < n$ and computes $b = a^2 \bmod n$.

2. At the same time, Bob chooses c in $\sqrt{n} < c < n$ and computes $d = c^2 \bmod n$.

3. Alice sends b to Bob and Bob sends d to Alice.

4. Alice receives d and solves $x^2 \equiv bd \pmod{n}$. (Note that this is possible because bd is a quadratic residue and she can compute its square roots since she knows the factors of n.) Let x_1 be one solution of this congruence.

5. At the same time, Bob tosses a fair coin and gets Heads or Tails, each with probability $1/2$. Bob sends H or T to Alice.

6. If Alice receives H, she sends a to Bob. If Alice receives T, she sends x_1 to Bob.

7. If Bob sent H to Alice, then he receives a from Alice and checks that $a^2 \equiv b \pmod{n}$. If Bob sent T to Alice, then he receives x_1 from Alice and checks that $x_1 \equiv bd \pmod{n}$.

Alice and Bob repeat steps 1 through 7 many (20 or 30) times.

If the check in step 7 is always okay, then Bob accepts that Alice knows the factorization of n.

But if Alice ever fails even one test, then Bob concludes that Alice is lying.

Why does this protocol work? If Alice really knows the factors of n, then she can compute all the required square roots by the methods of Chapter 7. If Alice doesn't know the factors of n, then she will not be able to compute general square roots modulo n. She could fake *one* of the two square roots, but not both. In this case, she could give Bob the correct square root only if she could guess which one he would request. If Bob really tosses a fair coin to decide which square root to request, then Alice would have to guess the outcome of each coin toss. There is less than one chance in a million that she could predict the outcome of twenty coin tosses.

Why does Bob not learn the factors of n? This is so because throughout the protocol, Bob learns only one square root of any single quadratic residue. He would have to know two different square roots (not satisfying $r_1 \equiv -r_2 \pmod{n}$) in order to apply Theorem 13.1.

20.5   Methods of Sharing Secrets

An important cipher key K must be protected from (a) accidental or malicious exposure (causing vulnerability) and (b) loss or destruction (causing inaccessibility).

Both problems may be alleviated by the use of shadows in a threshold scheme.

For $1 \leq t \leq w$, a (t, w) **threshold scheme** is a system of protecting a key K by breaking it into w **shadows** (pieces), $K_1, \ldots, K_w$, in such a way that (a) it is easy to compute K using knowledge of any t of the shadows K_i, but (b) it is impossible to compute K because of lack of information if one knows only $t - 1$ or fewer of the shadows K_i.

The w shadows are given to w users. Since at least t shadows are needed to find K, no group of fewer than t users can conspire to get the key.

At the same time, if a shadow is lost or destroyed, one can still compute K so long as at least t valid shadows remain.

The same mathematics can handle more complicated cases in which some people are more important than others. For example, suppose the secret key K opens a safe in a bank. Let the policy be that the safe can be opened by (a) any four tellers, or (b) a manager and two tellers, or (c) two managers, or (d) the bank president. Then one could use a $(4, w)$ threshold scheme, where w is large enough. Give each teller one shadow, each manager two shadows and the president four shadows. The safe can be opened whenever four shadows are available.

20.5.1   Secret Splitting

The special case of $t = w$ is easy to arrange. In this case the secret is split among w people and all w must get together to use the secret. This technique is called **secret splitting**. A trusted person or program, Tracy, prepares the w shadows. The first $w - 1$ people are given random bit strings of the same length as the secret. The last person receives the exclusive-or of the $w - 1$ random strings and the secret.

To reconstruct the secret the w people simply exclusive-or their strings. Each random string appears twice in the final exclusive-or and the secret appears once in it, so the exclusive-or of all w strings is the secret. If any $w - 1$ of the people try to compute the secret, they can't do it because they lack the missing bit string. Anything they could compute would be as random as the missing string.

Other forms of secret sharing can be done elegantly with number theory.

20.5.2   The Lagrange Interpolating Polynomial Scheme

Shamir [102] proposed a (t, w) threshold scheme based on Lagrange interpolating polynomials.

A polynomial of degree $t - 1$ is determined by its values at t distinct values of its argument. In numerical analysis it is shown that given t points $(i_1, K_{i_1}), \ldots, (i_t, K_{i_t})$ with different x coordinates i_j, there is a unique polynomial of degree $\leq t - 1$ passing through them. It is the Lagrange polynomial

$$h(x) = \sum_{s=1}^{t} K_{i_s} \prod_{\substack{j=1 \\ j \neq s}}^{t} \frac{(x - i_j)}{(i_s - i_j)}.$$

Indeed, it is easy to see that $h(x)$ is a polynomial of degree no more than $t - 1$, and when $x = i_j$ each term in the sum has a factor 0 in the product, except for the j-th term, which has the product equal to 1 and the value K_{i_j}, so that $h(i_j) = K_{i_j}$, as required.

Shamir's threshold scheme uses Lagrange polynomials modulo p.

The shadows come from a random polynomial of degree $t - 1$:

$$h(x) = (a_{t-1}x^{t-1} + \cdots + a_1 x + a_0) \bmod p$$

with constant term $a_0 = K$ and random numbers for $a_{t-1}, \ldots, a_1$.

All arithmetic is done modulo p, where p is a prime number greater than both K and w. Long keys can be broken into smaller blocks to avoid computing modulo a large prime.

Given the polynomial $h(x)$, the key K is easily computed by $K = h(0)$.

The w shadows are defined as the value of $h(x)$ at w distinct points. For example, one might let $K_i = h(i)$ for $1 \leq i \leq w$.

Given t shadows, $K_{i_1}, \ldots, K_{i_t}$, one may construct $h(x)$ as

$$h(x) = \sum_{s=1}^{t} K_{i_s} \prod_{\substack{j=1 \\ j \neq s}}^{t} \frac{(x - i_j)}{(i_s - i_j)} \bmod p.$$

If only $t - 1$ shadows are known, then, for any K_0, one could pick $(0, K_0)$ as the t-th point and compute a polynomial $h_0(x)$ with $K_0 = h(0)$. Hence, $t - 1$ shadows reveal nothing about K.

Example 20.1

Let $t = 3$, $w = 5$, $p = 13$, $K = 10$ and

$$h(x) = (6x^2 + 7x + 10) \bmod 13,$$

with random coefficients 6 and 7.

The five shadows are the values of $h(x)$ at $x = 1, 2, 3, 4, 5$:

$$K_1 = h(1) = (6 + 7 + 10) \bmod 13 = 10$$
$$K_2 = h(2) = (24 + 14 + 10) \bmod 13 = 9$$
$$K_3 = h(3) = (54 + 21 + 10) \bmod 13 = 7$$
$$K_4 = h(4) = (96 + 28 + 10) \bmod 13 = 4$$
$$K_5 = h(5) = (150 + 35 + 10) \bmod 13 = 0$$

We can recover $h(x)$ and $K = h(0)$ from any three of the shadows. For example, using K_1, K_3 and K_5 we have:

$$h(x) = 10\frac{(x-3)(x-5)}{(1-3)(1-5)} + 7\frac{(x-1)(x-5)}{(3-1)(3-5)} + 0\frac{(x-1)(x-3)}{(5-1)(5-3)} \bmod 13$$
$$= 10(x-3)(x-5)/8 + 7(x-1)(x-5)/(-4) \bmod 13$$
$$= 10(x-3)(x-5)5 + 7(x-1)(x-5)3 \bmod 13$$
$$= 50(x^2 - 8x + 15) + 21(x^2 - 6x + 5) \bmod 13$$
$$= 11(x^2 + 5x + 2) + 8(x^2 + 7x + 5) \bmod 13$$
$$= (19x^2 + 111x + 62) \bmod 13 = h(x).$$

20.5.3   The Asmuth and Bloom Threshold Scheme

Asmuth and Bloom [5] based their threshold scheme on the Chinese remainder theorem.

Let $K \geq 0$ be the key. Let $p, d_1, d_2, \ldots, d_w$ be integers such that $p > K$, $d_1 < d_2 < \cdots < d_w$, $\gcd(p, d_i) = 1$ for all i, $\gcd(d_i, d_j) = 1$ for all $i \neq j$, and $d_1 d_2 \cdots d_t > p d_{w-t+2} d_{w-t+3} \cdots d_w$.

The gcd requirements guarantee that the integers $p, d_1, d_2, \ldots, d_w$ are pairwise relatively prime. The last condition says that the product of the t smallest d_i's is greater than the product of p and the $t-1$ largest d_i's. Let $n = d_1 d_2 \cdots d_t$ be the product of the t smallest d_i's. Then n/p is greater than the product of any $t-1$ of the d_i's.

Let r be a random integer in the range $0 \leq r < n/p$. Write $K' = K + rp$. Then $0 \leq K' < n$. The w shadows are defined as $K_i = K' \bmod d_i$ for $i = 1, \ldots, w$.

To recover K, it suffices to find K' because $K = K' \bmod p$. If t shadows $K_{i_1}, \ldots, K_{i_t}$ are known, then by the Chinese remainder theorem, K' is known modulo $n_1 = d_{i_1} \cdots d_{i_t}$. Since $n_1 \geq n > K'$, the Chinese remainder theorem uniquely determines K'.

If only $t-1$ shadows $K_{i_1}, \ldots, K_{i_{t-1}}$ are known, then K' can only be known modulo $n_2 = d_{i_1} \cdots d_{i_{t-1}}$. Because $n/n_2 > p$ (the last condition above) and $\gcd(n_2, p) = 1$, the numbers x such that $x \leq n$ and $x \equiv K' \pmod{n_2}$ are nearly evenly distributed over all the congruence classes modulo p. Therefore, there is not enough information to determine K'.

Example 20.2

Let $K = 3$, $t = 2$, $w = 3$, $p = 5$, $d_1 = 7$, $d_2 = 9$ and $d_3 = 11$. Then $n = d_1 d_2 = 7 \cdot 9 = 63 > 5 \cdot 11 = p d_3$ as required.

We need to choose a random number between 0 and $(63/5)$, that is, between 0 and 12. Picking $r = 9$, we get

$$K' = K + rp = 3 + 9 \cdot 5 = 48.$$

The shadows are $K_1 = 48 \mod 7 = 6$, $K_2 = 48 \mod 9 = 3$ and $K_3 = 48 \mod 11 = 4$.

Given any two of the three shadows, we can compute K. Assume we know K_1 and K_3. Then $n_1 = d_1 d_3 = 7 \cdot 11 = 77$. The Chinese remainder theorem produces $K' \equiv 48 \pmod{77}$. Finally, $K = K' \mod p = 48 \mod 5 = 3$.

20.6 Blind Signatures

Suppose Bob uses RSA for signatures and has public keys n and e, and a private key d for this purpose. Alice wants Bob to sign a message M, but Alice doesn't want Bob to see the contents of M. Bob trusts Alice and agrees to provide a **blind signature** for M.

Normally, Bob would sign M by computing $M^d \mod n$ and giving this number to Alice. But if he did that, he could see what M says.

Instead, Alice chooses a random k in $1 < k < n$. She "blinds" (enciphers) M by computing $t = Mk^e \mod n$. To Bob, t looks like a random integer. He cannot compute M from t because he doesn't know k. Bob signs t as

$$t^d \equiv (Mk^e)^d \equiv M^d k^{ed} \equiv M^d k \pmod{n}.$$

After she leaves Bob with $t^d \mod n$, Alice "unblinds" (deciphers) the signed message by computing the multiplicative inverse $k^{-1} \mod n$ via the extended Euclidean algorithm and multiplying:

$$s = t^d k^{-1} \mod n \equiv (Mk^e)^d k^{-1} \equiv M^d k^{ed-1} \equiv M^d \pmod{n}.$$

Now Alice has Bob's blind signature $M^d \mod n$ for M, and Bob never saw M.

This protocol is the electronic equivalent of Alice sealing M inside an envelope with a piece of carbon paper, getting Bob to sign the outside of the envelope, so that the carbon paper copies his signature onto M, and then Alice opening the envelope later.

20.7 Exercises

1. Alice discovers a wonderful proof of the Riemann Hypothesis. She writes a manuscript containing her proof, but does not publish it immediately because she feels the world is not ready for the proof. However, she wants to be able to claim that she did it first in case someone else finds a proof later. What short string of letters or numbers could she publish in a classified advertisement in the *New York Times* to accomplish this goal?

2. Design a protocol for mental poker with three players.

3. The rules of draw poker allow players to discard (some) cards and replace them. Show that discarding and replacement can be done by a slight modification of the mental poker protocol.

4. Is there a way that either Alice or Bob could deliberately lose the coin tossing protocol? Explain your answer.

5. Can Bob deduce Alice's second message when she gives him both of her private RSA keys in Step 6 of the second oblivious transfer protocol?

6. Consider Shamir's Lagrange interpolating polynomial threshold scheme. Let $t = 4$, $p = 11$, $K = 7$ and $h(x) = (x^3 + 10x^2 + 3x + 7) \bmod 11$. Compute shadows for $x = 1, 2, 3, 4, 5, 6$ and 7. Reconstruct $h(x)$ from the shadows for $x = 1, 3, 5$ and 7.

7. Consider Asmuth and Bloom's key threshold scheme based on the Chinese remainder theorem. Let $t = 2$, $w = 4$, $p = 5$, $d_1 = 8$, $d_2 = 9$, $d_3 = 11$, $d_4 = 13$. Then $n = 8 \times 9 = 72$. Let $K = 3$ and $r = 10$, so that $K' = 53$. Compute the four shadows K_1, K_2, K_3 and K_4. Reconstruct K from K_1 and K_3.

8. Consider Asmuth and Bloom's key threshold scheme based on the Chinese remainder theorem. Suppose keys have three shadows, any two of which are enough to determine the key. Suppose the key is a non-negative integer less than $p = 5$, while the shadows are integers modulo $d_1 = 7$, $d_2 = 9$ and $d_3 = 11$. Determine the key K from the two shadows $K_1 = 0$ and $K_3 = 9$. Then find the second shadow K_2. Note that the two given shadows correspond to the *first* and *third* moduli 7 and 11.

9. A small bank has an electronic safe which may be opened by certain combinations of the president, the two managers and the five tellers. Policy dictates that the safe may be opened if and only if

   (a) the bank president decides to open it, or

   (b) the two managers both decide to open it, or

   (c) all five tellers decide to open it, or

   (d) one manager and three tellers decide to open it.

   Explain how you would choose the parameters and distribute the shadows of a Lagrange interpolation polynomial key threshold scheme to meet the requirements of this bank.

Chapter 21

Complicated Protocols

This chapter discusses several complicated protocols. Some of them use protocols from the previous chapter. See Chapters 5 and 6 of Schneier [100] for more about these protocols.

21.1 Contract Signing

Alice and Bob want to enter into a contract. They agree on it. Both want to sign, but neither wishes to sign unless the other signs as well.

The first protocol uses a trusted arbitrator, Tracy.

1. Alice signs a copy of the contract and mails it to Tracy.

2. Bob signs a copy of the contract and mails it to Tracy.

3. Tracy tells both Alice and Bob that the other has signed it.

4. Alice signs two copies of the contract and mails them to Bob.

5. Bob signs both copies, keeps one and mails the other to Alice.

6. Alice and Bob both tell Tracy that they each have a copy of the contract signed by both of them.

7. Tracy destroys the two copies of the contract that she has.

If Alice didn't sign the contract in Step 4, then Bob could get a copy she had signed from Tracy. Likewise, if Bob didn't sign it in Step 5, then Alice could get one he had signed from Tracy.

In the second protocol, Alice and Bob have no arbitrator, but are face-to-face in a room.

1. Alice signs the first letter of her name on two copies of the contract and hands them to Bob.

2. Bob signs the first letter of his name on both copies and hands them back to Alice.

3. Alice signs the second letter of her name on both copies and hands them back to Bob.

4. Bob signs the second letter of his name on both copies and hands them back to Alice.

5. This continues until both Alice and Bob have signed their entire names.

In the third protocol, Alice and Bob have no arbitrator and are not face-to-face. In this case, they exchange a series of signed messages of the form, "I agree that I am bound by the contract with probability p." Suppose that Alice doesn't want to be bound to the contract with a probability more than 2% higher than the probability Bob is bound to it. Suppose also that Bob doesn't want to be bound to the contract with a probability more than 3% higher than the probability Alice is bound to it.

1. Alice and Bob agree on a time by which the signing protocol should be completed.

2. Alice mails Bob a message with $p = 0.02$.

3. Bob mails Alice a message with $p = 0.05$.

4. Alice mails Bob a message with $p = 0.07$.

5. Bob mails Alice a message with $p = 0.10$.

6. These steps alternate until both have received messages with $p = 1$ or else the time in Step 1 has passed.

In case one of Alice and Bob stopped the protocol before the end, the other would take the last signed message to a judge, who would choose a random number between 0 and 1 and compare it to the p in the message to see whether the contract was valid.

In the fourth protocol, Alice and Bob have no arbitrator, are not face-to-face, and can't agree on the probabilities above.

1. Alice and Bob randomly select $2n$ Rijndael keys, in pairs.

2. Alice and Bob generate n pairs of messages: $L_i =$ "This is the left half of my signature." $R_i =$ "This is the right half of my signature." Each message also contains a digital signature of the contract and a time stamp. The contract is considered signed by a party if both L_i and R_i can be produced by the other party for some $1 \leq i \leq n$.

3. Alice and Bob encipher their message pairs using the $2n$ Rijndael pairs, the left message with the left key and the right message with the right key.

4. Alice and Bob mail each other their set of $2n$ enciphered messages, making it clear which is which.

5. For $1 \leq i \leq n$, Alice and Bob mail each other one of the keys in the i-th pair by oblivious transfer, omitting Step 6 for the moment. Now they each have one key in each pair, but neither knows which signature halves the other can read.

6. Both Alice and Bob decipher the messages they can, using the keys mailed to them. They check that the messages are valid.

7. Alice and Bob mail each other the first bit of all $2n$ Rijndael keys. They check the n first bits they already know.

8. Alice and Bob repeat Step 7 for the second, third, etc., bits, until all bits of all the Rijndael keys have been exchanged.

9. Alice and Bob decipher the remaining halves of the message pairs and the contract is signed.

10. Alice and Bob exchange the private RSA keys (Step 6 of the RSA version of the oblivious transfer protocol) and verify that the other has not cheated.

If Bob wanted to cheat, he could send garbage in Step 4 or 5, but Alice would notice this fraud in Step 6.

21.2 Secure Elections

Secure elections should have at least these properties:

1. Only registered voters can vote.

2. No person can vote more than once.

3. No one can determine for whom anyone else voted.

4. Every voter can make sure that his vote has been counted.

5. No person can duplicate any other person's vote.

6. No person can change any other person's vote undetected.

All of the following protocols use a Central Tabulating Facility, CTF. The following protocol satisfies Property 6, but not much else.

1. Each voter enciphers his vote with the public key of the CTF and mails it to the CTF.

2. The CTF deciphers the votes, tabulates them, and publishes the results.

There are many problems with this protocol. The CTF can't tell whether votes come from registered voters or whether registered voters vote more than once.

Protocol 2 is slightly better.

1. Each voter signs his vote with his private (RSA) key.

2. Each voter enciphers his signed vote with the CTF's public key and mails it to the CTF.

3. The CTF deciphers all votes, checks signatures, tabulates the votes and publishes the results.

This protocol satisfies Properties 1, 2 and 6: only registered voters can vote and no person can vote more than once. Also, no person can change any vote.

The problem is that the CTF knows who voted for whom. The voters must trust the CTF completely.

Protocol 3 solves many of these problems. The "blinding" mentioned in Step 2 refers to the blind signature protocol described in the previous chapter.

1. Each voter prepares ten sets of messages. Each set contains a valid vote for each possible outcome. Each message also contains a randomly generated twenty-digit number.

2. Each voter blinds each of these messages individually and mails them to the CTF.

3. The CTF checks its list to make sure the voter has not submitted his blinded votes previously in this election. It randomly chooses nine of the ten sets of votes and asks the voter to unblind these sets. The voter does and the CTF checks that these nine sets are properly formed. If they are, then it individually signs each message in the tenth set. It mails them back to the voter and stores his name in the list.

4. The voter unblinds the signed tenth set and is left with a set of all possible votes signed by the CTF. He can tell which is which.

5. The voter chooses one of the possible votes and enciphers it with the CTF's (second) public key. He mails it to the CTF.

6. The CTF deciphers the votes, checks the signatures, checks its list for a duplicate twenty-digit number, saves the twenty-digit number in the database and tabulates the votes. It publishes the results of the election, along with every twenty-digit number and the vote with it.

The blind signature ensures that votes are unique. No one can generate bogus votes or change votes of others because the CTF's private key is secret. A malicious CTF cannot determine how people voted. Each voter can confirm that his vote was tabulated correctly.

However, if the CTF can determine where the votes comes from, it can link votes with people. Even if it can't do this, it could still generate many signed valid votes and submit them itself. If Alice discovers that the CTF has changed her vote, she cannot prove this.

The next protocol has the six properties listed above and these two:

7. A voter can change his mind, that is, delete his old vote and cast a new one, within a certain time period.

8. If a voter finds that his vote is miscounted, he can correct the problem without hurting the secrecy of his ballot.

Protocol 4.

1. The CTF publishes a list of all eligible voters.

2. Before a deadline, each voter tells the CTF whether he intends to vote.

3. At this deadline, the CTF publishes a list of all eligible voters who intend to vote.

4. Each voter receives a unique twenty-digit number I, for example, by mental poker or blind signature protocol.

5. Each voter generates RSA keys n, e, d with encipher function E and decipher function D. If his vote is v, he mails the message $I, E(I; v)$ anonymously to the CTF.

6. The CTF acknowledges getting the vote by publishing $E(I, v)$.

7. Each voter mails the message I, d to the CTF.

8. The CTF deciphers the votes. It can do this since it has received the d for each vote, so it can compute $D(E(I; v)) = I; v$. When the election ends, it publishes the results and, for each different vote, the list of all $E(I; v)$ that contained that vote.

9. If a voter sees that his vote was not counted properly, he objects by mailing $I, E(I; v), d$ to the CTF.

10. If a voter wants to change his vote from v to v', he mails $I, E(I; v'), d$ to the CTF.

Steps 1 through 3 of Protocol 4 are preliminary to the actual voting. They reduce the ability of the CTF to add fraudulent votes.

If two voters get the same I in Step 4, the CTF discovers this in Step 5. It creates a new twenty-digit number I', chooses one of the two votes, and publishes $I', E(I; v)$.

The person who cast that vote recognizes it and votes again by repeating Step 5 with the new I'.

In Step 6, each voter can check that his vote is counted accurately. If not, he can prove this in Step 9.

One limited problem is that the CTF could make up fraudulent votes for people who respond in Step 2 but don't actually vote.

A more serious problem is that CTF could neglect to count a vote. Alice could claim that the CTF deliberately neglected her vote, while the CTF could claim that she never voted.

21.3   Electronic Cash

Electronic cash or **digital cash** is not a check, credit card or a debit card. They leave audit trails. It is as close to cash as can be.

Digital cash is anonymous and untraceable. It can be sent through computer networks. It can be used off-line, not connected to a bank. It is transferable. One can make change with it. It can be stolen. It can be spent only once. It would be used to pay for small things like tolls and food.

We do not achieve all of these goals, but we design electronic cash with many of these properties. In the next two sections we describe two quite different forms of digital cash.

21.3.1   Electronic Cash According to Chaum

We begin with a simple protocol for digital cash and follow with a series of more complicated protocols that fix the problems with the first one.

1. The bank gives Alice a note for $100 (like a money order or cashier's check) and subtracts $100 from Alice's bank account.

2. Alice spends the note with a merchant.

3. The merchant deposits the note in his bank account.

4. The merchant's bank clears the note with Alice's bank.

This protocol has many problems. As the note is electronic, it can be easily copied, so Alice could spend it twice. So could the merchant. Also, Alice could be traced if the bank remembered the serial number of the note.

We solve these problems one-by-one, following Chaum [26] and [27]. Protocol 1 solves the problem of Alice being traced by the bank and makes the money truly anonymous. The bank has RSA keys n, e, d, as usual.

Protocol 1.

1. Alice makes 100 anonymous money orders for $100 each. She blinds (enciphers) each one and gives them all to her bank.

2. The bank asks Alice to open 99 of the money orders, randomly chosen, and verifies that each one is a money order for $100. (If not so, Alice goes to jail.) To "open" the 99 money orders t, Alice tells the bank M and k for each. The bank verifies that each $t = Mk^e \bmod n$.

3. The bank put its blind signature on the last, unopened one, returns it to Alice and deducts $100 from her bank account.

4. Alice unblinds the money order (now signed by the bank) and spends it with a merchant.

5. The merchant verifies the bank's signature (which can be done without communication with bank) to make sure it is valid.

6. The merchant takes the money order to his bank, which verifies the signature and adds $100 to merchant's bank account.

Protocol 1 makes anonymous cash, but cash that can still be spent twice. Protocol 2 protects the bank from double spending, but doesn't catch the double spender.

Protocol 2.

1. Alice makes 100 anonymous money orders for $100 each. On each one she writes a random twenty-digit integer. She blinds each one and gives all to her bank.

2. The bank asks Alice to open 99 money orders, and verifies that each one is a money order for $100 and that all 99 twenty-digit integers differ. (If not, Alice goes to jail.) The bank puts its blind signature on the last money order, returns it to Alice and deducts $100 from her bank account.

3. Alice unblinds the money order (now signed by the bank) and spends it with a merchant. The merchant verifies the bank's signature to make sure it is valid.

4. The merchant takes the money order to his bank.

5. The merchant's bank verifies the signature, and checks in a database to make sure that a money order with the same twenty-digit integer has not been previously spent. If this has not happened, then it adds $100 to the merchant's bank account and records the twenty-digit number in the database used by all banks. But if the number is already in the database, then the bank doesn't accept the money order.

With Protocol 2, if Alice tries to spend the money order twice, or if the merchant tries to deposit it twice (in two different banks, say), the second bank will know and not accept it. Protocol 3 identifies the cheater.

Protocol 3.

1. Alice makes 100 anonymous money orders for $100 each. On each one she writes a random twenty-digit integer and 100 pairs of identity bit strings: $(I_{1L}, I_{1R}), \ldots, (I_{100L}, I_{100R})$. Each part is a bit-committed packet that Alice can be asked to open, and whose proper opening can easily be checked. Any pair, (I_{59L}, I_{59R}), say, reveals Alice's identity when the two strings are exclusive-or'ed together. But halves from different pairs, such as (I_{23L}, I_{81R}) or (I_{19L}, I_{61L}), do not reveal who Alice is. Alice blinds each money order and gives all 100 of them to her bank.

2. The bank asks Alice to open 99 money orders, and verifies the contents. If it finds an error, Alice goes to jail. The bank puts its blind signature on the last money order, returns it to Alice and deducts $100 from her bank account.

3. Alice unblinds the money order (now signed by the bank) and spends it with a merchant, who verifies the bank's signature to make sure it is valid.

4. The merchant asks Alice to randomly reveal either the left or right half of each of the 100 identity strings. (The merchant chooses the random numbers.) Alice reveals them.

5. The merchant takes the money order to his bank. The bank verifies the signature and checks the database for the twenty-digit number. If it is not found therein, the bank credits the merchant with $100 and records the money order in the database.

6. If the twenty-digit integer is already in the database, the bank does not accept the money order. It compares the 100 identity strings on the money order with those in the database. If they agree, that is, the same set of left halves has been opened, the bank knows that the merchant copied the money order. If they differ, a second merchant deposited the money order earlier and it was Alice who copied. Some of the 100 identity strings will have both halves revealed, so Alice can be identified. In that case, Alice goes to jail.

Protocol 3 is not transferable nor can one make change with this digital money. We did not achieve all of our goals. But Protocol 3 does some remarkable things anyway.

Can Alice cheat? She can copy her \$100 money order. It works the first time she spends it. But she gets caught the second time she spends it.

Can she create a money order with a bad identity string? There is one chance in 100 that she can, not worth going to jail.

Alice can't change the twenty-digit number or the identity strings, because then the bank's signature would no longer be valid.

Can the merchant cheat? No. If he tries to deposit the money order twice, he will be caught, and Alice will not be implicated.

Can the merchant and Alice conspire to spend the digital cash twice? No, because they can't change the twenty-digit number signed by the bank, so the bank will not have to pay the \$100 more than once.

Can Eve copy Alice's money order and spend it first? Yes. It is like cash. Even worse, if Alice didn't know that Eve copied it and spent it, then Alice would be caught when she spent it the first time. If Eve spent it twice (or even more times) before Alice spent it, then Alice would go to jail.

Eve could eavesdrop on communication between Alice and the merchant and deposit the money (as a merchant) before the merchant deposits it. When the merchant tries to deposit it, he will be found as a cheater.

Both Alice and the merchant must protect their digital cash as if it were cash. It must be enciphered when it is sent across the Internet.

The money order in Protocol 3 takes about a megabyte.

21.3.2 *Electronic Cash According to Brands*

Here is an alternate way of creating digital cash, due to Brands [15]. Trappe and Washington have an excellent treatment of this protocol in Chapter 9 of [115]. In addition to the properties of cash in the previous protocol, this one offers some protection against someone stealing the cash from Alice. If they do, they won't be able to spend it. The protocol sounds complicated, but keep in mind that it does everything that Protocol 3 above does, and more. It also takes far less storage than Protocol 3, only a few kilobytes.

We will call the unit of digital cash a "coin" in this protocol, rather than a "money order."

A central bank chooses a Sophie Germain prime q. This means that $p = 2q + 1$ is also prime. The prime p must be large enough so that no one can solve the discrete logarithm problem modulo p, say, $p > 2^{1000}$. The central bank also chooses a number g which is the square of a primitive root modulo p. Then g has order q modulo p, so $g^i \equiv g^j \pmod{p}$ if and only if $i \equiv j \pmod{q}$ by Theorem 6.15. The central bank chooses two random numbers k_1 and k_2, and computes $g_1 = g^{k_1} \bmod p$ and $g_2 = g^{k_2} \bmod p$. It makes public the numbers p, q, g, g_1 and g_2. The random numbers k_1 and k_2 are destroyed. It also publishes a standard hash function S, like SHA, and a standard way of applying it to the concatenation of four or five large integers, so that, given the numbers, everyone would compute the same message digest of them.

Each bank in the electronic cash system chooses a secret identity number, x, which it remembers, and publishes the three numbers $h = g^x \bmod p$, $h_1 = g_1^x \bmod p$ and $h_2 = g_2^x \bmod p$ on its Web page.

When Alice opens an account at a bank, she chooses a secret identity number u. She tells the bank the account number $I = g_1^u \bmod p$. She does not tell u to anyone, not even the bank. The bank stores I together with information identifying Alice. It sends $z_1 = (Ig_2)^x$ to Alice to use when she later creates coins.

Each merchant chooses an identification number m, which is registered with the bank, together with the merchant's name and address.

A coin consists of a 6-tuple (A, B, z, a, b, r) of integers modulo p or q, and takes about a kilobyte to store. When Alice wants to withdraw cash from her bank account she performs the following steps for each coin.

1. Alice identifies herself to the bank and tells it the value of the coin she wants to withdraw from her account.

2. The bank chooses a random number w modulo q, a new one for each coin, and sends $g_w = g^w \bmod p$ and $e = (Ig_2)^w \bmod p$ to Alice. The bank keeps w secret and remembers it until Step 4, where it is used again.

3. Alice chooses five secret random numbers $s \neq 0$, x_1, x_2, y_1 and y_2, all modulo q. She chooses different numbers for each coin. Alice computes

$$A \equiv (Ig_2)^s \bmod p,$$
$$B \equiv g_1^{x_1} g_2^{x_2} \bmod p,$$
$$z \equiv z_1^s \bmod p,$$
$$a \equiv g_w^{y_1} g^{y_2} \bmod p \text{ and}$$
$$b \equiv e^{sy_1} A^{y_2} \bmod p.$$

The case $A = 1$ is forbidden. This can happen only when $s = 0$, which is not allowed, or when $Ig_2 \equiv 1 \pmod{p}$, which is equivalent to $g_1^{-u} \equiv g_2 \pmod{p}$, and this would mean that Alice had solved a discrete logarithm problem modulo p when she chose u. But the prime p is so large that this can't happen. Alice computes $c = (y_1)^{-1} S(A, B, z, a, b) \bmod q$ and sends c to her bank.

4. The bank computes $c_1 = cx + w \bmod q$ and sends c_1 to Alice. The bank destroys w after this use. The bank deducts the value of the coin from Alice's account.

5. Alice computes $r = y_1 c_1 + y_2 \bmod q$ and the coin (A, B, z, a, b, r) is ready.

This may sound complicated, but Alice's wallet computer and the bank's computer can do it all in less than a second.

Now Alice takes the coin (A, B, z, a, b, r), and probably others, to the merchant. Her computer and the merchant's computer perform the following steps to let Alice spend the coin.

1. Alice gives the coin (A, B, z, a, b, r) to the merchant.

2. The merchant tests whether $g^r \equiv ah^{S(A,B,z,a,b)} \pmod{p}$ and $A^r \equiv bz^{S(A,B,z,a,b)} \pmod{p}$. If either congruence fails, the coin is invalid, and the merchant rejects the coin. If both congruences hold, the merchant computes $d = S(A, B, m, t)$, where t is a current time stamp. The merchant sends d to Alice.

3. Alice computes $r_1 = (dus + x_1) \bmod q$ and $r_2 = (ds + x_2) \bmod q$, where u is Alice's secret identity number she used to establish her bank account, and s, x_1, and x_2 are three of the secret random numbers she chose to generate this coin. Alice sends r_1 and r_2 to the merchant.

4. The merchant checks the congruence $g_1^{r_1} g_2^{r_2} \equiv A^d B \pmod{p}$. If so, the merchant accepts the coin and the transaction is complete. If not, the merchant rejects the coin because Alice stole it from someone else.

Some time after the transaction, the merchant tries to deposit the coin in the bank. He gives it to the bank, together with the triple (r_1, r_2, d). The bank performs these checks.

First, the bank checks whether the coin (A, B, z, a, b, r) is already stored in its database of used coins. If it is not there, the bank checks the same three congruences that the merchant checked earlier, namely,

$$g^r \equiv ah^{S(A,B,z,a,b)} \pmod{p},$$
$$A^r \equiv bz^{S(A,B,z,a,b)} \pmod{p}, \text{ and}$$
$$g_1^{r_1} g_2^{r_2} \equiv A^d B \pmod{p}.$$

If all three hold, the coin is valid and the bank credits it to the merchant's account. It enters the coin (A, B, z, a, b, r) and the triple (r_1, r_2, d) into its database of used coins.

If the coin (A, B, z, a, b, r) is already in the bank's database, then it has a triple (r_1', r_2', d') stored with it. The bank compares this triple with the triple (r_1, r_2, d) just submitted by the merchant. If the triples are the same, then the merchant must be submitting the same coin again because the merchant's identification number m and the time stamp t are hashed into d and Alice's secret u is in r_1. If the triples differ, and in particular if $r_2 \neq r_2'$, then the bank can compute Alice's secret identity number u from $u = (r_1 - r_1')(r_2 - r_2')^{-1} \pmod{q}$. Then it calculates $I = g_1^u \bmod p$ and learns Alice's identity. Alice goes to jail.

The difficulty of solving the discrete logarithm problem modulo p prevents most other possible fraud. The merchant cannot submit a coin twice, once with the triple (r_1, r_2, d) and once with a phony triple (r_1', r_2', d'), because he can't compute a second triple satisfying $g_1^{r_1'} g_2^{r_2'} \equiv A^{d'} B \pmod{p}$. For the same reason, one merchant cannot deposit a coin in the bank and also spend the same coin at another merchant. If someone stole a coin from Alice, they could not spend it because they could not compute r_1 and r_2 with $g_1^{r_1} g_2^{r_2} \equiv A^{d'} B \pmod{p}$, where d' was given to them by a merchant. Likewise, no one could forge a bogus coin because they could not compute a number r satisfying $g^r \equiv ah^{S(A,B,z,a,b)} \pmod{p}$ and $A^r \equiv bz^{S(A,B,z,a,b)} \pmod{p}$. A person working in the bank and knowing Alice's I could not forge a coin with Alice's identity because they don't know Alice's secret u, and that number is needed to produce r_1. If an evil merchant stole a coin and the triple (r_1, r_2, d) from a merchant before that merchant sent it to the bank, then the evil merchant could successfully deposit it first. This is a problem with real cash, too.

The transaction between Alice and the merchant is totally anonymous, just as it would be if real cash were used. Their computers merely exchange numbers. Alice's identity I is hidden in one of the numbers, but it cannot be extracted unless Alice spends the coin twice. The bank knows I. It could remember that w was a random number used to create a coin for Alice, although it is not supposed to store this number. It might even keep track of c and c_1. With all this knowledge, can it inspect incoming coins and tell which ones were spent by Alice? Only Alice knows the secret numbers s, x_1, x_2, y_1 and y_2. Therefore, A, B, z, a, b and r, which depend on these five secret random numbers are just six random numbers to anyone other than Alice.

21.4 Exercises

1. Explain why the protocols for contract signing work, if they do. What if one party has much faster computers than the other?

2. Explain why the protocols for elections work, if they do.

3. Make reasonable estimates for the sizes of the numbers used in Chaum's Protocol 3, and estimate the number of bytes Alice must carry from her bank to the merchant.

4. In Chaum's Protocol 3, why must Alice bit-commit the halves of the identity bit strings? After all, if she changed them, the bank's signature on them would not be valid.

5. Suppose a coin is created properly using the protocol of Brands. Prove that the three congruences checked by the merchant and the bank will hold.

6. Explain how someone who knows the bank's secret number x can create and spend valid coins without even having an account at the bank. What happens if he spends one of these coins twice?

7. In which digital cash protocols in this chapter can Alice and a Merchant conspire to cheat the Bank?

Chapter 22

Complete Systems

This chapter introduces two complete systems. See Chapters 14 and 15 of Stallings [114] for more about these systems.

22.1 Kerberos

Kerberos, developed by Project Athena at MIT, solves this problem: Assume an open distributed environment. Users at workstations wish to access services on various servers. Servers wish to restrict access to authorized users and be able to authenticate users' requests for service. A workstation cannot be trusted to identify its users correctly.

There are three threats:

1. A user may gain access to a workstation and pretend to be another user on that workstation.

2. A user may alter the network address of a workstation so that the requests from it appear to come from a different workstation.

3. A user may eavesdrop on exchanges and use a replay attack to gain entrance to a server or to disrupt operations.

Kerberos addresses these threats by providing a central authentication server, AS, to authenticate users to servers and servers to users. It requires a user to prove identity for each service invoked. It also requires that servers prove their identity to clients.

Kerberos uses only conventional cryptography (DES), no public key cryptography, and is supposed to be:

- Secure: No eavesdropper can impersonate a user.

- Reliable: No one can use any services unless permitted by Kerberos.

- Transparent: A user just types a password; all else is hidden.

- Scalable: It can support many clients and servers.

Version 4 was the first full version, and is still in use. Version 5 is the next full version.

Here is a simple authentication dialogue:

1. $C \to AS$: ID_C, P_C, ID_V
2. $AS \to C$: $Ticket$
3. $C \to V$: $ID_C, Ticket$

where $Ticket = E_{K_V}[ID_C, AD_C, ID_V]$.

AS = authentication server (Kerberos)
C = client
V = server
ID_C = identification of user on C
ID_V = identification of server V
P_C = password of user on C
AD_C = network address of C
K_V = secret key shared by AS and V.

The use of AD_C prevents ticket capture and reuse.

The ticket is valid only once and only from workstation C.

There are some problems with this simple dialogue:

1. A user on C must enter her password for each ticket, which is too many times. It would be better to make the ticket reusable.

2. The plaintext transmission of the password in Step 1. An eavesdropper could capture it.

These problems are solved by adding a Ticket Granting Server, TGS.

Once per user login session:

1. $C \to AS$: ID_C, ID_{TGS}.
2. $AS \to C$: $E_{K_C}[Ticket_{TGS}]$,

where $Ticket_{TGS} = E_{K_{TGS}}[ID_C, AD_C, ID_{TGS}, TS_1, LT_1]$.

Once per type of service (mail, print, login, etc.):

3. $C \to TGS$: $ID_C, ID_V, Ticket_{TGS}$.
4. $TGS \to C$: $Ticket_V$.

where $Ticket_V = E_{K_V}[ID_C, AD_C, ID_V, TS_2, LT_2]$.

Once per service session:

5. $C \to V$: $ID_C, Ticket_V$.

In 1 and 2, no password is sent over the network. Instead, in 2, C asks its user for a password (K_C) and uses it to decrypt the ticket.

The time stamps and lifetimes prevent reuse by an eavesdropper, unless he reuses it right away.

There are two problems with the dialogue above:

(a) The lifetime may be too long or too short. The TGS or V should be able to check that the person using the ticket is the same as the one to whom it was issued.

(b) Servers should have to prove their identity to users. Otherwise a bogus server could capture information from an unwary user and deny service.

These problems are solved by the following Kerberos 4 dialogue.

Once per user login session:

1. $C \rightarrow AS$: ID_C, ID_{TGS}, TS_1.

2. $AS \rightarrow C$: $E_{K_C}[K_{C,TGS}, ID_{TGS}, TS_2, LT_2, Ticket_{TGS}]$,

where $Ticket_{TGS} = E_{K_{TGS}}[K_{C,TGS}, ID_C, AD_C, ID_{TGS}, TS_2, LT_2]$.

Once per type of service:

3. $C \rightarrow TGS$: $ID_V, Ticket_{TGS}, Authenticator_{C,TGS}$,

where $Authenticator_{C,TGS} = E_{K_{C,TGS}}[ID_C, AD_C, TS_3]$.

4. $TGS \rightarrow C$: $E_{K_{C,TGS}}[K_{C,V}, ID_V, TS_4, Ticket_V]$,

where $Ticket_V = E_{K_V}[K_{C,V}, ID_C, AD_C, ID_V, TS_4, LT_4]$.

Once per service session:

5. $C \rightarrow V$: $Ticket_V, Authenticator_{C,V}$,

where $Authenticator_{C,V} = E_{K_{C,V}}[ID_C, AD_C, TS_5]$.

6. $V \rightarrow C$: $E_{K_{C,V}}[TS_5 + 1]$.

Now the lifetime is less important and can be made long enough, since knowledge of $K_{C,TGS}$ and $K_{C,V}$ prove the user is the grantee of the ticket.

In 6, V proves its identity to C.

Purdue University students Dole and Lodin broke Kerberos 4 a few years ago.

22.2 *Pretty Good Privacy*

Pretty Good Privacy, PGP, was written mostly by Phil Zimmermann.

He used the best available crypto algorithms as building blocks to create a system for enciphering both files and e-mail. It provides confidentiality and/or authentication. It is independent of operating system and machine. It has a small number of easy-to-use commands. It is freely available on the Internet. Authentication is provided by SHA signed by either RSA or DSS. Confidentiality is provided by encryption using either CAST-128, IDEA or Triple DES with a one-time key generated by the sender. PGP also provides ZIP compression, radix-64 conversion (for e-mail), as well as segmentation and reassembly of long messages.

The signature is generated before compression because:

(a) It is better to sign an uncompressed message so that you can store only the uncompressed message and signature for later verification. If you signed a compressed document, you would either have to store the compressed document or else recompress it at verification time.

(b) The ZIP compression algorithm is not deterministic. Different versions of ZIP produce different compressed files. Signing after compression would require the use of just one version of ZIP.

The message is enciphered after compression because the compressed message has less redundancy; so, its cryptanalysis is harder.

The random numbers for generating session keys come from the timing of the users' keystrokes.

Users may have more than one set of RSA keys (to change keys or to communicate with different sets of correspondents, say). Each public key is identified by its low-order 64 bits in messages sent to the recipient.

Each user of PGP has two data structures to hold keys: one for his own public/private key pairs and one to store the public keys of other users. These data structures are called the user's private-key ring and public-key ring.

The private keys are encrypted via a passphrase. SHA produces a 160-bit hash of the passphrase and 128 of these 160 bits are used as the key for CAST-128. Private keys are indexed either by their low-order 64 bits or by a userid.

The public keys are stored in a similar data structure, but which has additional fields for a time stamp and trust information.

Suppose Alice gets a public key for Bob from a source which has been compromised by Chuck, so that the key Alice thinks is Bob's really comes from Chuck. Then Chuck could send a message to Alice signed "Bob" and Alice would accept it as coming from Bob. Furthermore, Chuck could read any encrypted message from Alice to Bob.

One way to solve this problem would use X.509. PGP uses the notion of "trust" instead.

PGP provides a way for a public key to be "signed" by another public key. It also has a level of "trust" associated to each public key. The higher the level of trust, the stronger the binding of userid and key. A key that is signed by trusted keys is also trusted to a degree determined by number and degree of trust of the trusted keys.

The degrees of trust are: undefined, unknown user, usually not trusted to sign other keys, usually trusted to sign other keys, always trusted to sign other keys, and present in the secret key ring (ultimate trust).

If a user wishes to change one of his public keys or if he believes it has been compromised, then he widely disseminates a Key Revocation Certificate, signed by the associated private key.

22.3 Exercises

1. Find weaknesses in the Kerberos and PGP protocols.

Part III

Methods of Attack

Chapter 23

Direct Attacks

This section of the book describes various attacks on the ciphers and protocols mentioned in Part II. This section is not as long as it might have been since many attacks were already described when the algorithms were presented in Part II.

The attacks vary in the information known to the attacker, the computational power of the attacker and the attacker's goal. If the cryptanalyst knows only that the ciphertext is a string of bits, then it is hard to make any progress. We will assume that at least the type of cipher is known, in addition to some ciphertext. As explained in Chapter 1, this ciphertext-only attack is the most difficult. In case of a public-key cipher, we will assume that the public key data are also known.

It is an advantage for the cryptanalyst to know some pairs of plaintext and ciphertext. Often one can mount this known-plaintext attack by guessing some of the plaintext, such as a standard header.

The cryptanalyst may get even more help if he can choose plaintext and see the corresponding ciphertext. This chosen-plaintext attack is always available with a public-key cipher, and usually doesn't help much in that case.

This chapter introduces some direct attacks, which include the most obvious kinds of attack. In these attacks a direct assault is made on a secret key or message.

23.1 Try All Keys

This sounds dumb, and it is dumb, but occasionally it works. Most ciphers have so many possible keys that it would take too long to try all of them in a known-plaintext attack.

The Data Encryption Standard, DES, is a block cipher with 56-bit keys. One can build a special machine with many fast simple processors which can try all 2^{56} possible keys in a few hours.

If a defect in the key selection algorithm limits the number of possible keys, then one might be able to try all of the possible ones. This is how Bryn Dole and Steve Lodin [69] broke Kerberos 4.

Recall that Kerberos is a secret key network authentication protocol designed at MIT. See Section 22.1 for a description of the protocol. Its key distribution server and ticket granting server generate secret keys for the symmetric block cipher DES. A user who could guess these keys could intercept session keys, which are enciphered, and use them to access services without authorization. DES keys have 56 bits, and 2^{56} keys is too many for one to try all of them during the lifetime of a session key.

Kerberos 4 generates the 56-bit keys as follows. A pseudorandom number generator, PRNG, is seeded with a random 32-bit seed and called twice to produce two 32-bit random numbers. Every eighth bit is set as a parity bit. This 64-bit quantity is the 56-bit DES key for a session key. Since the key depends only on the 32-bit seed, the entropy of the 56-bit key K is only $H(K) = 32$. This is already a serious problem because one could try all 2^{32} seeds and test each DES key they yield in only a few hours on a workstation. But the lifetime of a typical session key is a few hours; so, this attack may or may not work.

But the situation is even worse. The 32-bit seed for the PRNG is formed as the exclusive-or of five random 32-bit numbers. They are:

1. the time-of-day in seconds since January 1, 1970,

2. the fractional part of the current time in microseconds,

3. the process ID of the Kerberos server process,

4. the cumulative count of session keys produced so far, and

5. the hostid of the machine on which Kerberos is running.

These five quantities have various amounts of entropy from 1 to 20 bits. The fractional part of the time has the most entropy. It is a random number modulo 1,000,000. Since the uncertain bits are always in the low-order 20 bit positions and since the five numbers are combined with exclusive-or, the entropy of the seed for the PRNG is only 20 bits. Therefore, one could generate and test all 2^{20} possible DES keys in a few seconds. This attack will succeed easily within the lifetime of the session key.

Rather than forming the exclusive-or of the five random numbers, the key generator should have computed a hash of their concatenation and chosen the bits of the session key from the bits of the message digest. Then every bit of randomness in the numbers would contribute to every bit of the session key. The PRNG, with its 32-bit seed, should not have been used at all.

This problem was corrected in Kerberos 5.

Another situation in which one can try all keys is when a human chooses a key. When a cipher program asks a user to type a password and then converts it into a key, the user often chooses an easy-to-remember word, like a word in a dictionary. This oversight happens frequently with login passwords. One can attack this error by trying all words in a dictionary. One can also try dictionary words spelled backwards, words with one letter capitalized, pairs of short words, foreign words, and strings of letters constructed from the letters in a user's name. One can guess a large percentage of login passwords this way.

One way to counter this problem is to have a program assign random passwords. But these are difficult to remember. Another approach is to use passphrases. These are easy to remember and the user can type the first letter of each word as he says the passphrase to himself. Another way of using passphrases is to have the user type the entire passphrase, let a program compute a message digest of it, and use part of the hash value as the key. Keep in mind that the rate of English is about one bit per letter. This translates into five or six normal words per 32 bits of key. A 64-bit key should have a passphrase of at least a dozen words to produce enough key entropy.

23.2 Factor a Large Integer

This attack might work for some public-key systems, like RSA or Rabin-Williams. Some day a polynomial-time integer factoring algorithm might be discovered. If that happens, these systems will be out of business.

Usually, the composite numbers n used as public keys are chosen so that they cannot be factored. Designers of these systems must be aware of all known integer factoring algorithms and choose keys that will make it impossible to factor n by any of these methods. One requirement is that the prime factors of n be large so that they cannot be found by trial division or the Pollard rho method. On the other hand, the larger n is, the slower the enciphering function will be, so n should be as small as possible, subject to being too hard to factor. This size requirement implies that n should have only two prime factors because if n had three prime factors, then one of them would have to be less than the cube root of n. If $n = pq$ is the product of two primes, they must be different because (i) prime powers are easy to detect and factor, and (ii) the cipher wouldn't work if $n = p^2$.

Thus, if we want n to be difficult to factor, we should choose $n = pq$ where the primes p and q are close to each other. But if they were too close, Fermat's difference of squares method could find them. To avoid this attack, choose $|q - p| > 10^{25}$.

The primes p and q should also have the property that all four of the numbers $p \pm 1$, $q \pm 1$ have at least one prime factor $> 10^{20}$. If either of the numbers $p \pm 1$ were 10^{10}-smooth, then p could be found by the Pollard $p - 1$ method or Williams' $p + 1$ method.

In order to avoid having p discovered by the elliptic curve method one would have to make certain that no integer N in the Hasse interval $p + 1 - 2\sqrt{p} < N < p + 1 + 2\sqrt{p}$ was 10^{10}-smooth, say. Of course, it is impossible to check this requirement. The best that one can do is make $p > 10^{80}$, say. The largest prime factor ever discovered by ECM has about 55 digits. This record increases slowly, but will take many years to pass 10^{80}.

The largest number n factored with the quadratic sieve algorithm has 135 decimal digits. If you choose $n = pq$ with $q > p > 10^{80}$, then n will be too large to factor with QS.

The general number field sieve has factored n with about 160 decimal digits, and this record is slowly increasing as faster computers are applied to factoring. To be safe from GNFS, one should choose $n > 10^{200}$, at least.

The special number field sieve has factored n with about 230 decimal digits, and by the time this is published, it will probably have factored an integer with more than 1024 bits, a common size specified for RSA cryptographic keys. Of course, the SNFS works only for integers with the special form $r^e - s$, where r and $|s|$ are small positive integers. Make sure that your RSA key n is not of this form. Use a program like this one to test candidate n.

```
for (e = 2 to 1000) {
        r = the nearest integer to the e-th root of n
        if (|r^e - n| < 1000) { print "n is bad"; exit }
        }
print "n is good"
```

Some cryptographic algorithms might be endangered by a sudden increase in the speed of computers due to new technology. Several computational paradigms loom on the horizon. Shor [109], [110] has shown how one may factor large integers using quantum computation. Shamir [104] proposed an optoelectronic processor for factoring integers via the quadratic or number field sieve. Paun et al. [80] explain how future computers might work via chemical reactions of DNA molecules. If any of these technologies succeed, we could factor integers and/or compute discrete logarithms much faster than we can with current machines. Only time will tell whether these new methods will work.

23.3   Solve a Discrete Logarithm Problem

This attack might apply to some private- or public-key systems, like Pohlig-Hellman or ElGamal, or to a key exchange protocol, like Diffie-Hellman. Some day faster algorithms for discrete logarithms might be discovered. If that happens, these systems may be out of business or one may have to use larger groups. The discrete logarithm problem is to solve for x in the equation $a^x = b$ in some group. The time to solve the problem depends mostly on the size of the group, although it might be easy for special a and b. Usually, the size of the group is chosen large enough so that the problem cannot be solved in reasonable time. But if it is chosen too large, then the algorithm speed will suffer. Designers of these systems must be aware of all known discrete logarithm algorithms and choose the size just large enough so that the problem cannot be solved by any of these methods.

Three types of groups are used in contemporary cryptography. The complexity of the discrete logarithm problem is quite different in the three types.

The first group of interest is the multiplicative group R_p of integers modulo p. This group appears in the Pohlig-Hellman and ElGamal ciphers, the ElGamal and Digital Signature Algorithms and the Diffie-Hellman key exchange protocol. Its discrete logarithm problem is a congruence, $a^x \equiv b \pmod{p}$. The parameter p is almost always prime in this application because the discrete logarithm problem is hardest in that case. The fastest algorithms for solving this problem are the index calculus method, which is similar to the quadratic sieve factoring algorithm, and the number field sieve, which is like the factoring algorithm of the same name. The complexity of these algorithms is comparable to that of the similar factoring algorithms when they factor a composite number of about the same size as p. This means that p should be larger than about 160 digits to be safe from the index calculus method and larger than about 200 digits to be safe from the number field sieve method. Work on these discrete logarithm algorithms has lagged work on the similar factoring algorithms. The record p's for solving discrete logarithms is quite a bit smaller than the largest hard numbers factored by these algorithms.

The second group used in cryptography is the group of points on an elliptic curve modulo a prime p. These groups have roughly p points. But they have no notion of smoothness, which is what makes the index calculus or number field sieve work. The product of smooth numbers in R_p is smooth—or congruent to a smooth number, which is all the algorithm wants. One might think of defining a point on an elliptic curve to be smooth if its coordinates are small compared to p. But then $P + Q$ need not be smooth when P and Q are smooth. Therefore, one must use other algorithms to solve the discrete logarithm problem on an elliptic curve. The other nontrivial algorithms include Shanks' baby-step-giant-step method and the rho and lambda methods of Pollard. All three of these algorithms have complexity $O(\sqrt{p})$, far slower than the index calculus and the number field sieve, whose complexities are $L(p)$ or better, where $L(x) = \exp(\sqrt{(\ln x)\ln\ln x})$. A rough comparison of $\sqrt{p}$ with $L(p)$ shows that the discrete logarithm problem for an elliptic curve modulo a 160-bit prime is a bit harder to solve than the problem in R_p modulo a 1024-bit prime. Both problems are too hard to solve with current algorithms and machines. With these parameter choices, encryption by ElGamal or Pohlig-Hellman would run about 100 times faster in an elliptic curve than in R_p.

We mentioned the third group commonly used in cryptography briefly at the end of Chapter 14. It is the multiplicative group of the field $\mathbf{F}_{2^n}$ with 2^n elements. This group is cyclic and arithmetic in it is fast on a binary computer. Coppersmith [30] found an algorithm for computing discrete logarithms in this field which works for n up to about 1000. It consists of a massive precomputation, which need be done only once for each n, and a fairly short main computation to find particular discrete logarithms.

23.4   Timing Attacks

These insidious attacks were discovered by Kocher [59] and apply to nearly all cryptographic algorithms whose execution time depends on the input value. This includes most of the ciphers, signature schemes and key exchange protocols discussed in this book.

In order to perform the attack, you must be able to observe a cipher program running on your computer and make precise measurements of the time it takes to run on various inputs. You must also know the input value and the parameters of the cryptographic algorithm other than the secret key. Someone with an account on the victim's machine and who could observe incoming packets could easily obtain the required information.

Let us use RSA as a simple example of a timing attack. The victim has modulus n, enciphering exponent e and deciphering exponent d. The latter is secret, while n and e are public. The victim receives ciphertext messages C and deciphers them by computing $M = C^d \bmod n$. Let $d = \sum_{i=0}^{k} b_i 2^i$ be the binary representation of d. The attacker records many ciphertexts C_j and the time t_j needed to decipher each. He deduces d one bit at a time, from b_0 to b_k. This is the order in which the bits are used in the fast exponentiation algorithm. Assume that the first r bits have been computed. We repeat the fast exponentiation algorithm from Chapter 6, but specialize it here to RSA decryption.

[Fast Exponentiation for RSA Deciphering]
Input: A modulus n, an exponent $d \geq 0$ and a ciphertext C.
Output: The value $M = C^d$.

```
e = d
M = 1
z = C
while (e > 0) {
          if (e is odd)  M = Mz mod n
          z = z² mod n
          e = ⌊e/2⌋
          }
return M
```

Let us suppose that the operation $M = Mz \bmod n$ takes longer for some pairs M, z than for other pairs and that the attacker can measure the execution time of the algorithm accurately enough to notice the difference. Because the first r bits of d have been computed, the attacker can perform the first r iterations of the while loop for input C_j and measure its time c_j precisely. The attacker can also measure the precise time d_j the operation $M = Mz \bmod$

n would take, if it were done. He knows whether this particular modular multiplication is fast or slow compared to the time for average pairs M, z. Using a formula from statistics, he can predict whether b_r is 0 or 1. He compares the two variances $v_1 = \mathbf{Var}(t_j - c_j)$ and $v_2 = \mathbf{Var}(t_j - c_j - d_j)$. If $v_1 > v_2$, the bit b_r is probably 1; but if $v_1 < v_2$, then b_r is probably 0. This works because of Part 4 of Theorem 2.6. For if the multiplication occurs, it is reasonable to assume that the time d_j it takes and the time $t_j - c_j - d_j$ for the part of the fast exponentiation after it are mutually independent, so

$$v_1 = \mathbf{Var}(t_j - c_j) = \mathbf{Var}(t_j - c_j - d_j) + \mathbf{Var}(d_j) > \mathbf{Var}(t_j - c_j - d_j) = v_2.$$

But if the multiplication does not occur, then the time d_j it takes and the time $t_j - c_j$ for the part of the fast exponentiation after it are mutually independent, so

$$v_2 = \mathbf{Var}(t_j - c_j - d_j) = \mathbf{Var}(t_j - c_j) + \mathbf{Var}(-d_j) > \mathbf{Var}(t_j - c_j) = v_1.$$

If a mistake is made, then no further significant differences between v_1 and v_2 will appear for larger r. In that case, the attacker will notice the error, back up and correct it. See [59] for more details.

The timing attack works against the Diffie-Hellman key exchange protocol provided Alice always uses the same random x_A, but different primitive roots g are used each time, and the modulus n is fixed.

23.5   Exercises

1. Read the source code of security programs, like the Secure Socket Layer, which generate random numbers. Estimate the entropy of these random numbers and find ways to predict them.

2. Eve routinely records all ciphertext messages sent from Alice to Bob. Bob uses the RSA cipher with public keys n_B and e_B. One day, Eve learns from another source that one of the *plaintext* messages Alice sent to Bob was not relatively prime to n_B (and it was not 0 either). Does this tidbit help Eve to decipher any of the ciphertext messages?

3. Alice was playing with Bob's RSA public keys and noticed that if she enciphered a message five times, she always got the message back. That is, $E_B(E_B(E_B(E_B(E_B(M))))) = M$ for every M she tried.

   a. If Alice intercepted a message C from Carol to Bob enciphered (once, of course) with Bob's public RSA keys, tell how she could decipher it.

   b. What mathematical property of Bob's RSA keys caused this weakness?

   c. Without expensive repeated enciphering of lots of messages, how could Bob have avoided this weakness when he chose his RSA keys?

(By "this weakness" I mean "enciphering any small number of times, not just five times, and returning to the plaintext.")

4. The timing attack could be thwarted if one could ensure that all modular multiplication operations, $Mz \bmod n$, take exactly the same time. Can you think of a way to make this happen?

Chapter 24

Exploiting an Error

In this chapter, we explore various ways an attacker might exploit an error made by the user(s) of a cipher or protocol, or even a hardware or software error. These errors may or may not relate to number theory. Many attacks on RSA in this chapter and the previous one are described in Boneh [12].

24.1 Key Management

We have seen how to choose keys for a number theoretic cipher sufficiently large so that no one can find them in a direct attack, by factoring a number or solving a discrete logarithm problem in some group. Although it requires some hard mathematics to be able to make these choices, this is the easy part. If you leave your passphrase written on a note on your terminal, it doesn't matter how large your keys are. If a knowledgeable attacker sees that note, he can use your cipher program to generate your key the same way the program generates it when you use it.

If your keys are stored enciphered in a file, and you type a passphrase to allow your cipher program to access them, then you must choose a passphrase that is hard to guess, and you must not write it down anywhere. What if you forget it? Then you will not be able to read your enciphered files or messages. If this matters to you, then you will have to write your passphrase, or at least the keys you cannot afford to lose, in some secure location, like on a sheet of paper kept in a safe. If you forget the combination to the safe, you can cut it open. If several people need to share secret keys, the methods of Section 20.5 might help.

When generating keys, you must create enough entropy to make all bits of the key as random as possible. It is not good enough to choose a 32-bit seed for a random number generator and use its output to produce your key. Even if your key has 160 bits, it will have no more than 32 bits of entropy. I once graded a student project which generated 512-bit primes for RSA keys. The primes it made were all $\equiv 1 \pmod{2^{480}}$. The extended-precision integer package used base 2^{30}, storing 30 bits per word. The student had formed random numbers rs000000000001 in this base as candidate large primes, putting all the randomness in the high-order 32 bits rs. Strong probable prime tests are especially fast for numbers p of this form because $p - 1$ is divisible by at least 480 twos. But that is no excuse for not making all fourteen digits random. And one should not call the system random number generator fourteen times to generate these digits, as there would be no more entropy in the prime than in the seed for the RNG. The high-order and low-order bits of a 512-bit random prime must be 1, but the remaining 510 bits must come from a source with at least 510 bits of entropy.

How do Alice and Bob exchange keys for secure communication over an insecure medium like the Internet? The best solution is for them to meet and trade keys before communicating. If that is not possible, then they should use the X.509 protocol to learn each other's public keys. They should verify the signatures certifying these keys, which are a part of the protocol. If Alice uses the Diffie-Hellman key exchange to establish a common private key with Bob over the Internet, then she will get a secret known only to her and to whomever is performing the other side of the protocol. The number theory behind this key exchange does not guarantee that it is Bob.

The protocols and algorithms in this book are secure only if the key remains secret. If the key is compromised, security is lost. If the stolen or revealed key was for a symmetric cipher, then Alice must change her key and hope for the best. But if it was a private key for an asymmetric cryptosystem, the damage is greater. Alice must quickly change her public key in every location that stores it and tell everyone who might use it about the new key. These messages should have a time stamp so that the recipient will know when the theft happened. Someone who discovered Alice's private key could read her enciphered mail, sign messages as Alice and literally become Alice. Each time you use someone's public key you should check to see whether it has been compromised and revoked.

24.2   Reuse of a Key

A cryptographic key should never be reused after it has been exposed. This is obvious for most ciphers.

We give a slightly less obvious example for RSA. Suppose Alice uses RSA modulus n, public encryption exponent e and private decryption exponent d. If d is exposed, Alice must change n because one can almost certainly factor n given e and d.

Note first that $ed \equiv 1 \pmod{\phi(n)}$ by construction of the RSA system. Therefore, $ed - 1 = k\phi(n)$ for some integer k. If we let $r = ed - 1$, then whenever $\gcd(a, n) = 1$ we have

$$a^r = a^{ed-1} = \left(a^{\phi(n)}\right)^k \equiv 1 \pmod{n} \tag{24.1}$$

by Euler's theorem. Note that r is even because $\phi(n)$ is even for $n > 2$.

Now write $r = 2^s d$ with d odd. Choose a random a in $1 < a < n - 1$. If $\gcd(a, n) > 1$, then n has been factored and we are done. Otherwise, compute $b_i = a^{2^i d} \bmod n$ for $0 \le i \le s$. We know that $b_s = a^r \bmod n = 1$ by Congruence (24.1). If for some $0 < i \le s$ we have $b_i = 1$ but $b_{i-1} \not\equiv \pm 1 \pmod{n}$, then $\gcd(b_{i-1} - 1, n)$ is a proper factor of n. If there is no such i, try a different random a. The reason this works is that $b_{i-1}^2 \equiv 1 \pmod{n}$, but $b_{i-1} \not\equiv \pm 1 \pmod{n}$, so $\gcd(b_{i-1} - 1, n)$ is a proper factor of n by Theorem 13.1. We can argue just as in the proof of Theorem 13.2 that each random a leads to a factorization of n with probability at least $1/2$.

Here is an example of the misuse of RSA. Suppose a trusted central authority chose a fixed modulus $n = pq$ for everyone to use. It would keep the primes p and q secret, of course, but n would be public. It would provide each user A with a pair of exponents e_A, d_A for encryption and decryption. Each e_A would be public, but only user A would know d_A. At first glance, this might seem reasonable since user B could not decipher a message enciphered with e_A because B does not know d_A. However, user B knows both e_B and d_B and, by the argument in the preceding paragraph, this is enough to factor n. Then user B could compute d_A from e_A, which is public, and read A's mail.

24.3 Bad Parameter Choice

There are many ways in which users may choose secret parameters that are easy for an attacker to guess or compute. As one simple example, suppose you must choose a random secret 100-digit integer. Suppose you do this by choosing a random 30-bit integer and multiplying it by 10^{91}. Then your 100-digit number will have at most 30 bits of entropy and will be easy to guess. Suppose you need a random secret 100-digit prime and construct it by choosing a random 30-bit integer, multiplying by 10^{91}, and selecting the first strong probable prime to base 2 greater than the product. Then your 100-digit prime will be insecure because it will have only 30 bits of entropy.

Here are more examples for RSA. The next attack is due to Hastad [52]. The smallest possible value for the public enciphering exponent is $e = 3$, and some implementations of RSA use this value of e. Sometimes this is done for communication between smart cards, which have relatively slow processors, and larger, faster computers. If the public key for the large machine has $e = 3$, then the slow smart card only has to raise M to the third power to encipher it. The message M might be a credit card number. Now suppose the smart card is used to make purchases at three different merchants, with public moduli n_1, n_2, n_3. An attacker who could observe the ciphertext transmitted during these three transactions would know

$$M^3 \bmod n_1, \quad M^3 \bmod n_2, \quad \text{and} \quad M^3 \bmod n_3$$

and, of course, the three public moduli. If the moduli were not relatively prime in pairs, then he could factor at least two of them and recover M by deciphering it the way the merchant's computer did it. If they were relatively prime in pairs, then the attacker could use the Chinese remainder theorem to determine $M^3 \pmod{(n_1 n_2 n_3)}$. But $0 < M < n_i$ for each i, and so $0 < M^3 < n_1 n_2 n_3$ and the attacker has found the actual value of M^3, just as Sun Tsu found the number of his soldiers. Then the attacker could find M by computing the cube root of M^3, for example, by Newton's method. This attack works not only for $e = 3$, but for any small e.

Users of RSA might wish to choose a small deciphering exponent d to speed decryption. Clearly, d must not be so small that one could guess it, but it must not even be as small as $n^{1/4}$, as shown by this theorem of Wiener [120], which uses continued fractions in its proof.

THEOREM 24.1 Small RSA deciphering exponents are bad
Let $n = pq$, where the primes p and q satisfy $q < p < 2q$. Let $ed \equiv 1 \pmod{\phi(n)}$, where $1 < e < \phi(n)$ and $1 < d < n^{1/4}/3$. An attacker given n and e can efficiently find d.

PROOF Since $ed \equiv 1 \pmod{\phi(n)}$, there is an integer k with $ed - k\phi(n) = 1$. Therefore,

$$\left| \frac{e}{\phi(n)} - \frac{k}{d} \right| = \frac{1}{d\phi(n)}.$$

This shows that k/d is an approximation to $e/\phi(n)$. The attacker does not know $\phi(n)$, but may use n as an approximation of it. Since $\phi(n) = n - p - q + 1$ and $p + q - 1 < 3\sqrt{n}$, we have $0 < n - \phi(n) < 3\sqrt{n}$. Thus,

$$\left| \frac{e}{n} - \frac{k}{d} \right| = \left| \frac{ed - k\phi(n) - kn + k\phi(n)}{nd} \right| = \left| \frac{1 - k(n - \phi(n))}{nd} \right| \leq \frac{3k\sqrt{n}}{nd} = \frac{3k}{d\sqrt{n}}.$$

But $k\phi(n) = ed - 1 < ed < d\phi(n)$, so $k < d < n^{1/4}/3$ and we find

$$\left| \frac{e}{n} - \frac{k}{d} \right| \leq \frac{n^{1/4}}{d\sqrt{n}} < \frac{1}{3d^2}.$$

This is a case of Inequality (13.1). It shows that k/d is such a close approximation to the fraction e/n, which is known to the attacker, that it must have the form A_i/B_i in the continued fraction expansion for e/n. The attacker computes the continued fraction expansion for e/n, which is fast, and checks each denominator B_i for being d, that is, he tests whether $M^{eB_i} \equiv M \pmod{n}$ for a few M. One of them must work. ∎

As this attack actually works for d slightly larger than $\sqrt[4]{n}$, one must choose d even larger to be safe. If $d > \sqrt{n}$, then the attack certainly doesn't work.

24.4 Partial Key Exposure

If some bits of a secret key are revealed, this reduces the key entropy. Then one might be able to discover it by trying all possible keys having the known bits. We give more examples for RSA.

The first result in this direction is due to Coppersmith [31].

THEOREM 24.2 One can find p given half its bits
Let $n = pq$ be a b-bit RSA modulus, so that the length of each of the primes p and q is about $b/2$ bits. One can efficiently factor n given either the $b/4$ most significant bits of p or the $b/4$ least significant bits of p.

See [31] for a proof. The next theorem, due to Boneh et al. [14], says that if e is small enough so that an attacker can perform $O(e \log e)$ operations, then an attacker can deduce d from just a few of its bits.

THEOREM 24.3 You can find d from its least significant bits
Let n be an RSA modulus of b bits. Let e and d be the enciphering and deciphering exponents. Given the $b/4$ least significant bits of d, an attacker can find d in $O(e \log e)$ steps.

PROOF Write $n = pq$. Since $ed \equiv 1 \pmod{\phi(n)}$, there is an integer k so that

$$ed - k(n - p - q + 1) = ed - k\phi(n) = 1.$$

As $d < \phi(n)$, we have $0 < k \leq e$. Multiply the equation by p, replace q by n/p and reduce modulo $2^{b/4}$ to get

$$(ed)p - kp(n - p + 1) + kn \equiv p \pmod{2^{b/4}}.$$

In this congruence, the attacker knows n, e and the $b/4$ least significant bits of d. Therefore, he knows the value of $ed \bmod 2^{b/4}$. For each k, the congruence is a quadratic equation in p. For each of the e possible values of k, the attacker solves the quadratic congruence using the methods of Chapter 7 and obtains some candidate values for $p \bmod 2^{b/4}$. One can show that there are no more than $e \log_2 e$ candidate values of $p \bmod 2^{b/4}$ in total. For each of these values, the attacker runs the algorithm of Theorem 24.2 and tests whether the output p actually divides n. This shows that n will be factored after at most $O(e \log e)$ steps. ∎

24.5 Computer Failure

Computers occasionally make hardware or software errors. If an error happens during computation of a cryptographic function, a key may be leaked. We give an example involving the RSA signature scheme. In it, the attacker will be able to factor the modulus n, and so be able to sign messages with it, just like the legitimate user.

Suppose $n = pq$ and the encryption and decryption exponents are e and d. The public information is n and e, while p, q and d are private. The message to be signed is M; its signature is $S = M^d \bmod n$.

We assume the trick described in Section 18.1 is used to speed signature generation. This trick computes $S_p = M^d \bmod p$ and $S_q = M^d \bmod q$ and combines them with the Chinese remainder theorem to form $S = aS_p + bS_q$. Suppose an error causes S_p to be incorrect. Suppose the attacker has the correct signature S as well as the incorrect one S' formed by the Chinese remainder theorem using the wrong S_p and the correct S_q. Then $S \equiv S_p \pmod{p}$, $S \equiv S_q \pmod{q}$, $S' \not\equiv S_p \pmod{p}$, $S' \equiv S_q \pmod{q}$ and we have $\gcd(S - S', n) = q$.

A. Lenstra found that this attack works even if the attacker only knows S' and M, but not S. We have $M = S^e \bmod n$. In case the error occurred in computing S_p, we have $M \equiv (S')^e \pmod{q}$ but $M \not\equiv (S')^e \pmod{p}$. Therefore, $\gcd(M - (S')^e, n) = q$.

Boneh et al. [13] describe the attack above and also an attack on RSA signatures not using the trick, that is, in which $S = M^d \bmod n$ is computed directly. That attack is less likely, as it requires the attacker to see several faulty signatures each with one bit flipped somewhere during the computation.

24.6 Exercises

1. What is the flaw in the following solution to a key exposure problem? Bob accidentally reveals his private RSA key d. Because so many people already know his public modulus n and enciphering exponent e, he decides to keep them. He chooses two new secret primes with product n and uses them and the old e to compute a new deciphering exponent d'.

2. In the RSA cipher, each user has a public modulus n, a public key e, and a private key d. Suppose Bob accidentally reveals his private key d. Because it takes so long to generate large primes, Bob decides to keep his old modulus n, which is public anyway, and just create a new e and d. After creating them, he makes the new e public and keeps the new d secret. Is this choice of RSA parameters safe? Explain your answer.

3. Eve notices that both Alice and Bob use the RSA cipher with the same modulus n, although they have different public encryption exponents e_A and e_B, which happen to be relatively prime. Eve learns through one of her agents that Chuck has just sent the same plaintext M to both Alice and Bob via RSA, and she intercepts the two ciphertexts C_A and C_B. Explain how Eve can read M, given C_A, C_B and public data. Can Eve factor n easily? Explain your answer.

4. The 1988 X.509 text says this about choosing secure RSA keys: "It must be ensured that $e > \log_2 n$ to prevent attack by taking the e-th root modulo n to discover the plaintext." The reason given is incorrect, although it is a good idea to satisfy the inequality. What is wrong with the reason, and what is the real reason?

5. Alice uses the ElGamal signature scheme with common public modulus p and primitive root g. Alice has a secret x and publishes $y = g^x \bmod p$. She uses a secret random number k to sign one message M. Since k is used only once, she discards the piece of paper on which she had written k. The next day Tom, Alice's trash collector, finds the sheet in her garbage can and realizes its significance. Tom learns M and Alice's signature for it from Eve, who collects all of Alice's signed messages as her hobby. Explain how Tom can forge Alice's signature (with different random k's) on arbitrary messages.

6. Alice and Bob use the elliptic curve ElGamal public key cipher for their secret communication. One day, Bob tosses a coin and sends Alice the enciphered result. Knowing only public data and that the plaintext is either "Heads" or "Tails," can Eve tell which plaintext it is from the ciphertext she has intercepted?

7. Alice and Bob use the coin-tossing protocol from Section 20.3. Alice chooses 100-digit primes p and q as her secret. Because of a defect in his random number generator, Bob always chooses x in $\sqrt{n} < x < n/1000$. Does this fault matter? Can either Alice or Bob almost always win the coin toss? Explain your answer. Would the same fault in the random number generator affect the zero-knowledge proof protocol?

8. Alice and Bob use the elliptic curve Diffie-Hellman key exchange protocol to choose random Rijndael keys. The elliptic curve group is public and has order N near 2^{160}. Because of a defect in her random number generator, the low-order 100 bits of Alice's random x_A are always one, but the high-order 60 bits are really random. Eve is aware of this defect because she has studied the source code of Alice's random number generator. Eve records all messages passing between Alice and Bob. Eve has a computer powerful enough to perform about 2^{30} elliptic curve group additions in a reasonable time. Explain how Eve can compute, with high probability, the Rijndael keys chosen by Alice and Bob.

Chapter 25

Active Attacks

This chapter considers some active attacks in which the attacker takes some action to fool the victim, causing an error, obtaining a key, or introducing his own devious message which others accept as authentic.

25.1 Force a User to Make a Mistake

For example, the mistake might expose a flaw in the cryptographic algorithm or it might make a user reveal part of a key.

Here is an attack on a flaw in one standard zero-knowledge proof protocol.

Alice knows n, p and q, where p and q are large primes and $n = pq$. Bob knows n but not p or q. Alice wants to convince Bob that she knows the factors p, q of n. But she does not want to reveal the factors to him.

1. Alice chooses a in $\sqrt{n} < a < n$ and computes $b = a^2 \bmod n$.

2. At the same time, Bob chooses c in $\sqrt{n} < c < n$ and computes $d = c^2 \bmod n$.

3. Alice sends b to Bob and Bob sends d to Alice.

4. Alice receives d and solves $x^2 \equiv bd \pmod{n}$. (Note that this is possible because bd is a quadratic residue and she can compute its square root because she knows the factors of n.) Let x_1 be one solution of this congruence.

5. At the same time, Bob tosses a fair coin and gets Heads or Tails, each with probability 1/2. Bob sends H or T to Alice.

6. If Alice receives H, she sends a to Bob. If Alice receives T, she sends x_1 to Bob.

7. If Bob sent H to Alice, then he receives a from Alice and checks that $a^2 \equiv b \pmod{n}$. If Bob sent T to Alice, then he receives x_1 from Alice and checks that $x_1 \equiv bd \pmod{n}$.

Alice and Bob repeat steps 1 through 7 many (20 or 30) times.

If the check in step 7 is always okay, then Bob accepts that Alice knows the factorization of n.

But if Alice ever fails even one test, then Bob concludes that Alice is lying.

Bob could cheat by finding the factors of n as follows:

Bob skips Step 2. In Step 3, he waits until he receives b from Alice. He computes $d = b^3 \bmod n$, and sends this d to Alice. Note that d is a quadratic residue modulo n, by Theorem 20.1. In Step 5, Bob sends T to Alice. In Step 6, Alice sends x_1 to Bob, and x_1 is a solution to the congruence $x^2 \equiv bd \equiv b^4 \pmod{n}$. Bob already knows one solution, b^2, to this congruence. If $x_1 \equiv \pm(b^2) \pmod{n}$, then Bob learns nothing when he receives x_1. But if Alice sends Bob one of the other two square roots of $b^4 \bmod n$ as x_1 in Step 6, then Bob can factor n by taking a greatest common divisor, by Theorem 13.1.

Alice could prevent this by checking whether $d = b^3 \bmod n$. But Bob could mount a similar attack with $d = b^5 \bmod n$, $d = b^7 \bmod n$, etc., and there are too many possibilities for Alice to check.

One way for Alice to avoid this trap is to wait for d to arrive from Bob before she sends b to him. (Could Alice cheat in a similar way then? I'm not sure.) A better way to avoid the problem is for both Alice and Bob to bit-commit to their quadratic residues before sending them.

25.2   Man-in-the-Middle Attacks

In these attacks, an attacker stands between Alice and Bob. Communication between them passes through the attacker's computer. He may modify parts of a message and pass it on to the intended recipient. The result of this devious action is that the attacker learns something about their communication or else fools one of them into accepting as authentic a bogus message supposedly from the other.

Here is a simple example of the **man-in-the-middle attack**. Suppose Alice wishes to communicate securely with Bob. She initiates the conversation by sending Bob a plaintext message with her public key. She expects Bob to reply with his public key and that they will use their public keys to talk secretly. However, Mike intercepts Alice's first message, replaces her public key with his and forwards the message to Bob. Bob receives a message that claims to be from Alice. It contains a number which is supposed to be her public key. He replies to Alice with a message containing his public key. Mike intercepts this message, replaces Bob's public key with his own and forwards the message to Alice. Then Alice and Bob communicate, but they both unknowingly encipher all their messages with Mike's public key. Mike intercepts all their messages, deciphers them, reenciphers them with the appropriate public key and forwards the messages to the intended recipient. Mike could even change the messages if he wished. The clever number theory behind public key cryptography is worthless against this attack.

This attack would work even if the public keys of Alice and Bob were stored in a public database, so long as Mike could intercept their queries to the database. Mike would intercept Alice's request for Bob's public key and reply with his own public key. He would do the same with Bob's query for Alice's public key and the attack would continue as before. The attack would be the same if Alice and Bob stored their public keys on their home pages. Mike would intercept their browser requests and reply with his own public key.

The X.509 protocol prevents the man-in-the-middle attack. In it, each user's public key is signed by a trusted authority, Tracy, perhaps through a hierarchy of trusted signatures. The signed record contains the identity of the public key owner in addition to the public key. When Alice requests Bob's public key and receives a number in reply, that number comes with a certificate signed by Tracy that this number is Bob's public key. Mike can intercept Alice's request and Bob's reply, but if he replaces Bob's public key with his own, either the signature won't match the one on the signed message or else the message will identify the public key as Mike's—not Bob's.

If Alice and Bob used PGP, they would receive each other's public key from several sources, each having a level of "trust" or confidence. Mike would have to compromise all of these sources to perform a man-in-the-middle attack. And Alice and Bob would likely be informed by PGP that there was little confidence in each other's public key, so they would realize that a man-in-the-middle attack could be happening.

Another type of man-in-the-middle attack involves replaying a message recorded earlier, in the hope that the recipient will believe it was just sent. For example, suppose Mike creates a phony person called Bob and arranges for Bob to sell a valuable diamond to Alice. Alice pays for the diamond by a bank transfer. Mike records the bank transfer, which is enciphered, of course, and replays it several times to Bob's bank. This causes Bob's bank to credit the payment to Bob's account several times. Mike withdraws the money from Bob's account, which Mike created, and retires in Brazil. Banks require several forms of identification when you open an account, or even access it, because of attacks like this one. Furthermore, bank transfer orders contain time stamps and sequence numbers to prevent this type of attack. After the first bank transfer, Bob's bank would notice that the others were copies of the first one, and not credit the account multiple times. Whether Alice received her diamond is another matter.

25.3   Birthday Attacks

Hash functions $h(M)$ may be subject to birthday attacks if the message digest is too short. Suppose M is a contract that Alice and Bob want to sign. Alice prepares M and sends it to Bob to sign. Both agree that signing $h(M)$ electronically binds the contract. Alice prepares two contracts, M which is fair to Bob, and M' which favors Alice. Let m be the length in bits of the message digest produced by h. In each contract, Alice finds $m/2 + 5$ places where the text could read in either of two ways, such as changing active voice to passive. She hashes all $2^{m/2+5}$ versions of both messages and looks for a match $h(M') = h(M)$ between versions of the two messages. By Theorem 2.5, this is likely to happen because there are 2^m possible message digests and more than $2^{m/2}$ versions of each message. She sends M and $h(M)$ to Bob. Alice and Bob sign $h(M)$. Later, Alice produces M' and Bob's signature on $h(M') = h(M)$. Bob finds that he has signed a contract favorable to Alice. This attack can be prevented by choosing m large enough so that Alice can't compute $2^{m/2+5}$ hash values.

25.4   Subliminal Channels

Several cryptographic algorithms have **subliminal channels**. These are covert ways that an attacker can send a second message hidden within a normal message. A simple example of this is the bit string formed by the parity of the number of letters in each word of a message. A carefully constructed, innocent sounding message may contain another message hidden in this bit string. As another example, a message may be hidden in certain bits of a digitized photo. Some cryptographic algorithms known to have subliminal channels are schemes that choose random numbers used just once, such as the Digital Signature Algorithm and the signature algorithms of ElGamal and Ong-Schnorr-Shamir.

Here is a subliminal channel for the DSA. Alice is a spy who sends many plaintext messages to her contact Bob. Alice's employer reviews these messages to ensure she is not divulging any secrets. All of the messages are signed by the DSA. Alice and Bob secretly agree on a prime p_1, different from any parameter of the DSA. When Alice signs an innocuous message M, she hides a subliminal bit in it. If she wants to send a 1 bit, she tries different random numbers k until the r parameter of the signature is a quadratic residue modulo p_1. For a 0 bit, she makes r a quadratic nonresidue modulo p_1. Since half of the r are quadratic residues and half are quadratic nonresidues, she doesn't have to try many random k to do this.

Bob checks each DSA signature to be sure the message came from Alice. Alice's employer could make the same check and find nothing amiss. Bob would read the subliminal bit in each message by evaluating the Legendre symbol (r/p_1).

In fact, Alice could send several subliminal bits per message. Suppose Bob and Alice agree on j secret primes $p_1, \ldots, p_j$. Then j subliminal bits could be sent per message, with the i-th bit being 0 or 1 according as r is a quadratic residue or nonresidue modulo p_i. On average, Alice would have to try 2^j values of k to get an r with the required properties, so she can't make j too big or her employer might wonder why her DSA was so slow. She could choose $j = 16$ and send two bytes per message.

An evil implementer of DSA could use the same subliminal channel to leak Alice's 160-bit private key. Mike sells low-cost DSA chips with a 14-bit subliminal channel using fourteen primes that only he knows. When the chip signs a message, the user must supply it with her private 160-bit key x. Alice buys a DSA chip from Mike and uses it to sign messages. The chip breaks the 160-bit key x into sixteen ten-bit pieces. It chooses a random four-bit number e and sends the subliminal message consisting of e and the e-th piece of x. Mike observes the signature and deduces ten bits of Alice's private key x. After seeing many of Alice's signatures, he will learn most of the ten-bit pieces of x. When he knows all but one or two of the pieces he can compute the remaining bits by trying all possibilities. Then Mike can forge Alice's signature on messages. Even if Alice or someone else knew that Mike was stealing DSA private keys this way, they could not prove it unless they knew Mike's fourteen secret primes.

25.5   *Exercises*

1. Bob is a bit naive about using RSA. He created a public modulus n and a public encryption exponent e and made them public. He knew that he should keep his private decryption exponent d secret. For a long time no one sent him RSA-enciphered messages. Finally, he received a ciphertext message C from Alice. He deciphered this to get the plaintext M, a rather personal note. Evil Eve has intercepted C and would love to know what Alice said in M. Eve makes Bob feel guilty that his RSA system was used only once. In response, he agrees to decrypt any message sent to him, so long as it is not C, and return the answer to the sender. Eve sends him the ciphertext $(7^e C \bmod n)$. Bob deciphers this message, puzzles over the apparently meaningless "plaintext," but sends the latter to Eve as agreed. Explain how this trick allows Eve to read M.

2. Alice and Bob use the Diffie-Hellman key exchange protocol to establish a Rijndael key for their daily chat. Explain how Eve can mount a man-in-the-middle attack and relay all messages between them so that she can read them, but they won't know that anything is wrong.

3. Write a contract or find an existing one. Find forty places in the text where you could express something in two different ways without changing the meaning.

4. Find a subliminal channel in the ElGamal signature scheme.

5. In the DSA subliminal channel, suppose Mike has acquired 25 of Alice's signatures produced by the chip he sold her. What is the probability that these signatures reveal to him at least fourteen of the sixteen pieces of Alice's private key x?

6. Alice and Bob use this simplified version of the zero-knowledge proof protocol for Alice to convince Bob she knows the factorization of a huge integer n. Both Alice and Bob know n.

   (a) Bob chooses a random x in $\sqrt{n} < x < n$ and sends $y = x^2 \bmod n$ to Alice.

   (b) Alice computes the four square roots of y modulo n, picks one, r, say, and sends it to Bob.

   (c) Bob checks whether $y = r^2 \bmod n$. If this fails, Alice does not know the factorization of n.

   Alice and Bob repeat the three steps thirty times, but stop if the equation fails in Step (c).

   a. Explain why Bob should believe that Alice knows the factors of n if the check in Step (c) is valid all thirty times.

   b. Explain why this is not a zero-knowledge proof protocol.

   c. Suppose that no value of y is repeated during the thirty repetitions. Eve knows n and observes the numbers y and r passed between Alice and Bob, but cannot see their private data, such as x. Can Eve use this information to convince Chuck that she, Eve, knows the factorization of n using the same protocol? Could Eve convince him if she and Chuck use the full, unsimplified zero-knowledge proof protocol instead? Explain your answers.

   d. Same question as part c., except that one y value was repeated.

References

[1] L. M. Adleman, C. Pomerance, and R. S. Rumely. On distinguishing prime numbers from composite numbers. *Ann. of Math.*, 117:173–206, 1983.

[2] W. Alford, A. Granville, and C. Pomerance. There are infinitely many Carmichael numbers. *Ann. of Math.*, 139:703–722, 1994.

[3] W. R. Alford, A. Granville, and C. Pomerance. On the difficulty of finding reliable witnesses. In L. Adleman and M.-D. Huang, editors, *Algorithmic Number Theory, Proc. ANTS-I, Ithaca, NY*, volume 887 of *Lecture Notes in Computer Science*, pages 1–16, Springer-Verlag, Berlin, 1994.

[4] W. R. Alford and C. Pomerance. Implementing the self-initializing quadratic sieve on a distributed network. In A. van der Poorten, I. Shparlinski, and H. G. Zimmer, editors, *Number Theoretic and Algebraic Methods in Computer Science*, pages 163–174, Moscow, 1993.

[5] C. Asmuth and J. Bloom. A modular approach to key safeguarding. *IEEE Trans. on Info. Theory*, IT-29(2):208–210, 1983.

[6] A. O. L. Atkin and F. Morain. Elliptic curves and primality proving. *Math. Comp.*, 61:29–68, 1993.

[7] E. Bach. *Analytic Methods in the Analysis and Design of Number-Theoretic Algorithms*. The MIT Press, Cambridge, Massachusetts, 1985.

[8] E. Bach and J. Shallit. *Algorithmic Number Theory, Volume I: Efficient Algorithms*. The MIT Press, Cambridge, Massachusetts, 1996.

[9] R. Baillie and S. S. Wagstaff, Jr. Lucas pseudoprimes. *Math. Comp.*, 35:1391–1417, 1980.

[10] T. H. Barr. *An Invitation to Cryptology*. Prentice-Hall, Upper Saddle River, New Jersey, 2002.

[11] L. Blum, M. Blum, and M. Shub. A simple unpredictable pseudo-random number generator. *SIAM J. Comput.*, 15:364–383, 1986.

[12] D. Boneh. Twenty years of attacks on the RSA cryptosystem. *Amer. Math. Soc. Notices*, 46:202–213, 1999.

[13] D. Boneh, R. A. DeMillo, and R. J. Lipton. On the importance of checking cryptographic protocols for faults. In *Advances in Cryptology— EUROCRYPT '97*, volume 1223 of *Lecture Notes in Computer Science*, pages 37–51, Springer-Verlag, Berlin, 1997.

[14] D. Boneh, G. Durfee, and Y. Frankel. An attack on RSA given a frac-

tion of the private key bits. In *Advances in Cryptology—ASIACRYPT '98*, volume 1514 of *Lecture Notes in Computer Science*, pages 25–34, Springer-Verlag, Berlin, 1998.

[15] S. Brands. Untraceable off-line cash in wallets with observers. In *Advances in Cryptology—CRYPTO '93*, volume 773 of *Lecture Notes in Computer Science*, pages 302–318, Springer-Verlag, Berlin, New York, 1994.

[16] J. Brillhart, M. Filaseta, and A. M. Odlyzko. On an irreducibility theorem of A. Cohn. *Canad. J. Math.*, 33:1055–1059, 1981.

[17] J. Brillhart, D. H. Lehmer, and J. L. Selfridge. New primality criteria and factorizations of $2^m \pm 1$. *Math. Comp.*, 29:620–647, 1975.

[18] John Brillhart, D. H. Lehmer, J. L. Selfridge, Bryant Tuckerman, and S. S. Wagstaff, Jr. *Factorizations of $b^n \pm 1$, $b = 2, 3, 5, 6, 7, 10, 11, 12$ up to high powers*. Amer. Math. Soc., Providence, Rhode Island, Third edition, 2002. Electronic book available at http://www.ams.org/online_bks/conm22.

[19] J. Buhler, H. W. Lenstra, Jr., and C. Pomerance. Factoring integers with the number field sieve. In A. K. Lenstra and H. W. Lenstra, Jr., editors, *The Development of the Number Field Sieve*, volume 1554 of *Lecture Notes in Mathematics*, pages 50–94, Springer-Verlag, Berlin, New York, 1993.

[20] D. A. Burgess. A note on the distribution of residues and non-residues. *J. London Math. Soc.*, 38:253–256, 1963.

[21] M. Burrows, M. Abadi, and R. M. Needham. A logic of authentication. Technical Report 39, Digital Equipment Corporation Systems Research Center, Palo Alto, CA, 1989.

[22] M. Burrows, M. Abadi, and R. M. Needham. A logic of authentication. *ACM Trans. on Computer Systems*, 8:18–36, February 1990.

[23] E. Canfield, P. Erdős, and C. Pomerance. On a problem of Oppenheim concerning "factorisatio numerorum." *J. Number Theory*, 17:1–28, 1983.

[24] J. Cassels. Diophantine equations with special reference to elliptic curves. *J. London Math. Soc.*, 41:193–291, 1966.

[25] S. Cavalar. *On the number field sieve integer factorisation algorithm*. PhD thesis, Leiden University, June 2002.

[26] D. Chaum. Security without identification: Transaction systems to make Big Brother obsolete. *Communications of the ACM*, 28:1030–1044, 1985.

[27] D. Chaum, A. Fiat, and M. Naor. Untraceable electronic cash. In *Advances in Cryptology—CRYPTO '88*, pages 319–327, Springer-Verlag, Berlin, New York, 1990.

[28] H. Cohen. *A Course in Computational Algebraic Number Theory*. Springer-Verlag, New York, 1996.

[29] H. Cohen and H. W. Lenstra, Jr. Primality testing and Jacobi sums. *Math. Comp.*, 42:297–330, 1984.

[30] D. Coppersmith. Fast evaluation of logarithms in fields of characteristic

two. *IEEE Trans. on Info. Theory*, 30:587–594, 1984.

[31] D. Coppersmith. Small solutions to polynomial equations, and low exponent RSA vulnerabilities. *J. Cryptology*, 10:233–594, 1997.

[32] J.-M. Couveignes. Computing a square root for the number field sieve. In A. K. Lenstra and H. W. Lenstra, Jr., editors, *The Development of the Number Field Sieve*, volume 1554 of *Lecture Notes in Mathematics*, pages 95–102, Springer-Verlag, Berlin, New York, 1993.

[33] R. Crandall and C. Pomerance. *Prime Numbers: A Computational Perspective*. Springer-Verlag, New York, 2001.

[34] J. A. Davis and D. B. Holdridge. Factorization using the quadratic sieve algorithm. In D. Chaum, editor, *Advances in Cryptology—CRYPTO '83*, pages 103–113, Plenum Press, New York, 1984.

[35] N. G. de Bruijn. On the number of positive integers $\leq x$ and free of prime factors $> y$. *Proc. Kon. Ned. Akad. Wet.*, A54:50–60, 1951.

[36] D. E. Denning. *Cryptography and Data Security*. Addison-Wesley, Reading, Massachusetts, 1983.

[37] D. E. Denning and G. M. Sacco. Timestamps in key distribution protocols. *Communications of the ACM*, 24:533–536, 1981.

[38] M. Deuring. Die Typen der Multiplikatorenringe elliptischer Funktionenkörper. *Abh. Math. Sem. Hansischen Univ.*, 14:197–272, 1941.

[39] K. Dickman. On the frequency of numbers containing prime factors of a certain relative magnitude. *Ark. Mat., Astronomi och Fysik*, 22A, 10:1–14, 1930.

[40] W. Diffie and M. Hellman. New directions in cryptography. *IEEE Trans. on Info. Theory*, IT-22(6):644–654, 1976.

[41] C. Ding, G. Xiao, and W. Shan. *The Stability Theory of Stream Ciphers, Lecture Notes in Computer Science 561*. Springer-Verlag, New York, 1991.

[42] R.-M. Elkenbracht-Huizing, P. L. Montgomery, R. D. Silverman, R. K. Wackerbarth, and S. S. Wagstaff, Jr. *The number field sieve on many computers*, volume 19 of *CRM Proceedings and Lecture Notes*, pages 81–85. Centre de Researches Mathématiques, 1999.

[43] W. Feller. *An Introduction to Probability Theory and Its Applications*. John Wiley, New York, 1957.

[44] W. F. Friedman. *Elements of Cryptanalysis*. Aegean Park Press, Laguna Hills, California, 1976.

[45] C. F. Gauss. *Disquisitiones Arithmeticae*. Yale University Press, New Haven, English edition, 1966.

[46] J. L. Gerver. Factoring large numbers with a quadratic sieve. *Math. Comp.*, 41:287–294, 1983.

[47] S. Goldwasser and J. Kilian. Almost all primes can be quickly certified. In *Proc. Eighteenth Annual ACM Symp. on the Theory of Computing (STOC), Berkeley, May 28-30, 1986*, pages 316–329. ACM, 1986.

[48] S. W. Golumb. *Shift Register Sequences*. Holden-Day, San Francisco, California, 1967.

[49] D. Gordon. Discrete logarithms in $GF(p)$ via the number field sieve. *SIAM J. Discrete Math.*, 16:124–138, 1993.

[50] J. Hadamard. Résolution d'une question relative aux déterminants. *Bull. Sci. Math. (2)*, 17:235–246, 1893.

[51] G. H. Hardy and E. M. Wright. *An Introduction to the Theory of Numbers*. Clarendon Press, Oxford, England, Fifth edition, 1979.

[52] J. Hastad. Solving simultaneous modular equations of low degree. *SIAM J. Comput.*, 17:336–341, 1988.

[53] I. N. Herstein. *Topics in Algebra*. Xerox, Lexington, Second edition, 1975.

[54] L. S. Hill. Cryptography in an algebraic alphabet. *Amer. Math. Monthly*, 36:306–312, 1929.

[55] D. Kahn. *The Codebreakers*. Macmillan Co., New York, 1967.

[56] D. E. Knuth. *The Art of Computer Programming, Volume 2, Seminumerical Algorithms*. Addison-Wesley, Reading, Massachusetts, Second edition, 1981.

[57] D. E. Knuth and L. Trabb Pardo. Analysis of a simple factorization algorithm. *Theoretical Computer Science*, 3:321–348, 1976.

[58] N. Koblitz. *A Course in Number Theory and Cryptography*. Springer-Verlag, New York, 1987.

[59] P. C. Kocher. Timing attacks on implementations of Diffie-Hellman, RSA, DSS, and other systems. In *Advances in Cryptology—CRYPTO '96*, volume 1109 of *Lecture Notes in Computer Science*, pages 104–113, Springer-Verlag, Berlin, New York, 1996.

[60] M. Kraitchik. *Théorie des nombres, Tome II*. Gauthiers-Villars, Paris, France, 1926.

[61] A. M. Legendre. *Théorie des nombres, Tome I*. Gauthiers-Villars, Paris, France, 1830.

[62] D. H. Lehmer. A photo-electric number sieve. *Amer. Math. Monthly*, 40:401–406, 1933.

[63] D. H. Lehmer. An announcement concerning the delay line sieve DLS 127. *Math. Comp.*, 20:645–646, 1966.

[64] A. K. Lenstra and H. W. Lenstra, Jr. *The Development of the Number Field Sieve*. Springer-Verlag, New York, 1993.

[65] A. K. Lenstra and M. S. Manasse. Factoring with two large primes. *Math. Comp.*, 63:785–798, 1994.

[66] H. W. Lenstra, Jr. Primality testing algorithms (after Adleman, Rumley and Williams). In *Seminar Bourbaki 33 (1980/81), Lecture Notes in Mathematics*, volume 901, pages 243–258, Springer-Verlag, Berlin, 1981.

[67] H. W. Lenstra, Jr. Factoring integers with elliptic curves. *Ann. of Math.*, 126:649–673, 1987.

[68] P. Leyland, A. K. Lenstra, B. Dodson, A. Muffett, and S. S. Wagstaff, Jr. MPQS with three large primes. In *Algorithmic Number Theory, Proceedings ANTS 2002*, volume 2369 of *Lecture Notes in Computer Science*, pages 448–462. Springer-Verlag, 2002.

[69] S. Lodin, B. Dole, and E. H. Spafford. Misplaced trust: Kerberos 4 random session keys. In *Proc. of Internet Society Symposium on Network and Distributed System Security*, pages 60–70. The Internet Society, February 1997.

[70] J. L. Massey and J. K. Omura. Method and apparatus for maintaining the privacy of digital messages conveyed by public transmission. U. S. Patent # 4,567,600, 28 January 1986.

[71] P. Mihăilescu. Cyclotomy primality proving—recent developments. In *Algorithmic Number Theory (Portland, OR, 1998), Lecture Notes in Computer Science*, volume 1423, pages 99–110, Springer-Verlag, Berlin, 1981.

[72] G. Miller. Riemann's hypothesis and tests for primality. *J. Comput. System Sci.*, 13:300–317, 1976.

[73] L. Monier. Evaluation and comparison of two efficient probabilistic primality testing algorithms. *Theoret. Comput. Sci.*, 12:97–108, 1980.

[74] P. Montgomery. Square roots of products of algebraic numbers. In W. Gautschi, editor, *Mathematics of Computation 1943–1993*, volume 48 of *Proc. Symp. Appl. Math.*, pages 567–571. Amer. Math. Soc., 1994.

[75] P. Montgomery. A block Lanczos algorithm for finding dependencies over $GF(2)$. In A. J. Menezes and S. A. Vanstone, editors, *Advances in Cryptology—EUROCRYPT '95*, volume 921 of *Lecture Notes in Computer Science*, pages 106–120, Springer-Verlag, Berlin, 1995.

[76] M. A. Morrison. A note on primality testing using Lucas sequences. *Math. Comp.*, 29:181–182, 1975.

[77] M. A. Morrison and J. Brillhart. A method of factoring and the factorization of F_7. *Math. Comp.*, 29:183–205, 1975.

[78] I. Niven, H. S. Zuckerman, and H. L. Montgomery. *An Introduction to the Theory of Numbers.* John Wiley, New York, Fifth edition, 1991.

[79] A. M. Odlyzko. Discrete logarithms in finite fields and their cryptographic significance. In T. Beth, N. Cot, and I. Ingemarsson, editors, *Advances in Cryptology—EUROCRYPT '84*, volume 209 of *Lecture Notes in Computer Science*, pages 224–313, Springer-Verlag, Berlin, 1985.

[80] G. Paun, G. Rozenberg, and A. Salomaa. *DNA Computing: New Computing Paradigms.* Springer-Verlag, New York, 1998.

[81] R. Peralta. A quadratic sieve on the n-dimensional cube. In *Advances in Cryptology—CRYPTO '92*, volume 740 of *Lecture Notes in Computer Science*, pages 324–332, Springer-Verlag, Berlin, New York, 1993.

[82] R. G. E. Pinch. The pseudoprimes to 10^{13}. In *Algorithmic Number Theory (Leiden, 2000)*, volume 1838 of *Lecture Notes in Computer Science*, pages 459–473. Springer-Verlag, Berlin, 2000.

[83] S. Pohlig and M. Hellman. An improved algorithm for computing logarithms over $\mathbf{GF}(p)$ and its cryptographic significance. *IEEE Trans. on Info. Theory*, IT-24(1):106–110, 1978.

[84] J. M. Pollard. Theorems on factorization and primality testing. *Proc. Cambridge Philos. Soc.*, 76:521–528, 1974.

[85] J. M. Pollard. A Monte Carlo method for factorization. *Nordisk Tidskr. Informationsbehandling (BIT)*, 15:331–335, 1975.

[86] J. M. Pollard. Monte Carlo methods for index computation (mod p). *Math. Comp.*, 32:918–924, 1978.

[87] C. Pomerance. Analysis and comparison of some integer factoring algorithms. In H. W. Lenstra, Jr. and R. Tijdeman, editors, *Computational Methods in Number Theory, Part 1*, volume 154 of *Math. Centrum Tract*, pages 89–139, CWI, Amsterdam, 1982.

[88] C. Pomerance. The quadratic sieve factoring algorithm. In T. Beth, N. Cot, and I. Ingemarsson, editors, *Advances in Cryptology— EUROCRYPT '84*, volume 209 of *Lecture Notes in Computer Science*, pages 169–182, Springer-Verlag, Berlin, 1985.

[89] C. Pomerance, J. L. Selfridge, and S. S. Wagstaff, Jr. The pseudoprimes to $25 \cdot 10^9$. *Math. Comp.*, 35:1003–1026, 1980.

[90] V. R. Pratt. Every prime has a succinct certificate. *SIAM J. Comput.*, 4:214–220, 1975.

[91] G. E. Purdy. A high security log-in procedure. *Communications of the ACM*, 17:442–445, August 1974.

[92] M. Rabin. Digitized signatures and public-key functions as intractable as factoring. Technical Report LCS/TR-212, M.I.T. Lab for Computer Science, 1979.

[93] M. Rabin. Probabilistic algorithms for testing primality. *J. Number Theory*, 12:128–138, 1980.

[94] V. Ramaswami. The number of positive integers $< x$ and free of prime divisors $> x^c$, and a problem of S. S. Pillai. *Duke Math. J.*, 16:99–109, 1949.

[95] J. A. Reeds and P. J. Weinberger. File security and the UNIX crypt command. *AT&T Tech. J.*, 63:1673–1683, October, 1984.

[96] H. Riesel. *Prime Numbers and Computer Methods of Factorization.* Birkhäuser, Boston, Massachusetts, Second edition, 1994.

[97] R. L. Rivest, A. Shamir, and L. Adleman. A method for obtaining digital signatures and public-key cryptosystems. *Comm. A. C. M.*, 21(2):120–126, 1978.

[98] N. Robbins. *Beginning Number Theory.* Wm. C. Brown, Dubuque, Iowa, 1993.

[99] K. H. Rosen. *Elementary Number Theory and Its Applications.* Addison-Wesley, Reading, Massachusetts, Second edition, 1988.

[100] B. Schneier. *Applied Cryptography.* Wiley, New York, Second edition, 1996.

[101] R. Schoof. Elliptic curves over finite fields and the computation of square roots mod p. *Math. Comp.*, 44:483–494, 1985.

[102] A. Shamir. How to share a secret. *Communications of the ACM*, 24:612–613, 1979.

[103] A. Shamir. *On the Generation of Cryptographically Strong Pseudorandom Sequences.* Dept. of Applied Math., The Weizmann Institute of Science, Rehovat, Israel, 1981.

[104] A. Shamir. Factoring large numbers with the TWINKLE device (extended abstract). In Ç. Koç and C. Paar, editors, *Cryptographic Hardware and Embedded Systems, First International Workshop, CHES '99, Worcester, MA*, volume 1717 of *Lecture Notes in Computer Science*, pages 2–12, Springer-Verlag, New York, 1999.

[105] D. Shanks. Class number, a theory of factorization, and genera. In *1969 Number Theory Institute, Stony Brook, N.Y.*, volume 20 of *Proc. Sympos. Pure Math.*, pages 415–440. Amer. Math. Soc., 1971.

[106] C. E. Shannon. A mathematical theory of communication. *Bell Syst. Tech. J.*, 27:379–423, 623–656, 1948.

[107] C. E. Shannon. Communication theory of secrecy systems. *Bell Syst. Tech. J.*, 28:656–715, 1949.

[108] C. E. Shannon. Predilection and entropy of printed English. *Bell Syst. Tech. J.*, 30:50–64, 1951.

[109] P. Shor. Algorithms for quantum computation: discrete logarithms and factoring. In *Proceedings of the Thirty-Fifth Annual Symposium on the Foundations of Computer Science*, pages 124–134, 1994.

[110] P. Shor. Polynomial-time algorithms for prime factorization and discrete logarithms on a quantum computer. *SIAM Review*, 41:303–332, 1999.

[111] J. Silverman. *The Arithmetic of Elliptic Curves.* Springer-Verlag, New York, 1986.

[112] R. D. Silverman and S. S. Wagstaff, Jr. A practical analysis of the elliptic curve factoring algorithm. *Math. Comp.*, 61:445–462, 1993.

[113] J. W. Smith and S. S. Wagstaff, Jr. An extended precision operand computer. In *Proc. of the Twenty-First Southeast Region ACM Conference*, pages 209–216, ACM, 1983.

[114] W. Stallings. *Cryptography and Network Security, Principles and Practice.* Prentice Hall, Upper Saddle River, New Jersey, Third edition, 2003.

[115] W. Trappe and L. C. Washington. *Introduction to Cryptography with Coding Theory.* Prentice Hall, Upper Saddle River, New Jersey, 2002.

[116] P. C. van Oorschot and M. J. Wiener. Parallel collision search with cryptanalytic applications. *Journal of Cryptology*, 12:1–28, 1999.

[117] S. S. Wagstaff, Jr. Prime numbers with a fixed number of one bits or zero bits in their binary representation. *Experimental Math.*, 10:267–273, 2001.

[118] S. S. Wagstaff, Jr. and J. W. Smith. Methods of factoring large integers. In D. V. Chudnovsky, G. V. Chudnovsky, H. Cohn, and M. B. Nathanson, editors, *Number Theory, New York, 1984-1985*, volume 1240 of *Lecture Notes in Mathematics*, pages 281–303, Springer-Verlag, New York, 1987.

[119] W. Weaver. *Lady Luck, the Theory of Probability.* Anchor Books, Gar-

den City, New York, 1963.

[120] M. J. Wiener. Cryptanalysis of short RSA secret exponents. *IEEE Trans. on Info. Theory*, 36:553–558, 1990.

[121] H. C. Williams. Primality testing on a computer. *Ars Combinatoria*, 32:127–185, 1978.

[122] H. C. Williams. A modification of the RSA public-key encryption procedure. *IEEE Trans. on Info. Theory*, IT-26(6):726–729, 1980.

[123] H. C. Williams. A $p+1$ method of factoring. *Math. Comp.*, 39:225–234, 1982.

[124] H. C. Williams. *Edouard Lucas and Primality Testing*, volume 22 of *Canadian Mathematics Society Series of Monographs and Advanced Texts*. John Wiley & Sons, New York, 1998.

[125] H. C. Williams and R. Holte. Some observations on primality testing. *Math. Comp.*, 32:905–917, 1978.

[126] H. C. Williams and J. S. Judd. Determination of the primality of N using factors of $N^2 \pm 1$. *Math. Comp.*, 30:157–172, 1976.

[127] H. C. Williams and J. S. Judd. Some algorithms for primality testing using generalized Lehmer functions. *Math. Comp.*, 30:867–886, 1976.

[128] H. C. Williams and C. D. Patterson. A report on the University of Manitoba sieve unit. *Congressus Numerantium*, 37:85–98, 1983.

[129] T. Y. C. Woo and S. S. Lam. Authentication for distributed systems. *Computer*, 25(1):39–52, January 1992.

[130] T. Y. C. Woo and S. S. Lam. "Authentication" revisited. *Computer*, 25(3):10, March 1992.

[131] C. Zhang. *An extension of the Dickman function and its application*. PhD thesis, Purdue University, June 2002.

Index

J